Selected Papers from the 2018 41st International Conference on Telecommunications and Signal Processing (TSP)

Selected Papers from the 2018 41st International Conference on Telecommunications and Signal Processing (TSP)

Special Issue Editors

Norbert Herencsar
Francesco Benedetto
Jorge Crichigno

MDPI • Basel • Beijing • Wuhan • Barcelona • Belgrade

MDPI

Special Issue Editors

Norbert Herencsar
Brno University of Technology
Czech Republic

Francesco Benedetto
University of "Roma TRE"
Italy

Jorge Crichigno
University of South Carolina
USA

Editorial Office
MDPI
St. Alban-Anlage 66
4052 Basel, Switzerland

This is a reprint of articles from the Special Issue published online in the open access journal *Applied Sciences* (ISSN 2076-3417) from 2018 to 2019 (available at: https://www.mdpi.com/journal/applsci/special_issues/tsp)

For citation purposes, cite each article independently as indicated on the article page online and as indicated below:

LastName, A.A.; LastName, B.B.; LastName, C.C. Article Title. *Journal Name* **Year**, *Article Number*, Page Range.

ISBN 978-3-03921-040-4 (Pbk)
ISBN 978-3-03921-041-1 (PDF)

Contents

About the Special Issue Editors vii

Norbert Herencsar, Francesco Benedetto and Jorge Crichigno
Special Issue "Selected Papers from the 2018 41st International Conference on
Telecommunications and Signal Processing (TSP)"
Reprinted from: *Appl. Sci.* **2019**, *9*, 2056, doi:10.3390/app9102056 1

G.Baldini, R. Giuliani and G. Steri
Physical Layer Authentication and Identification of Wireless Devices Using the
Synchrosqueezing Transform
Reprinted from: *Appl. Sci.* **2018**, *8*, 2167, doi:10.3390/app8112167 6

**Eylem Erdogan, Sultan Aldırmaz Çolak, Hakan Alakoca, Mustafa Namdar, Arif Basgumus
and Lutfiye Durak-Ata**
Interference Alignment in Multi-Hop Cognitive Radio Networks under Interference Leakage
Reprinted from: *Appl. Sci.* **2018**, *8*, 2486, doi:10.3390/app8122486 25

Tomas Horvath, Petr Munster, Vaclav Oujezsky and Josef Vojtech
Activation Process of ONU in EPON/GPON/XG-PON/NG-PON2 Networks
Reprinted from: *Appl. Sci.* **2018**, *8*, 1934, doi:10.3390/app8101934 38

David Kubanek, Todd J. Freeborn, Jaroslav Koton Jan Dvorak
Validation of Fractional-Order Lowpass Elliptic Responses of $(1 + \alpha)$-Order Analog Filters [†]
Reprinted from: *Appl. Sci.* **2018**, *8*, 2603, doi:10.3390/app8122603 56

**Jan Mucha, Jiri Mekyska, Zoltan Galaz, Marcos Faundez-Zanuy, Karmele Lopez-de-Ipina,
Vojtech Zvoncak, Tomas Kiska, Zdenek Smekal, Lubos Brabenec and Irena Rektorova**
Identification and Monitoring of Parkinson's Disease Dysgraphia Based on Fractional-Order
Derivatives of Online Handwriting [†]
Reprinted from: *Appl. Sci.* **2018**, *8*, 2566, doi:10.3390/app8122566 73

**Zoltan Galaz, Jiri Mekyska, Tomas Kiska, Vojtech Zvoncak, Jan Mucha,
Zdenek Smekal, Ilona Eliasova, Martina Mrackova, Milena Kostalova, Irena Rektorova,
Marcos Faundez-Zanuy, Jesus B. Alonso-Hernandez and Pedro Gomez-Vilda**
Changes in Phonation and Their Relations with Progress of Parkinson's Disease
Reprinted from: *Appl. Sci.* **2018**, *8*, 2339, doi:10.3390/app8122339 91

David Luengo, David Meltzer, and Tom Trigano
An Efficient Method to Learn Overcomplete Multi-Scale Dictionaries of ECG Signals
Reprinted from: *Appl. Sci.* **2018**, *8*, 2569, doi:10.3390/app8122569 109

Martin Kolařík, Radim Burget, Václav Uher, KamilŘíha, Malay Kishore Dutta
Optimized High Resolution 3D Dense-U-Net Network for Brain and Spine Segmentation
Reprinted from: *Appl. Sci.* **2019**, *9*, 404, doi:10.3390/app9030404 127

Anamaria Radoi and Corneliu Burileanu
Retrieval of Similar Evolution Patterns from Satellite Image Time Series [†]
Reprinted from: *Appl. Sci.* **2018**, *8*, 2435, doi:10.3390/app8122435 144

Xiangyu Liu, Hui Xu, Yinan Wang, Yingqiang Dai, Nan Li and Guiqing Liu
Calibration for Sample-And-Hold Mismatches in M-Channel TIADCs Based on Statistics
Reprinted from: *Appl. Sci.* **2019**, *9*, 198, doi:10.3390/app9010198 165

About the Special Issue Editors

Norbert Herencsar, Associate Professor, (S'07–M'12–SM'15) received his M.Sc. and Ph.D. degrees in Electronics & Communication and Teleinformatics from Brno University of Technology (BUT), Brno, Czech Republic, in 2006 and 2010, respectively. In 2013 and 2014, he was Visiting Researcher at Bogazici University and Dogus University, both in Istanbul, Turkey. Since March 2019, he has been Visiting Professor at the University of Calgary, Canada. He has been Associate Professor at the Department of Telecommunications of BUT since 2015, and has collaborated on numerous research projects supported by the Czech Science Foundation since 2006. Currently, he is an MC Member and the Science Communications Manager of the COST Action CA15225. Dr. Herencsar is the author of 82 articles published in SCIE peer-reviewed journals and about 120 papers published in the proceedings of international conferences. His research interests include analog electronics, current-mode circuits, and fractional-order systems synthesis. Since 2010, Dr. Herencsar has been Deputy Chair of the International Conference on Telecommunications and Signal Processing (TSP) and organizing or TPC member of the AFRICON, ELECO, I2MTC, ICUMT, IWSSIP, SET-CAS, MWSCAS, and ICECS conferences. In 2016, he was the General Co-Chair of the COST/IEEE-CASS Seasonal Training School in Fractional-Order Systems. He has been the General Co-Chair of the TSP since 2017. Since 2011, he has contributed as Guest Co-Editor of several journal Special Issues in *AEÜ—Int. Journal of Electronics and Communications, Radioengineering*, and *Telecommunication Systems*. Since 2014, he has served as Associate Editor of the *Journal of Circuits, Systems and Computers (JCSC)*, *IEEE Access, IEICE Electronics Express (ELEX)*, and as Editorial Board Member of Radioengineering as well as *Fractal and Fractional* since 2017 and 2018, respectively. Since 2015, he has served in the IEEE Czechoslovakia Section Executive Committee as SP/CAS/COM Joint Chapter Chair. Dr. Herencsar is also a Senior Member of the IACSIT and IRED and a Member of the IAENG, ACEEE, and RS.

Francesco Benedetto, Associate Professor, was born in Rome, Italy, on August 4th, 1977. He received his Dr. Eng. degree in Electronic Engineering from the University of ROMA Tre, Rome, Italy, in May 2002, and Ph.D. degree in Telecommunication Engineering from the University of Roma Tre, Rome, Italy, in April 2007. In 2007, he was a Research Fellow at the Department of Applied Electronics at the Third University of Rome. In 2008, he became an Assistant Professor of Telecommunications at the Third University of Rome (2008–2012, Applied Electronics Dept.; 2013–present, Economics Dept.). In 2017, Dr. Benedetto was unanimously awarded the National Academic Qualification for this position, and since 2019, he has been teaching the course "Elements of Telecommunications" (formerly *Signals and Telecommunications*) as part of the Computer Engineering degree, the course "Software Defined Radio" as part of the Laurea Magistralis in Information and Communication Technologies, and the course "Lab. of Statistical Analysis Systems" as part of the Laurea Magistralis in the School in Economics and Business Studies. Since the academic year 2013/2014, Dr. Benedetto has also been in charge of the course "Cognitive Communications" as part of the Ph.D. degree in Applied Electronics at the Department of Engineering, University of Roma Tre. He is a Senior Member of the Institution of Electrical and Electronic Engineers (IEEE), and an active member of the following IEEE Societies: IEEE Standard Association, IEEE Young Professionals, IEEE Software Defined Networks, IEEE Communications, IEEE Signal Processing, IEEE Vehicular Technology, and also a member of CNIT (Italian Inter-Universities Consortium for Telecommunications). Dr. Benedetto's research interests are in the field of ground-penetrating radar and software, and cognitive radio and digital signal and image processing for telecommunications and economics, code acquisition, and synchronization of 3G mobile communication systems and multimedia communications. Since 2016, he has been the Chair of the IEEE 1900.1 working group on "Definitions and Concepts for Dynamic Spectrum Access: Terminology Relating to Emerging Wireless Networks, System Functionality, and Spectrum Management". He is the Leader of the WP 3.4 on "Development of Advanced GPR Data Processing Technique" of the European COST Action TU1208—Civil Engineering Applications of Ground Penetrating Radar. He is an Editor of the IEEE SDN NEWSLETTER, was the General Chair of the Series of International Workshops on Signal Processing for Secure Communications (SP4SC-2014, -2015, and -2016), and the Lead Guest Editor of the Special Issue on Advanced Ground-Penetrating Radar Signal Processing Techniques for the *Signal Processing Journal* (Elsevier). Dr. Benedetto has also served as a Reviewer for *IEEE Transactions*, the *IET* (formerly IEE) *Proceedings*, *EURASIP*, and several Elsevier journals. He was a TPC Member for several IEEE international conferences and symposia in the same fields.

Jorge Crichigno is currently Associate Professor at the Department of Integrated Information Technology (IIT), University of South Carolina (USC), since January of 2018. He has also served as a Research Associate at the Electrical Engineering Department, University of South Florida, and at the Florida Center for Cybersecurity since 2016. Dr. Crichigno's research focuses on the practical implementation of Science DMZs. This includes the design and implementation of high-speed switched networks, TCP optimization, and experimental evaluation of congestion control algorithms such as BBR, HTCP, and Cubic. Dr. Crichigno's work has been funded by the U.S. National Science Foundation (NSF) and other agencies. In this regard, he has led the design, implementation, and testing of several high-speed networks for big data transfers. These include the 100 Gbps research network at USC to move terabyte-scale data to national laboratories (e.g., Argonne, Fermi, Oak Ridge, Savannah River, Los Alamos) and the U.S. national network of supercomputer centers, XSEDE. Dr. Crichigno's work also includes the design and implementation of overlay (emulated) networks used for training and research on high-throughput networks for big science data transfers. He is the Principal Investigator of the NSF-funded project "Cyberinfrastructure Expertise on High-Throughput Networks for Big Science Data Transfers."

applied
sciences

MDPI

Editorial

Special Issue "Selected Papers from the 2018 41st International Conference on Telecommunications and Signal Processing (TSP)"

Norbert Herencsar [1,*], Francesco Benedetto [2] and Jorge Crichigno [3]

[1] Department of Telecommunications, Brno University of Technology, Technicka 3082/12,
     616 00 Brno, Czech Republic
[2] Signal Processing for Telecommunications and Economics Laboratory (SP4TE), University of "Roma TRE",
     via Vito Volterra 62, 00146 Rome, Italy; francesco.benedetto@uniroma3.it
[3] College of Engineering and Computing, University of South Carolina, Innovation Center,
     550 Assembly Street, Suite 1300, Columbia, SC 29208, USA; jcrichigno@cec.sc.edu
*   Correspondence: herencsn@feec.vutbr.cz

Received: 13 May 2019; Accepted: 14 May 2019; Published: 18 May 2019

Dear Readers,

This Special Issue contains a series of excellent research works on telecommunications and signal processing; selected from the 2018 41st International Conference on Telecommunications and Signal Processing (TSP), which was held during 4–6 July 2018, in Athens, Greece. The Conference was organized in cooperation with the IEEE Region 8 (Europe, Middle East and Africa), IEEE Greece Section, IEEE Czechoslovakia Section, and IEEE Czechoslovakia Section SP/CAS/COM Joint Chapter by seventeen universities, from the Czech Republic, Hungary, Turkey, Taiwan, Japan, Slovak Republic, Spain, Bulgaria, France, Slovenia, Croatia, and Poland, for academics, researchers, and developers, and it serves as a premier annual international forum to promote the exchange of the latest advances in telecommunication technology and signal processing. The aim of the conference is to bring together both novice and experienced scientists, developers, and specialists, to meet new colleagues, collect new ideas, and establish new cooperation between research groups from universities, research centers, and private sectors worldwide. It is our great pleasure to introduce a collection of 10 selected high-quality research papers and let us briefly introduce the published works in this Special Issue.

In the first paper of this Special Issue [1], written by G. Baldini et al., authors address the problem of authentication and identification of wireless devices using their physical properties derived from their radio frequency (RF) emissions. This technique is based on the concept that small differences in the physical implementation of wireless devices are significant enough and they are carried over to the RF emissions to distinguish wireless devices with high accuracy. The technique can be used both to authenticate the claimed identity of a wireless device or to identify one wireless device among others. In the literature, this technique has been implemented by feature extraction in the 1D time domain, 1D frequency domain or also in the 2D time frequency domain. This paper describes the novel application of the synchrosqueezing transform to the problem of physical layer authentication. The idea is to exploit the capability of the synchrosqueezing transform to enhance the identification and authentication accuracy of RF devices from their actual wireless emissions. An experimental dataset of 12 cellular communication devices is used to validate the approach and to perform a comparison of the different techniques. The results described in this paper show that the accuracy obtained using 2D synchrosqueezing transform (SST) is superior to conventional techniques from the literature based in the 1D time domain, 1D frequency domain or 2D time frequency domain.

In the next paper [2], E. Erdogan et al. examine the interference alignment (IA) performance of a multi-input multi-output (MIMO) multi-hop cognitive radio (CR) network in the presence of multiple primary users. In the proposed architecture, it is assumed that linear IA is adopted at the secondary network to alleviate the interference between primary and secondary networks. By doing so, the secondary source can communicate with the secondary destination via multiple relays without causing any interference to the primary network. Even though linear IA can suppress the interference in CR networks considerably, interference leakages may occur due to a fast fading channel. To this end, the authors focus on the performance of the secondary network for two different cases: (i) The interference is perfectly aligned; (ii) the impact of interference leakages. For both cases, closed-form expressions of outage probability and ergodic capacity are derived. The results, which are validated by Monte Carlo simulations, show that interference leakages can deteriorate both system performance and the diversity gains considerably.

In the paper [3], T. Horvath et al. present a numerical implementation of the activation process for gigabit and 10 gigabit next generation and Ethernet passive optical networks (PONs). The specifications are completely different because gigabit PON (GPON), next generation PON (XG-PON) and next generation PON Stage 2 (NG-PON2) were developed by the International Telecommunication Union, whereas Ethernet PON was developed by the Institute of Electrical and Electronics Engineers. The speed of an activation process is the most important in a blackout scenario because end optical units have a timer after expiration transmission parameters are discarded. Proper implementation of an activation process is crucial for eliminating inadvisable delay. An optical line termination chassis is dedicated to several GPON (or other standard) cards. Each card has up to eight or 16 GPON ports. Furthermore, one GPON port can operate with up to 64/128 optical network units (ONUs). The results indicate a shorter duration activation process (due to a shorter frame duration) in Ethernet-based PON, but the maximum split ratio is only 1:32 instead of up to 1:64/128 for gigabit PON and newer standards. An optimization improves the reduction time for the GPON activation process with current physical layer operations and administration and maintenance messages and with no changes in the transmission convergence layer. The activation time was reduced from 215 ms to 145 ms for 64 ONUs.

In the paper [4] by D. Kubanek et al., fractional-order transfer functions to approximate the passband and stopband ripple characteristics of a second-order elliptic lowpass filter are designed and validated. The necessary coefficients for these transfer functions are determined through the application of a least squares fitting process. These fittings are applied to symmetrical and asymmetrical frequency ranges to evaluate how the selected approximated frequency band impacts the determined coefficients using this process and the transfer function magnitude characteristics. MATLAB simulations of $(1 + \alpha)$ order lowpass magnitude responses are given as examples with fractional steps from $\alpha = 0.1$ to $\alpha = 0.9$ and compared to the second-order elliptic response. Further, MATLAB simulations of the $(1 + \alpha) = 1.25$ and 1.75 using all sets of coefficients are given as examples to highlight their differences. Finally, the fractional-order filter responses were validated using both SPICE simulations and experimental results using two operational amplifier topologies realized with approximated fractional-order capacitors for $(1 + \alpha) = 1.2$ and 1.8 order filters.

The next paper [5] by J. Mucha et al. deals with Parkinson's disease (PD) dysgraphia, which affects the majority of PD patients and is the result of handwriting abnormalities mainly caused by motor dysfunctions. Several effective approaches to quantitative PD dysgraphia analysis, such as online handwriting processing, have been utilized. In this study, authors aim to deeply explore the impact of advanced online handwriting parameterization based on fractional-order derivatives (FD) on the PD dysgraphia diagnosis and its monitoring. For this purpose, 33 PD patients and 36 healthy controls from the PaHaW (PD handwriting database) are used. Partial correlation analysis (Spearman's and Pearson's) was performed to investigate the relationship between the newly designed features and patients' clinical data. Next, the discrimination power of the FD features was evaluated by a binary classification analysis. Finally, regression models were trained to explore the new features' ability to assess the progress and severity of PD. These results were compared to a baseline, which is based on conventional online handwriting features. In comparison with the conventional parameters, the FD handwriting features correlated more significantly with the patients' clinical characteristics and provided a more accurate assessment of PD severity (error around 12%). On the other hand, the highest classification accuracy (ACC = 97.14%) was obtained by the conventional parameters. The results of this study suggest that utilization of FD in combination with properly selected tasks (continuous and/or repetitive, such as the Archimedean spiral) could improve computerized PD severity assessment.

In the paper [6], Z. Galaz et al. focus on hypokinetic dysarthria, which is associated with PD, affects several speech dimensions, including phonation. Although the scientific community has dealt with a quantitative analysis of phonation in PD patients, a complex research revealing probable relations between phonatory features and progress of PD is missing. Therefore, the aim of this study is to explore these relations and model them mathematically to be able to estimate progress of PD during a two-year follow-up. Authors enrolled 51 PD patients who were assessed by three commonly used clinical scales. In addition, eight possible phonatory disorders in five vowels were quantified. To identify the relationship between baseline phonatory features and changes in clinical scores, a partial correlation analysis was performed. Finally, XGBoost models to predict the changes in clinical scores during a two-year follow-up were trained. For two years, the patients' voices became more aperiodic with increased microperturbations of frequency and amplitude. Next, the XGBoost models were able to predict changes in clinical scores with an error in range 11–26%. Although some significant correlations between changes in phonatory features and clinical scores were identified, they are less interpretable. This study suggests that it is possible to predict the progress of PD based on the acoustic analysis of phonation. Moreover, it recommends utilizing the sustained vowel /i/ instead of /a/.

In the paper [7], D. Luengo et al. describe an efficient method to construct an overcomplete and multi-scale dictionary for sparse electrocardiogram (ECG) representation using waveforms recorded from real-world patients. The ECG was the first biomedical signal for which digital signal processing techniques were extensively applied. By its own nature, the ECG is typically a sparse signal, composed of regular activations (QRS complexes and other waveforms, such as the P and T waves) and periods of inactivity (corresponding to isoelectric intervals, such as the PQ or ST segments), plus noise and interferences. Unlike most existing methods (which require multiple alternative iterations of the dictionary learning and sparse representation stages), the proposed approach learns the dictionary first, and then applies a fast sparse inference algorithm to model the signal using the constructed dictionary. As a result, the introduced method is much more efficient from a computational point of view than other existing algorithms, thus becoming amenable to dealing with long recordings from multiple patients. Regarding the dictionary construction, first all the QRS complexes were located in the training database, then authors computed a single average waveform per patient, and finally the most representative waveforms (using a correlation-based approach) as the basic atoms that were resampled to construct the multi-scale dictionary were selected. Simulations on real-world records from Physionet's PTB database show the good performance of the proposed approach.

In the work [8], written by M. Kolařík et al., a fully automatic method for high resolution 3D volumetric segmentation of medical image data using modern supervised deep learning approach is presented. Authors introduce 3D Dense-U-Net neural network architecture implementing densely connected layers. It has been optimized for graphic process unit accelerated high resolution image processing on currently available hardware (Nvidia GTX 1080ti). The method has been evaluated on MRI brain 3D volumetric dataset and computed tomography (CT) thoracic scan dataset for spine segmentation. In contrast with many previous methods, the approach is capable of precise segmentation of the input image data in the original resolution, without any pre-processing of the input image. It can process image data in 3D and has achieved accuracy of 99.72% on MRI brain dataset, which outperformed results achieved by human expert. On lumbar and thoracic vertebrae CT dataset it has achieved the accuracy of 99.80%. The architecture proposed in this paper can also be easily applied to any task already using U-Net network as a segmentation algorithm to enhance its results. Complete source code was released online under open-source license.

Technological evolution in the remote sensing domain has allowed the acquisition of large archives of satellite image time series (SITS) for Earth Observation. In this context, the need to interpret Earth Observation image time series is continuously increasing and the extraction of information from these archives has become difficult without adequate tools. In the paper [9], *A. Radoi* and *C. Burileanu* propose a fast and effective two-step technique for the retrieval of spatio-temporal patterns that are similar to a given query. The method is based on a query-by-example procedure whose inputs are evolution patterns provided by the end-user and outputs are other similar spatio-temporal patterns. The comparison between the temporal sequences and the queries is performed using the Dynamic Time Warping alignment method, whereas the separation between similar and non-similar patterns is determined via Expectation-Maximization. The experiments, which are assessed on both short and long SITS, prove the effectiveness of the proposed SITS retrieval method for different application scenarios. For the short SITS, two application scenarios, namely the construction of two accumulation lakes and flooding caused by heavy rain were considered. For the long SITS, a database formed of 88 Landsat images was used, and authors showed that the proposed method is able to retrieve similar patterns of land cover and land use.

In the last paper [10], X. Liu et al. discuss the time-interleaved analog-to-digital converter (TIADC), which is a good option for high sampling rate applications. However, the inevitable sample-and-hold (S/H) mismatches between channels incur undesirable error and then affect the TIADC's dynamic performance. Several calibration methods have been proposed for S/H mismatches which either need training signals or have less extensive applicability for different input signals and different numbers of channels. This paper proposes a statistics-based calibration algorithm for S/H mismatches in M-channel TIADCs. Initially, the mismatch coefficients are identified by eliminating the statistical differences between channels. Subsequently, the mismatch-induced error is approximated by employing variable multipliers and differentiators in several Richardson iterations. Finally, the error is subtracted from the original output signal to approximate the expected signal. Simulation results illustrate the effectiveness of the proposed method, the selection of key parameters and the advantage to other methods.

In summary, this Special Issue contains a series of excellent research works on telecommunications and signal processing. This collection of 10 papers is highly recommended and believed to be interesting, inspiring, and motivating readers in their further research.

Acknowledgments: We would like to thank all authors, the many dedicated referees, the editor team of *Applied Sciences*, and especially Xiaoyan Chen (Managing Editor) for their valuable contributions, making this special issue a success.

Conflicts of Interest: The authors declare no conflict of interest.

References

1. Baldini, G.; Giuliani, R.; Steri, G. Physical Layer Authentication and Identification of Wireless Devices Using the Synchrosqueezing Transform. *Appl. Sci.* **2018**, *8*, 2167. [CrossRef]
2. Erdogan, E.; Çolak, S.A.; Alakoca, H.; Namdar, M.; Basgumus, A.; Durak-Ata, L. Interference Alignment in Multi-Hop Cognitive Radio Networks under Interference Leakage. *Appl. Sci.* **2018**, *8*, 2486. [CrossRef]
3. Horvath, T.; Munster, P.; Oujezsky, V.; Vojtech, J. Activation Process of ONU in EPON/GPON/XG-PON/NG-PON2 Networks. *Appl. Sci.* **2018**, *8*, 1934. [CrossRef]
4. Kubanek, D.; Freeborn, T.J.; Koton, J.; Dvorak, J. Validation of Fractional-Order Lowpass Elliptic Responses of $(1 + \alpha)$-Order Analog Filters. *Appl. Sci.* **2018**, *8*, 2603. [CrossRef]
5. Mucha, J.; Mekyska, J.; Galaz, Z.; Faundez-Zanuy, M.; Lopez-de Ipina, K.; Zvoncak, V.; Kiska, T.; Smekal, Z.; Brabenec, L.; Rektorova, I. Identification and Monitoring of Parkinson's Disease Dysgraphia Based on Fractional-Order Derivatives of Online Handwriting. *Appl. Sci.* **2018**, *8*, 2566. [CrossRef]

6. Galaz, Z.; Mekyska, J.; Zvoncak, V.; Mucha, J.; Kiska, T.; Smekal, Z.; Eliasova, I.; Mrackova, M.; Kostalova, M.; Rektorova, I.; et al. Changes in Phonation and Their Relations with Progress of Parkinson's Disease. *Appl. Sci.* **2018**, *8*, 2339. [CrossRef]

7. Luengo, D.; Meltzer, D.; Trigano, T. An Efficient Method to Learn Overcomplete Multi-Scale Dictionaries of ECG Signals. *Appl. Sci.* **2018**, *8*, 2569. [CrossRef]

8. Kolařík, M.; Burget, R.; Uher, V.; Říha, K.; Dutta, M.K. Optimized High Resolution 3D Dense-U-Net Network for Brain and Spine Segmentation. *Appl. Sci.* **2019**, *9*, 404. [CrossRef]

9. Radoi, A.; Burileanu, C. Retrieval of Similar Evolution Patterns from Satellite Image Time Series. *Appl. Sci.* **2018**, *8*, 2435. [CrossRef]

10. Liu, X.; Xu, H.; Wang, Y.; Dai, Y.; Li, N.; Liu, G. Calibration for Sample-And-Hold Mismatches in M-Channel TIADCs Based on Statistics. *Appl. Sci.* **2019**, *9*, 198. [CrossRef]

applied
sciences

MDPI

Article

Physical Layer Authentication and Identification of Wireless Devices Using the Synchrosqueezing Transform

Gianmarco Baldini *, Raimondo Giuliani and Gary Steri

European Commission, Joint Research Centre, 21027 Ispra, Italy; raimondo.giuliani@ec.europa.eu (R.G.); gary.steri@ec.europa.eu (G.S.)
* Correspondence: gianmarco.baldini@ec.europa.eu; Tel.: +39-0332-78-6618

Received: 9 October 2018; Accepted: 2 November 2018; Published: 6 November 2018

Abstract: This paper addresses the problem of authentication and identification of wireless devices using their physical properties derived from their Radio Frequency (RF) emissions. This technique is based on the concept that small differences in the physical implementation of wireless devices are significant enough and they are carried over to the RF emissions to distinguish wireless devices with high accuracy. The technique can be used both to authenticate the claimed identity of a wireless device or to identify one wireless device among others. In the literature, this technique has been implemented by feature extraction in the 1D time domain, 1D frequency domain or also in the 2D time frequency domain. This paper describes the novel application of the synchrosqueezing transform to the problem of physical layer authentication. The idea is to exploit the capability of the synchrosqueezing transform to enhance the identification and authentication accuracy of RF devices from their actual wireless emissions. An experimental dataset of 12 cellular communication devices is used to validate the approach and to perform a comparison of the different techniques. The results described in this paper show that the accuracy obtained using 2D Synchrosqueezing Transform (SST) is superior to conventional techniques from the literature based in the 1D time domain, 1D frequency domain or 2D time frequency domain.

Keywords: authentication; identification; security; wireless communication; machine learning

---

1. Introduction

The authentication of wireless devices can be implemented using various approaches. Historically, authentication has been implemented using information known by the device (cryptographic information) or owned by the device (a SIM card). Another form of authentication is based on what a device is (i.e., the physical properties). A well-known example is biometric authentication such as scanning of the human eye iris used to prove the authenticity of a person. This approach has disadvantages and advantages, which are well known in literature [1]. A known advantage is that the intrinsic information of a device is difficult to clone, to steal or remove from a device. The disadvantages are that the extraction of the information can be more difficult to achieve or it can provide a statistical-only confirmation of authenticity (e.g., biometrics matching to a certain level of accuracy), rather than a precise confirmation as it is obtained with cryptographic means (e.g., the signing of a message with a key). In this paper, we investigate the identification and authentication of wireless devices using the physical properties (physical layer authentication) in their transmission components, which generate specific analogical artifacts in their RF emissions. This paper is an extended version of our paper published in the 2018 41st International Conference on Telecommunications and Signal Processing (TSP) [2].

Physical layer authentication is relevant when the authentication based on cryptographic means is difficult to achieve, either due to the limitations of IoT devices or because the context does not support an efficient distribution of cryptographic materials (e.g., keys and certificates). In the case of the Internet of Things, the authors in [3] highlighted that, although the design of authentication mechanisms based on cryptography is always desirable, it may not be applicable to several IoT scenarios because it may imply high computational cost and/or always connected trusted entities. We propose an authentication mechanism for wireless devices, which can be an alternative to cryptography or can complement and strengthen it (i.e., multi-factor authentication).

The effectiveness of this identification and authentication technique has been demonstrated in the literature in various settings and propagation conditions, and it has different names: Radiometric Identification (RAI) [4], Special Emitter Identification (SEI) [5] or Radio Frequency DNA (RF-DNA) [6], because RF fingerprints resemble the DNA of human beings. It is due to small differences in the material and the composition of the electronic circuits used for wireless transmission, which are represented in the RF signal over the air. These differences are usually not relevant to hamper the correct functioning of wireless services, but they are significant enough to identify the model or the electronic device itself uniquely [7]. The differences in the wireless transmission components, which become embedded in the RF signal, are usually stable during the transmission time (even if environment and aging effects have been reported), and they are not strongly related to the transmitted content. In many cases, RF fingerprinting in ideal wireless propagation conditions (i.e., high signal to noise radio or low fading effects) can provide a very high authentication accuracy of wireless devices. This technique requires the selection of features and classification algorithms, which are both accurate and time effective. This is often a design trade-off, because the application of sophisticated features and algorithms may require a longer processing time than the application of simple features and algorithms, even if the former provide a better identification accuracy. RAI has been applied to a large variety of electronic devices and wireless standards including WiFi [8], ZigBee [9], WiMAX [6] and Global System for Mobile Communications (GSM) in [10]. An analysis of existing literature in this area is reported in Section 2.

In the rest of this paper, the terms identification and verification are used in a manner consistent with the sources in literature: Authentication is the process of confirming the claimed identity of a wireless device. In this case, the RF fingerprints of a wireless device, which claims to be the device A, are compared to the previously recorded (e.g., after the product phase and before marked deployment) fingerprints of the device A using the techniques described in this paper. Authentication (also called verification in other sources) is based on a binary classification. Identification is the process where the recognition system determines a wireless device's identity by comparing the device fingerprints with reference fingerprint templates for all known devices in the test set. Identification requires a one-to-many comparison and multi-classification algorithms.

A potential application scenario is where a wirelessly-connected central node can accept data only from authenticated wireless devices, but the computing capabilities of the wireless devices are not sufficient to support cryptographic-based authentication or the cryptographic material (e.g., private keys) in the wireless device cannot be adequately protected for cost reasons [3]. In this scenario, before the deployment of the wireless device, its wireless signals are analyzed and recorded by the central node in order to compute the RF fingerprints [7,8]. In a subsequent phase, the authentication is performed using the approach described in this paper. In this scenario, the authentication accuracy must be maximized to reduce the number of false alarms, and the processing time must be minimized. Another scenario is the fight against the distribution of counterfeit electronic products, where the RF fingerprints can be used to distinguish between counterfeit and proper products because the fingerprints will be different in counterfeit products of the same model [11].

A significant challenge for researchers both for identification and authentication is the definition of features or signal representations, which can be used to detect the differences and authenticate the wireless devices. A common strategy is to extract statistical features from the RF signal and then use a machine learning algorithm to classify the obtained set of features and correlate them to the identity of the wireless device. There is an extensive literature on the selection of different statistical features for RAI including variance, entropy, skewness, kurtosis and others [8,12].

Our contribution: Following the recent trend of using 2D time frequency domain representations of the signal emitted by a wireless devices for the purpose of RAI, in this paper, we apply the SST algorithm to the problem of physical layer authentication and identification. In particular, we use a Wavelet Synchrosqueezed Transform (WSST) based on the Continuous Wavelet Transform (CWT). In the rest of this paper, such a transform will be called Wavelet Synchrosqueezed Transform (WSST). The WSST algorithm has been applied as a time frequency analysis tool for different kinds of nonlinear signals, such as the vibration signal in [13], for the detection of frequency shifting of earthquake damaged structures in [14] and to Time Frequency Domain (ECG) signal analysis in [15], but it has not been applied to the problem of physical layer authentication to the knowledge of the authors. In a similar way to the approach adopted by the other authors, this paper makes a comparison of the performance of WSST to methods based on the 1D time domain, 1D frequency domain and 2D Short Time Fourier Transform (STFT). The performance is evaluated on an experimental dataset of RF emissions transmitted by 12 wireless devices (i.e., GSM mobile phones) collected by the authors in a test bed environment.

As mentioned before, this paper is an extended version of our paper published in the 2018 41st International Conference on Telecommunications and Signal Processing (TSP) [2]. The following improvements and extensions have been made:

- A more extensive review of the related work in the literature for the problem of physical layer authentication (e.g., RAI, SEI or RF-DNA) and on the application of WSST.
- In the initial paper, only the identification problem was analyzed. In this paper, we also evaluate the verification/authentication problem.
- In the initial paper, only the K nearest neighbor with K = 1 was used to compare the performance of WSST with the other representations. In this paper, the authors have compared the results from different machine learning algorithms.
- A more extensive analysis and optimization of the hyperparameters of WSST and machine learning algorithms is performed in this paper.

Structure of this paper: The structure of the paper is the following: Section 2 provides a review of the related work on physical layer authentication. Section 3 provides a definition of the WSST. Section 4 provides a description of the methodology used to collect the RF signals and the test bed. Section 5 provides the experimental results and the related analysis where a comparison of different statistical features and machine learning algorithms is performed. In the first part, the results for the identification of wireless devices are provided. The second part presents the results for the verification or authentication of a wireless device. Finally, Section 6 concludes this paper.

2. Related Work

The concept of RAI is not new, as it was first proposed in the military domain to detect and identify hostile sources of RF emissions like radar systems, and it can be considered part of SIGnals INTelligence (SIGINT) or Measurement and Signature Intelligence (MASINT) [16].

More recently, the progress in electronic equipment for RF signal collection and analysis allowed the use of RAI in non-military contexts.

In some initial works [8,12], physical layer authentication was performed by analyzing the RF signal in space in the time domain or the frequency domain and by extracting statistical features or other signal characteristics. The RF devices used in the test were consumer mass market devices based on WiFi standards. The classification was performed both for the amplitude and phase components in the time domain by exploiting the non-content parts of the bursts defined in the wireless standard and transmitted by the device. The parts of the burst not related to the content must be used because the content (e.g., data, voice) can introduce a bias. In other words, the classification could be performed on the content rather than the physical properties of the device itself. Two outcomes already appeared from these studies and other similar studies [17]. The first is that bursts are usually composed of a transient element and a steady element (e.g., the preamble). Then, a design choice appears in the design of the physical layer authentication process. As described in [17,18], a transient signal can be described as a short signal (typically lasting a few microseconds) that occurs during transmitter power-on. It is noted in [17,18] that the capture and digitization of the transient signal requires very high oversampling rates and sophisticated and expensive receiver architectures. In contrast to the transient signal, the steady-state signal can be much longer than the transient part of the burst, thus providing more information for classification purposes. The choice on which element should be used for classification depends on the wireless standard and the test bed equipment. This aspect appears in this paper, as well, where an empirical analysis has been performed on the entire burst (i.e., the non-content portion) to identify the more suitable element. In this paper, the transient is proven to provide the best performance.

Recent papers have shown that other representations of the signal can be more effective for physical layer authentication than the specific time domain or frequency domain. 2D time frequency representations have been recently used for radiometric identification in [6], where a joint time frequency Gabor Transform (GT) and Gabor–Wigner Transform (GWT) features have been used for WiMAX wireless devices. The assessments in [6] show that Gabor-based RF-DNA fingerprinting is much more effective than either 1D time domain or frequency domain methods.

As described before, WSST has not been applied (until this paper) to the problem of physical layer authentication or identification, but it has been used in other contexts. We also note that the concept of the physical layer authentication of electronic devices is not only limited to RF devices, but it can also be applied to other components like MEMS [19].

The authors have used the synchrosqueezing transform method in [20] to detect gearbox fault signals in wind turbines. The paper presents an improved diagnosis method for wind turbines via the combination of the synchrosqueezing transform and local mean decomposition. In the area of geophysics, the authors in [21] have compared different time frequency techniques, and they have highlighted the advantages for interpretations for seismic signals in the areas of speech signals and volcanic tremors. The authors in [14] have applied synchrosqueezing to interpretations of seismic signals. The results of the paper shows that synchrosqueezing outperforms other time frequency transforms like the Gabor–Wigner transform, Wigner–Ville distribution and S-transform. Both [14,21] used synchrosqueezing based on CWT (i.e., WSST). In particular [14] showed that synchrosqueezing outperforms its CWT basis. These results support the choice by the authors of this paper to use a synchrosqueezing based on CWT. To summarize: WSST provides better frequency localization and good time support in comparison with the 2D Time Frequency Domain (TFD) such as Wigner–Ville distribution (WVD) and GWT used in [6]. In relation to Empirical Mode Decomposition (EMD) used in [22] as part of Hilbert–Huang Transform (HHT), EMD lacks solid mathematical foundations, though it is attractive due to its simplicity and effectiveness, as discussed in [23]. Extensions of the synchrosqueezing algorithm in combination with other time frequency representations apart from CWT are also possible. In [24], the authors developed the Synchrosqueezing Generalized S-Transform (SSGST) for the analysis of field seismic data. Similar approaches can be used for future extensions of this paper.

As described before, WSST has not been applied until this paper to physical layer authentication or identification, which is the novelty of this paper.

3. Definition of the Wavelet Synchrosqueezing Transform

The SST is a time frequency analysis method. It is a special case of the reallocation method whose aim is to "sharpen" a time frequency representation by allocating its value to a different point in the time frequency plane [25]. This reassignment compensates for the spreading effects caused by the mother wavelet, and it is performed only in the frequency direction, thus preserving the time resolution of the signal.

The starting point in the application of the synchrosqueezing algorithm in this paper is the continuous wavelet transform of the input signal from which instantaneous frequencies are extracted. After the extraction, an instantaneous frequency value is reassigned to a single value at the centroid of the CWT time frequency region. This final part corresponds to the squeezing of the CWT, which results in a sharpened output.

Therefore, following the description above, the synchrosqueezing algorithm can be summarized as three main steps. The first one is the application of the CWT to the original signal s, which is given by:

$$W_s(a,b) = \int s(t)a^{-\frac{1}{2}}\bar{\psi}\left(\frac{t-b}{a}\right)dt, \tag{1}$$

where $\psi(t)$ is the mother wavelet function and the bar denotes the complex conjugate, a is the scale factor and b the translation.

The extraction of the instantaneous frequencies from W_s, the second step, is done using a phase transform proportional to the first derivative of the CWT. Therefore, given a phase transform ω_s, the instantaneous frequencies can be expressed by [25]:

$$\omega_s(a,b) = -i(W_s(a,b)^{-1})\frac{\partial W_s(a,b)}{\partial b} \tag{2}$$

Finally, the resulting wavelet coefficients containing the same instantaneous frequencies can be combined. This corresponds to the application of the SST that for a given set of wavelet coefficients $W_s(a,b)$ is expressed as follows:

$$SST(\omega_1,b) = \sum W_s(a_k,b)a^{-3/2}(\Delta a)_k, \tag{3}$$

where $(\Delta a)_k = a_k - a_{k-1}$ and ω_1 are the frequency bins. Since the SST inherits the invertibility property of the CWT, the signal can be reconstructed. In this paper, the focus is not on the signal reconstruction.

The WSST is applied to each of the bursts collected from the wireless devices as described in Section 4.2.

4. Materials and Methods

4.1. Materials

The material used in the experiment are the following:

- Twelve wireless devices (i.e., GSM mobile phones) of 4 different brands (Sony Experia, HTC One, Samsung S5 and Apple iPhone): three phones were used for each of the four models.
- An OpenBTS software was used to activate the GSM communication from each of 12 wireless Devices Under Test (DUT) and to generate the signal in space.
- A Universal Software Radio Peripheral (USRP) type N200 receiver (RX) configured with a sampling rate of 1 MHz is used to collect the signal in space from each of the 12 transmitting wireless devices. The wireless devices were linked with a GSM base station implemented using OpenBTS running on a USRP N200. The base station and digitizer were fully disciplined and synchronized using a Global Positioning System (GPS) receiver with a Global Positioning System Disciplined Oscillator (GPSDO). To support repeatability and stability, the same USRP digitizer, as well as the same base station were used for all tests. All tests were performed after a minimum half hour lock after the Global Navigation Satellite System (GNSS) receiver was properly synchronized

on at least four satellites. In the Software-Defined Radio (SDR), the signal is received and down-converted using a WBX, flexible frequency front-end compatible with the USRP with a passing bandwidth of 40 MHz and tuning capabilities from 20 MHz to 2 GHZ. Then, the signal is digitally down-converted by the built-in Digital Down Converter (DDC) employing half band and Cascaded Integrator Comb (CIC) decimators from 100 MHz to 1 MHz.

A summary of the parameters and settings used for the collection of the signal in space is provided in Table 1.

Table 1. Parameters for signal collection.

Sampling frequency	1 MS/s IQ
Sample recording time	60 s
Downlink frequency	935.2 MHz
Uplink frequency	890.2 MHz
Synchronization	GPS only, using GPSDO (min 4 satellites, min 30 min lock)
Distance between DUT and RX	0.84 m
USRP gain	5
GSM arfcn	1
OpenBTS version	3.1.3

4.2. Methodology

The overall methodology for the classification of the wireless devices using the WSST is shown in Figure 1.

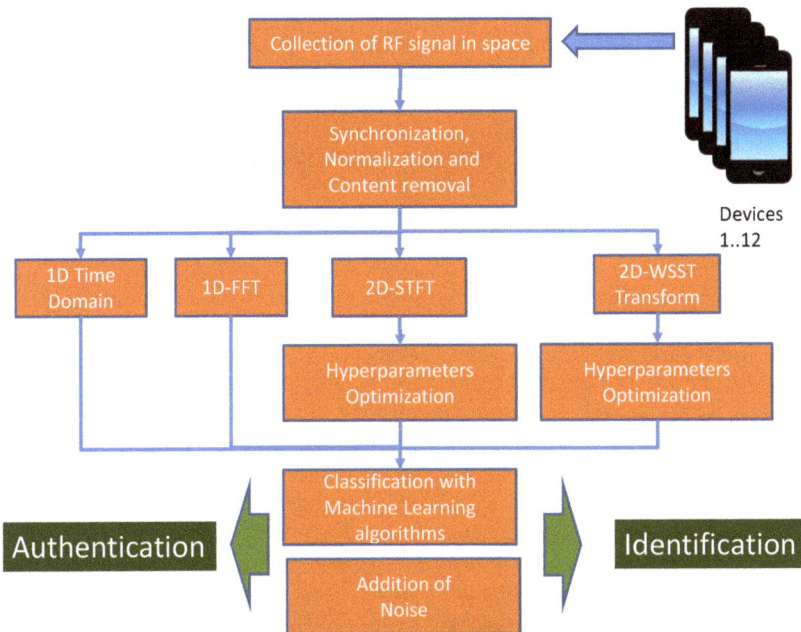

Figure 1. Overall methodology used in this paper for the identification and authentication of the wireless devices.

The methodology to generate the fingerprints from the RF signals consisted of the following steps:

1. Each of the 12 wireless devices (i.e., GSM mobile phones) were activated, and they started to transmit in a controlled environment where a specific transmission channel is used.
2. The signal in space from the wireless devices was collected using the SDR USRP type N200 receiver with the configuration described in the previous section.
3. The real-valued signal samples were sampled directly in In-phase and Quadrature components (IQ) format and then synchronized and normalized offline to extract the burst of traffic associated with each payload. For each wireless device, a set of 800 bursts was processed for a total of $800 \times 12 = 9600$ bursts.
4. From each burst, the content (payload data associated with the voice communication) was removed. In this way, each burst has only the transients and the preamble, which is the same for all the bursts and all the devices. After the removal, each burst is around 130 samples in length. An image of the normalized magnitude of the GSM bursts after synchronization, normalization and content removal is presented in Figure 2, where the differences among wireless devices can be seen especially near the transients. We note that the granularity of the digitized signal (i.e., number of samples for each burst) used for identification is quite inferior to the granularity of the datasets used by other authors [6,12,18], where a very high identification accuracy is obtained. This is intentional because the objective of this paper is to show that the application of WSST provides a better performance than conventional techniques from the literature in difficult datasets like the one used in this paper.
5. WSST was applied to each of the bursts recorded in the test bed. A representation of the burst is shown in Figure 3 for the Morlet mother wavelet, the scale factor $a = 10$ and the entire GSM burst.
6. Different machine learning algorithms are used for classification to implement identification and authentication: Support Vector Machine (SVM), K Nearest Neighbor (KNN) and decision trees. A 10-fold method was used for all the machine learning algorithms. Each collection of statistical fingerprints is divided into ten blocks. Nine blocks from each device are used for training, and one block is held out for classification. The training and classification process is repeated ten times until each of the ten blocks has been held out and classified. Thus, each block of statistical fingerprints is used once for classification and nine times for training. Final cross-validation performance statistics are calculated by averaging the results of all folds.
7. Optimization of the hyperparameters: In the application of WSST, the scale factor a from Equation (1) is used as a hyperparameter. The window size both for WSST and STFT is set to 10 because this is roughly the size of the transient of the burst. Each of the machine learning algorithms can be optimized on the basis of specific parameters (e.g., K index for the KNN algorithm). The optimization of these parameters is described in detail in Section 5.
8. Metrics definition: For identification, the overall identification accuracy is used as a metric to evaluate the performance of the identification. The accuracy is defined as the sum of the True Positives (TP)s and True Negatives (TN)s divided by the number of all samples. To show the relevance of False Positives (FP) and False Negatives (FN) in the final results, a confusion matrix is also provided. For verification and authentication, the adopted metrics are the Receiver Operative Characteristics (ROC) and the Equal Error Rate (EER), which is the point on the ROC where false positive and false negative rates are equal. The value of the X axis is used to determine the EER in this paper.
9. Impact of noise. Additive White Gaussian Noise (AWGN) is added to the original data sample to simulate the presence of noise in the environment. This is a common practice in the literature [26,27] to evaluate the performance of the classification algorithm in terms of identification accuracy for different values of Signal Noise Ratio (SNR).

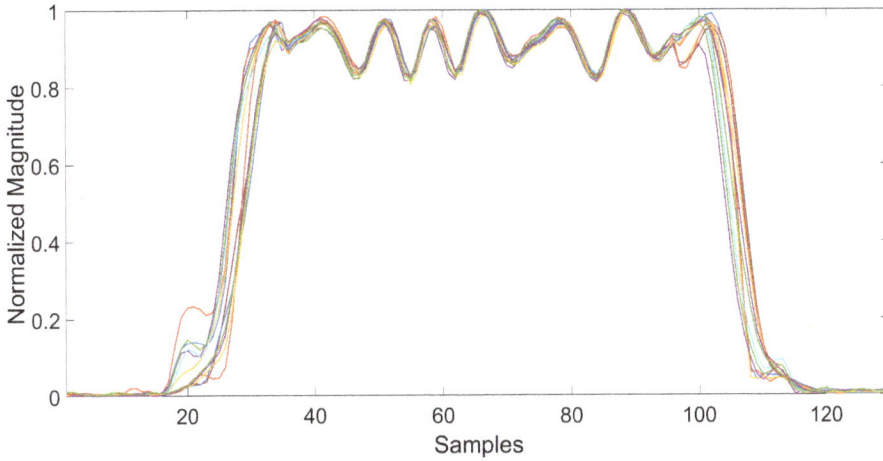

Figure 2. Normalized magnitude of 12 bursts (one for each wireless device).

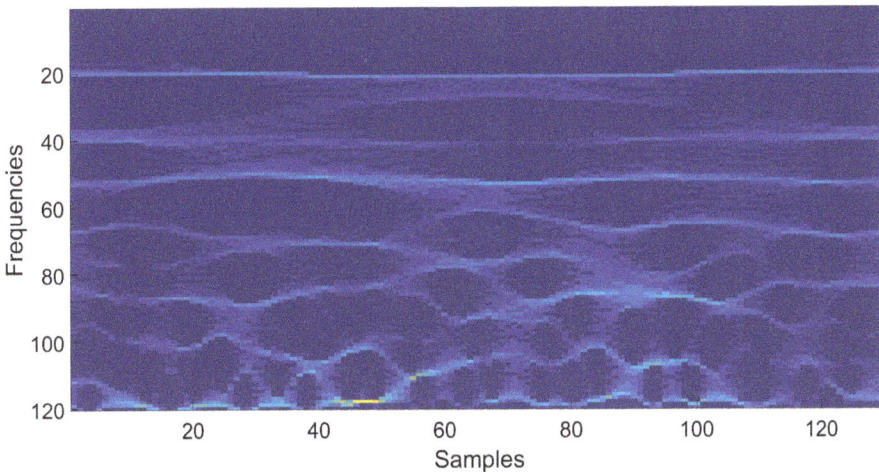

Figure 3. Spectrogram of the synchrosqueezing transform of the digitized signal from the wireless devices.

5. Discussion of the Results

5.1. Identification

In this section, the identification problem is evaluated. The first sub-section is focused on the optimization of the hyperparameters. The second section evaluates the performance in the presence of AWGN.

5.1.1. Optimization of the Hyperparameters

The WSST has different parameters (e.g., degrees of freedom), which can be optimized for the specific problem to be addressed. In this paper, we choose the two following parameters for our analysis: (a) the number of octaves and (b) the mother wavelet: Morlet or bump. Another parameter is the identification of a specific segment of the digital representation of the signal, which can be more appropriate for classification. Signals captured from the wireless devices are usually represented as bursts, which are repeated in time, and they are usually composed by a transient portion and a steady portion. The digital representation of the GSM burst signal used in this work is provided in Figure 2. As described in the related work, some papers focus only on the transient phase of the digital signal, while other papers focus on the steady part. In this paper, the optimal segment is determined in an empirical way. In addition, the WSST can be applied to the entire complex digital output from RF emissions or only the magnitude (see Figure 2) and phase components in the time domain. Then, we have four degrees of freedom, which must be evaluated to determine the optimal values. Because the analysis of all four degrees of freedom (1 magnitude or phase, 2 segment, 3 scale factor, 4 base wavelet) will be too complex to pursue (the optimization should be conducted in a four-dimensional space), a piecewise approach is adopted. In the first step, the segment and the scale factor a from Equation (1) are optimized. A simple KNN neighbor algorithm with K = 1 is used to avoid the process of the optimization of the hyperparameters of the machine learning algorithm.

The next step is to optimize the scale value a and the segment of the burst at the same time. In comparison to the preliminary paper [2], where the optimization was performed with a specific scale factor value ($a = 10$), in this paper, we conduct a more extensive analysis on the bidimensional space of the WSST scale factor and the segment index.

In this paper, the WSST representation is divided into 10 segments along the X axis (e.g., the samples of the burst). The reason for this choice is based on two reasons: (1) the visual analysis of Figure 2, which shows that the strongest deviations from the ideal shape of the GSM burst are in the transient region, and (2) the results of [28], which showed an improved accuracy using transients rather than the steady portion of the GSM burst.

The result is shown in Figure 4, where it can be seen that the optimal segment is three, which is consistent with the original paper [2], while the optimal scale factor is 20, which provides a slighter improvement to the values of 10 adopted in the original paper. We note that the choice of the segment is the most relevant element to contribute to the performance.

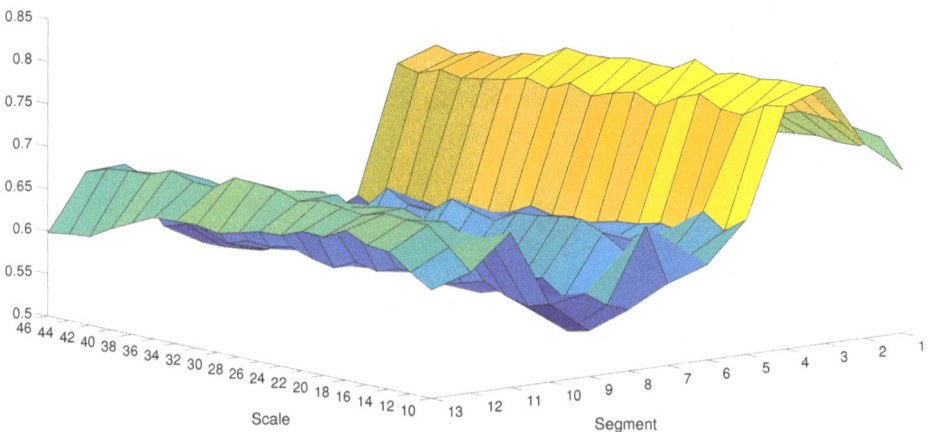

Figure 4. Accuracy for different values of segment Id and the scale factor value.

Then, a comparison of the Morlet and bump wavelets was performed, and the results are shown in the bar graph of Figure 5. The evaluation was performed using the KNN machine learning algorithm for different values of K (from one to 10). The result showed that the accuracy obtained using the Morlet wavelet in WSST was significantly better than the bump wavelet for all the considered values of K. The results are consistent with the initial findings of [2].

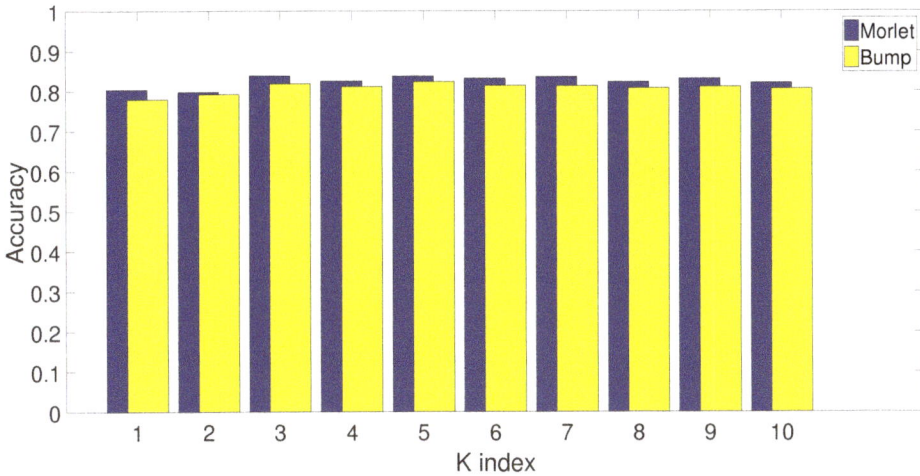

Figure 5. Comparison of the performance accuracy with KNN using Morlet and bump wavelets for different values of K.

A comparison of the amplitude and phase components of the digitized signal was performed, and the results are shown in the bar graph of Figure 6. The evaluation was performed using the KNN machine learning algorithm for different values of K. The results show that the best accuracy was obtained using the magnitude component in the time domain. The results are consistent for all the considered values of K, and they also confirm the initial findings of [2].

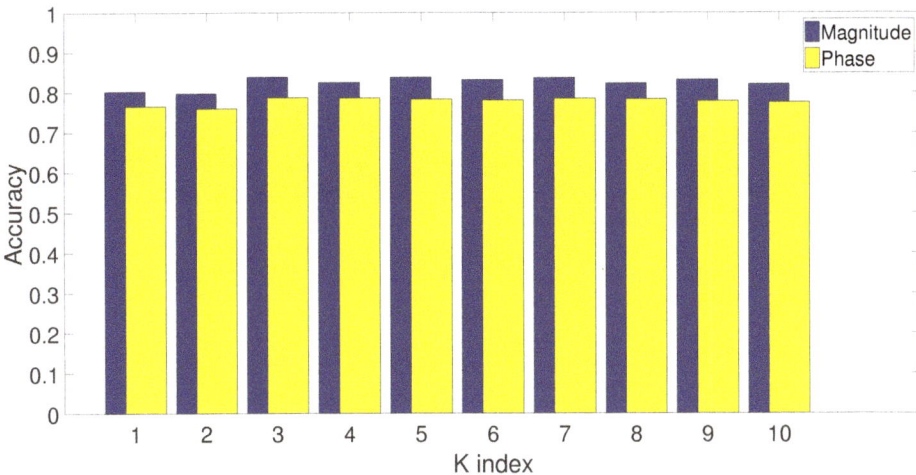

Figure 6. Comparison of the performance accuracy with KNN for the magnitude and phase components of the signal in the time domain for different values of K.

Finally, the optimization of the hyperparameters of the machine learning algorithms was performed. In this paper, we compared the results of three different machine learning algorithms: SVM (using a Radial Basis Function (RBF) kernel), KNN and decision trees.

The optimization of the SVM was performed for two hyperparameters: the scaling factor of the RBF kernel and the penalty factor C [29]. A grid approach was adopted to calculate the hyperparameters in the range of 2^1 to 2^{12} for both hyperparameters. The optimization process was applied to each of the 10 folds, and the results were averaged. The results of the optimization analysis are shown in Figure 7 for a specific fold (where the specific fold result was the same of the average value), where the optimal value is highlighted with a black circle.

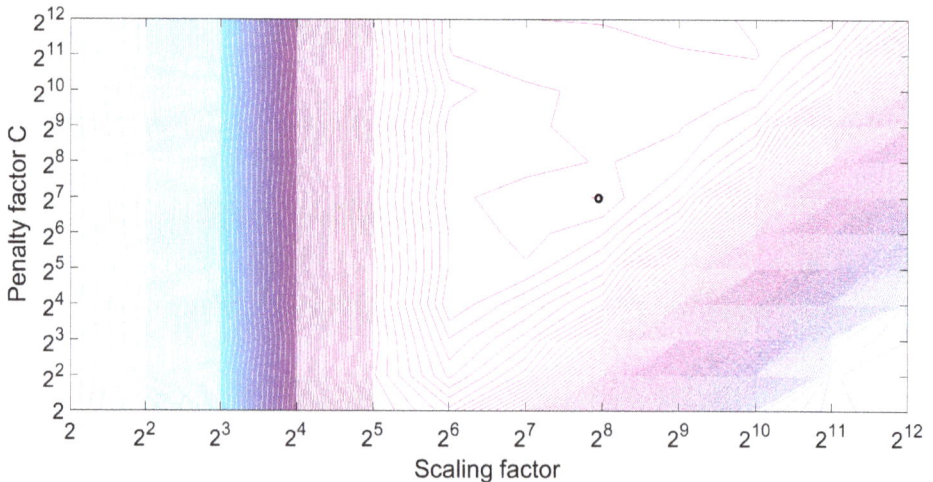

Figure 7. Optimization of the scaling factor and the penalty factor for SVM.

For WSST and KNN, the optimization results can also be seen from the previous Figures 5 and 6 where K (K = 3) for the KNN algorithm is the optimal value. In the application of the decision tree algorithm, the hyperparameter to optimize is the maximum number of splits for each branch. The optimal value is 12 for WSST.

The optimization process was repeated for the other representations of the digital signal (time domain, frequency domain and STFT) by averaging the results from each fold. The summary of the optimal values is shown in Table 2 for all the representations and for all the machine learning algorithms. The optimal values and machine learning algorithms were used in the subsequent sections of this paper.

Table 2 shows that the technique based on WSST significantly outperformed the other representations of the digital signal (time domain, frequency domain and STFT) for all the used machine learning algorithms. In other words, the superior performance of WSST was proven in a consistent way regardless of the machine learning algorithm.

Table 2 also provides the computation time requested by each machine learning algorithm and for each representation. The times were also based on averaging the results as written above. The computation time was expressed as the ratio with the smallest computation time (i.e., decision tree with time domain representation). The computation time is another metric (together with the identification accuracy) that a user can evaluate to decide the best technique in the practical deployment of the physical layer authentication approach described in this paper. It can be seen that the WSST required a larger computation time than the other techniques, while the most efficient technique was the time domain representation. The results from Table 2 were obtained on the experimental dataset where the RF signal was collected by the SDR in Line Of Sight (LOS) conditions and for high values of SNR. In the next subsection, we evaluate the performance of the WSST-based technique in the presence of AWGN for different values of SNR.

Table 2. Optimal values for the hyperparameters of the machine learning algorithms with the related identification accuracy and computing time ratio.

Machine Learning Algorithm	Optimal Values	Identification Accuracy	Computing Time Ratio
WSST	===	===	===
SVM	$C = 2^7, \gamma = 2^8$	**0.9236**	10
KNN	$K = 3$	0.8388	8.57
Decision Tree	$N_s = 12$	0.8225	7.28
STFT	===	===	===
SVM	$C = 2^7, \gamma = 2^9$	**0.8503**	4.64
KNN	$K = 17$	0.706	4.285
Decision Tree	$N_s = 18$	0.753	4
1D Frequency domain (magnitude component)	===	===	===
SVM	$C = 2^{12}, \gamma = 2^8$	0.753	4.35
KNN	$K = 3$	**0.7558**	1.785
Decision Tree	$N_s = 18$	0.7352	1.428
1D Time domain (magnitude component)	===	===	===
SVM	$C = 2^{12}, \gamma = 2^8$	0.7558	3.64
KNN	$K = 9$	**0.7910**	1.1
Decision Tree	$N_s = 17$	0.7265	1

5.1.2. Comparison of the WSST-Based Approach with Other Signal Representations in the Presence of AWGN

In the preliminary paper [2], an initial comparison was performed for different values of SNR using KNN as a machine learning algorithm with K = 1. The result has shown that the WSST-based techniques outperformed the other techniques for medium and higher values of SNR. The robustness of the identification algorithm in the presence of noise is an analysis commonly performed in the literature [26,27], to evaluate the impact of attenuation due to the propagation loss or the presence of obstacles (e.g., walls).

In this paper, we present an updated version of the comparison of the performance for the different representations, using the optimized hyperparameters from Table 2. The result is shown in Figure 8, where it can be seen that WSST still outperformed the other techniques for high values of SNR, while STFT had a better performance for lower values of SNR. The black line identifies the results from the preliminary paper [2] using the WSST technique. We note that the optimization of the parameters provided a significant improvement for all the values of SNR.

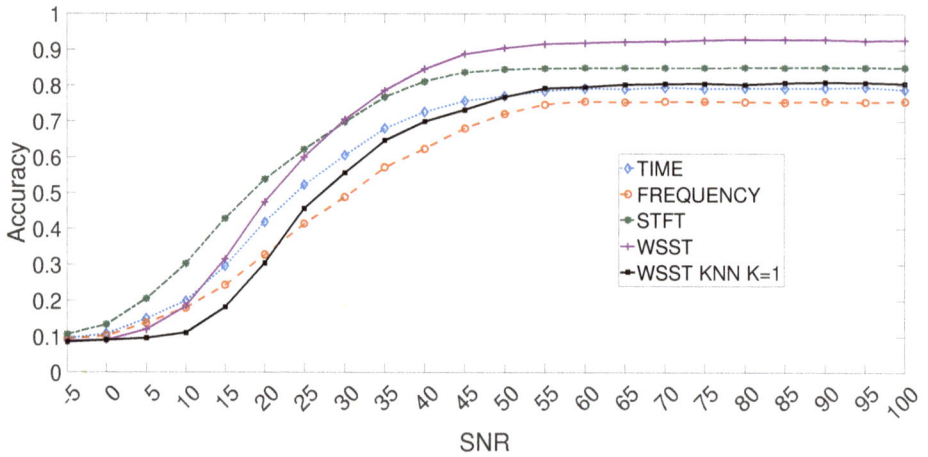

Figure 8. Accuracy for different values of SNR (expressed in dB) for 1D TD, 1D FD, 2D STFT and 2D WSST using the optimal selection of hyperparameters defined in Table 2. The black line is inserted to show the comparison with the application of the WSST technique with KNN (K = 1).

Finally, the confusion matrices were calculated. Two figures are presented: The original confusion matrix from [2] is presented in Figure 9, where KNN with K = 1 was used. A new confusion matrix is provided in Figure 10, which is based on the optimal values reported in Table 2: SVM with $C = 2^7$, $\gamma = 2^8$. Both confusion matrices have been calculated at SNR = 50. The lower accuracy for the last three phones (10 to 12) was due to the strong similarity of the RF emissions of the Apple (i.e., iPhone) models.

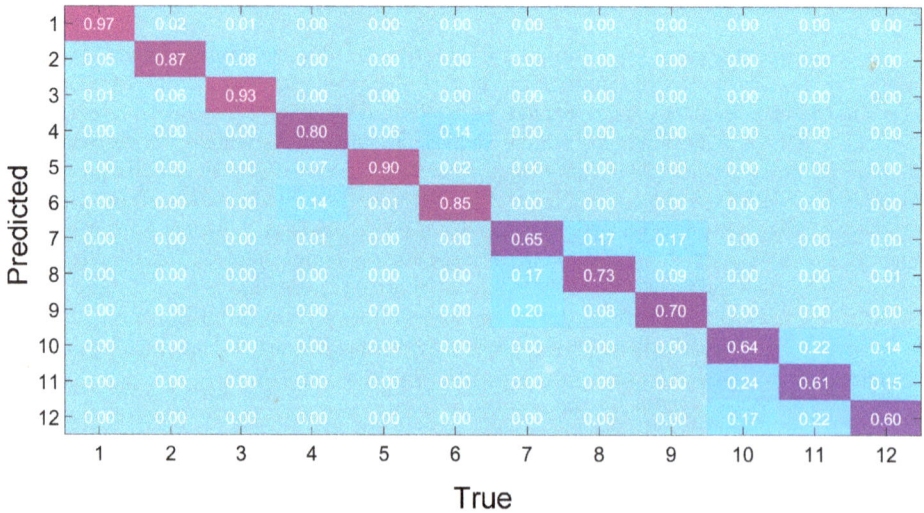

Figure 9. Confusion matrix for WSST calculated with the KNN machine learning algorithm with K = 1 at SNR = 50 dB.

	1	2	3	4	5	6	7	8	9	10	11	12
1	98.25	1.375	0	0	0	0	0	0	0	0	0	0
2	1.75	93	5.375	0	0	0.125	0	0	0	0.125	0	0
3	0	5.625	94.75	0	0	0	0	0	0	0	0	0
4	0	0	0	94.875	4	4.125	0	0	0	0	0	0
5	0	0	0	2.125	96	0.625	0	0	0	0	0	0
6	0	0	0	3.125	0	95.125	0	0	0	0	0	0
6	0	0	0	0	0	0	82	11.375	11.875	0	0	0
8	0	0.125	0	0	0	0.125	8.25	82.25	4	0	0.125	0
9	0	0	0	0	0	0	9.75	6.375	84.25	0	0	0
10	0	0	0	0	0	0	0	0	0	84.5	6.625	10.75
11	0	0	0	0	0	0	0	0	0	5.625	84.75	7.25
12	0	0	0	0	0	0	0	0	0	9.75	8.5	82.125

Predicted (vertical axis) / *True* (horizontal axis)

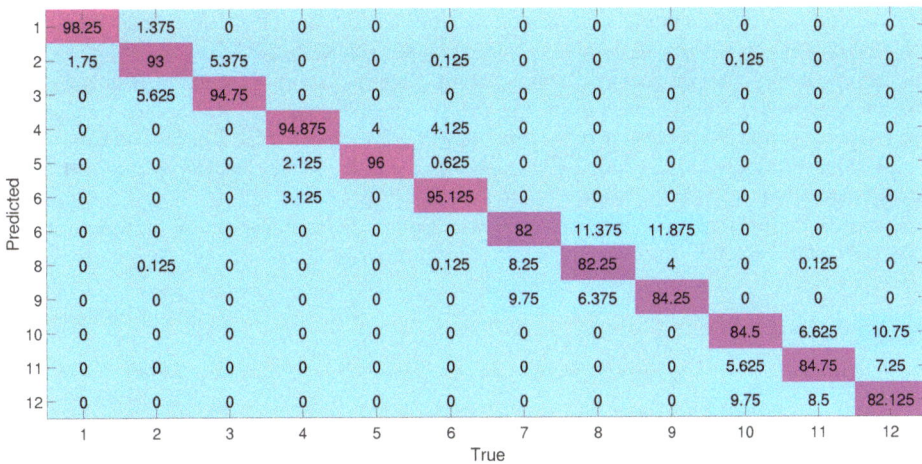

Figure 10. Confusion matrix for WSST calculated with the optimized values of Table 2 and SNR = 50 dB.

5.2. Authentication

In this section, we investigate the performance of WSST in comparison to the other techniques for the problem of authentication. The confusion matrix provided in Figure 9 shows that the final three mobile phones were the most difficult to distinguish. Then, in this section, we focus on the authentication of two of these phones (Phone 10 and Phone 12).

As for the previous results, we used the optimal values of WSST from the previous section and the SVM machine learning algorithm identified in Table 2.

Figure 11 shows the results for the EER metric for the authentication of Phone 10 against Phone 12. The scenario is that Phone 12 claimed to be Phone 10 and the EER was used to measure how well the algorithm was able to authenticate Phone 10. Figure 11 shows the comparison of the performance of WSST against the other representations. The results are consistent with the analysis on the identification accuracy: the WSST-based approach provided a higher authentication accuracy (lower EER for medium/high values of SNR) in comparison to the other representations. For lower values of SNR, the other techniques performed better than WSST, which was consistent with the findings of the previous section where STFT performed better than WSST at lower SNR. On the other side, a low accuracy limits the practical use of this technique. Then, for low values of SNR, all the techniques would have a limited use (as also reported in the literature), and filtering techniques should be used to remove the presence of noise [7].

Figure 12 shows the ROCs for the WSST-based approach with SVM for different values of SNR. The results are consistent with the previous Figure 11 because higher values of SNR generated ROC curves that showed a better authentication.

5.3. Authentication of Unknown Devices

In the previous section, we provided the results with a closed set of devices used both in training and testing where a 10-fold cross-validation was applied. This section deals with the identification of unknown devices, which were not in the training set: What if unknown devices, which are not used in training, tried to identify or authenticate themselves? To address this question, we have analyzed two cases: (a) when the unknown device is of the same model of some of the devices in the training set; in this case, the algorithm should predict that it is of a specific model, but not of the other models; (b) when the unknown device is of a different model from the models already present in the training set. In this case, the algorithm should determine that it is completely unknown. To implement the first case (a), we have first decreased the initial training set to 11 devices. Device 12 (i.e., an Apple iPhone) has been removed from the training set and then tested against the training model built with the other 11 devices, where nine devices belonged to three different models (i.e., three Sony Experia devices, three HTC One devices and three Samsung S5 devices) and two devices belonged to the same iPhone model. The expected result was that the classification algorithm should predict that the unknown phone did not belong to the three models Sony Experia, HTC One and Samsung S5. It should predict that it was an iPhone model. On the other side, the classification algorithm should also predict that the unknown device was not one of the two specific iPhones (called iPhone 1 and iPhone 2 in the rest of this section). In an ideal case, the classification results should indicate a random choice (50% probability) that the unknown device was one of the two iPhones.

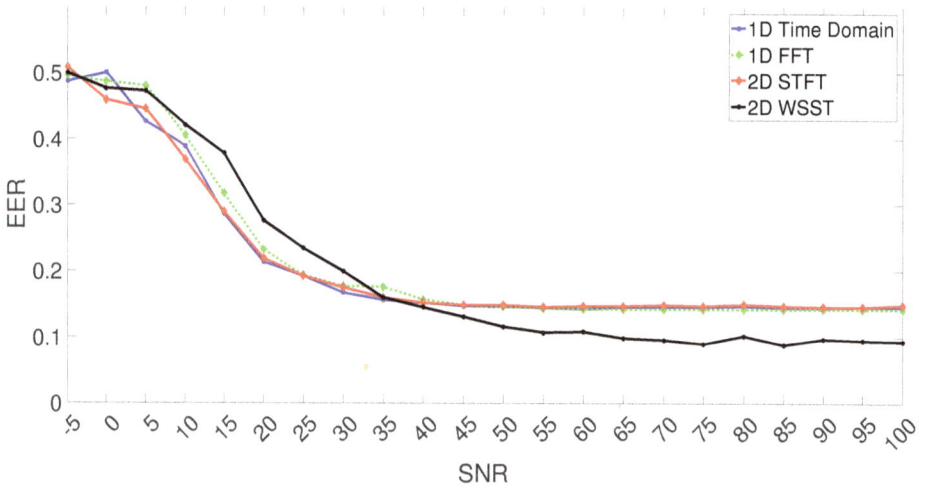

Figure 11. Evaluation of the authentication performance in the presence of noise between Device 10 and Device 12 using EER for the different techniques. SNR is expressed in dB.

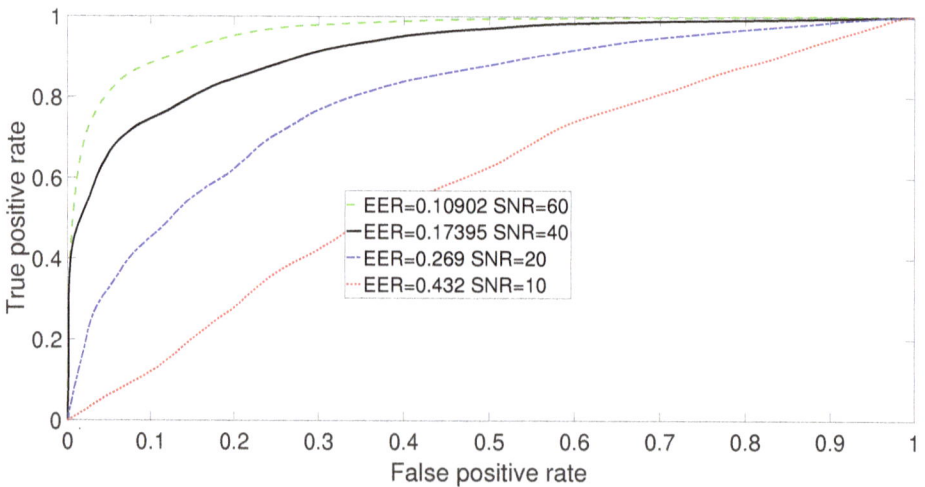

Figure 12. ROCs for the authentication performance between Device 10 and Device 12 for specific values of SNR. SNR is expressed in dB.

The results for Case (a) are shown in Table 3 for all the different representations. The SVM algorithm has been used in this case. For completeness, the classification has been repeated for different values of SNR. The predicted percentage in the first column shows the predictions against the devices of the three models (Sony Experia, HTC One and Samsung S5), and the second and third columns show the predicted percentage against the iPhone 1 and iPhone 2.

Table 3. Predicted percentage with an unknown device.

Technique	Predicted Percentage for the Three Models	Predicted Percentage for iPhone 1	Predicted Percentage for iPhone 2
SNR = 100	===	===	===
WSST	0	0.51	0.49
STFT	0	0.72	0.28
FFT	0	0.57	0.43
TIME	0	0.77	0.23
SNR = 10	===	===	===
WSST	0.45	0.18	0.37
STFT	0.3	0.21	0.49
FFT	0.44	0.18	0.38
TIME	0.26	0.26	0.48
SNR = 0	===	===	===
WSST	0.8	0.09	0.11
STFT	0.71	0.1	0.19
FFT	0.77	0.1	0.13
TIME	0.69	0.15	0.16

The results confirm the initial assumptions: for high values of SNR, the algorithm successfully predicted that the unknown device was not one of the three models (i.e., Sony Experia, HTC One and Samsung S5) as the predictions were zeros for all the techniques (first column in Table 3). The predicted percentage was not zero for the iPhone 1 and iPhone 2 as the algorithm predicted that the unknown device was of type iPhone. We note that the WSST-based technique provided the best predictions, because the predicted percentage was almost equally divided between iPhone 1 and iPhone 2. In other words, the algorithm did not associate the unknown device to a specific known device of the same model. The other techniques (STFT and TIME) predicted that the unknown device was more similar to iPhone 1, which was an inaccurate prediction. The FFT-based technique provided similar results (but slightly worse) to the WSST-based technique. In the presence of noise (low SNR values), the prediction degraded significantly, as expected from the results in Section 5.1.2. For SNR = 0, the algorithm was not able to provide accurate predictions for all techniques. In Case (b), the training set was composed only of the nine devices of the three models (i.e., Sony Experia, HTC One and Samsung S5), and the algorithm was tested against each of the iPhone devices. We obtained a predicted percentage of 0.112 for iPhone 1, 0.131 for iPhone 2 and 0.1 for iPhone 3, which shows that the algorithm recognized them as unknown devices in comparison to the training set.

6. Conclusions

This paper has presented the novel application of WSST to the physical layer authentication and identification of wireless devices for an experimental dataset based on the collection of RF emissions of 12 wireless devices (e.g., GSM mobile phones). The dataset is particularly challenging because the RF emissions has been collected with a low sample rate (1 MHz). This paper has performed an analysis on the application of WSST both for the problem of identification and authentication. The analysis includes the evaluation of the performance in the presence of AWGN. In both cases, the application of WSST outperforms STFT and the time domain and the frequency domain representation for medium and high SNR values. The results are consistent for different machine learning algorithms. An extensive analysis of the hyperparameters for the application of WSST has been implemented.

Author Contributions: Conceptualization, G.B., R.G. and G.S. Methodology, G.B. Software, G.B., G.S. Validation, G.S. Resources, R.G. Data curation, R.G. Writing, original draft preparation, G.B. Writing, review and editing, G.B., R.G. and G.S. Project administration, G.B. Funding acquisition, G.B.

Funding: This research has been supported by the EC H2020-IOT-2016-2017 (H2020-IOT-2017) Program under Grant Agreement 780139 for the SEcuRe and safe Internet Of Things (SerIoT) Research and Innovation Action.

Conflicts of Interest: The authors declare no conflict of interest. The founding sponsors had no role in the design of the study; in the collection, analyses or interpretation of the data; in the writing of the manuscript; nor in the decision to publish the results.

Abbreviations

The following abbreviations are used in this manuscript:

AWGN	Additive White Gaussian Noise
CMOS	Complementary Metal Oxide Semiconductor
CWT	Continuous Wavelet Transform
DDC	Digital Down Converter
ECG	Electro-CardioGram
EER	Equal Error Rate
EMD	Empirical Mode Decomposition
FAR	False Accept Rate
FN	False Negatives
FP	False Positives
FRR	False Reject Rate
GNSS	Global Navigation Satellite System
GSM	Global System for Mobile Communications
GT	Gabor Transform
GWT	Gabor–Wigner Transform
HHT	Hilbert–Huang Transform
ISM	Industrial, Scientific and Medical
ISO	International Organization for Standardization
JPEG	Joint Photographic Experts Group
KNN	K Nearest Neighbor
PNU	Pixel Non-Uniformity
PRNU	Photo-Response Non-Uniformity noise
PUF	Physical Unclonable Functions
RAI	Radiometric Identification
RF	Radio Frequency
RF-DNA	Radio Frequency DNA
ROC	Receiver Operative Characteristics
SDR	Software-Defined Radio
SNR	Signal to Noise Ratio
SST	Synchrosqueezing Transform
STFT	Short Time Fourier Transform
SVM	Support Vector Machine
TAR	True Accept Rate
TD	Time Domain
TFD	Time Frequency Domain
TN	True Negatives
TP	True Positives
UMTS	Universal Mobile Telecommunications System

USRP   Universal Software Radio Platform
WSST   Wavelet Synchrosqueezed Transform
WVD   Wigner–Ville distribution

References

1. Sandhu, R.; Samarati, P. Authentication, access control, and audit. *ACM Comput. Surv.* **1996**, *28*, 241–243. [CrossRef]
2. Baldini, G.; Steri, G.; Giuliani, R. Synchrosqueezing Transform Based Methodology for Radiometric Identification. In Proceedings of the 2018 41st International Conference on Telecommunications and Signal Processing (TSP), Athens, Greece, 4–6 July 2018; pp. 1–5.
3. Restuccia, F.; D'Oro, S.; Melodia, T. Securing the Internet of Things in the Age of Machine Learning and Software-defined Networking. *IEEE Internet Things J.* **2018**. [CrossRef]
4. Brik, V.; Banerjee, S.; Gruteser, M.; Oh, S. Wireless device identification with radiometric signatures. In Proceedings of the 14th ACM International Conference on Mobile Computing and Networking, San Francisco, CA, USA, 14–19 September 2008; pp. 116–127.
5. Huang, G.; Yuan, Y.; Wang, X.; Huang, Z. Specific Emitter Identification Based on Nonlinear Dynamical Characteristics. *Can. J. Electr. Comput. Eng.* **2016**, *39*, 34–41. [CrossRef]
6. Reising, D.R.; Temple, M.A.; Oxley, M.E. Gabor-based RF-DNA fingerprinting for classifying 802.16 e WiMAX mobile subscribers. In Proceedings of the 2012 International Conference on Computing, Networking and Communications (ICNC), Maui, HI, USA, 30 January–2 February 2012; pp. 7–13.
7. Xu, Q.; Zheng, R.; Saad, W.; Han, Z. Device Fingerprinting in Wireless Networks: Challenges and Opportunities. *IEEE Commun. Surv. Tutor.* **2016**, *18*, 94–104. [CrossRef]
8. Suski, W.C., II; Temple, M.A.; Mendenhall, M.J.; Mills, R.F. Using spectral fingerprints to improve wireless network security. In Proceedings of the IEEE GLOBECOM 2008-2008 IEEE Global Telecommunications Conference, New Orleans, LO, USA, 30 November–4 December 2008; pp. 1–5.
9. Bihl, T.J.; Bauer, K.W.; Temple, M.A.; Ramsey, B. Dimensional reduction analysis for Physical Layer device fingerprints with application to ZigBee and Z-Wave devices. In Proceedings of the Military Communications Conference, MILCOM 2015, Tampa, FL, USA, 26–28 October 2015; pp. 360–365. [CrossRef]
10. Reising, D.R.; Temple, M.A.; Mendenhall, M.J. Improved wireless security for GMSK-based devices using RF fingerprinting. *Int. J. Electron. Secur. Digit. Forensics* **2010**, *3*, 41–59. [CrossRef]
11. Lakafosis, V.; Traille, A.; Lee, H.; Gebara, E.; Tentzeris, M.M.; DeJean, G.R.; Kirovski, D. RF Fingerprinting Physical Objects for Anticounterfeiting Applications. *IEEE Trans. Microw. Theory Tech.* **2011**, *59*, 504–514. [CrossRef]
12. Klein, R.; Temple, M.A.; Mendenhall, M.J.; Reising, D.R. Sensitivity Analysis of Burst Detection and RF Fingerprinting Classification Performance. In Proceedings of the 2009 IEEE International Conference on Communications, Dresden, Germany, 14–18 June 2009; pp. 1–5. [CrossRef]
13. Li, C.; Liang, M. Time–frequency signal analysis for gearbox fault diagnosis using a generalized synchrosqueezing transform. *Mech. Syst. Signal Process.* **2012**, *26*, 205–217. [CrossRef]
14. Kumar, R.; Sumathi, P.; Kumar, A. Synchrosqueezing Transform-Based Frequency Shifting Detection for Earthquake-Damaged Structures. *IEEE Geosci. Remote Sens. Lett.* **2017**, *14*, 1393–1397. [CrossRef]
15. Wu, H.T.; Chan, Y.H.; Lin, Y.T.; Yeh, Y.H. Using synchrosqueezing transform to discover breathing dynamics from ECG signals. *Appl. Computat. Harmonic Anal.* **2014**, *36*, 354–359. [CrossRef]
16. Dudczyk, J.; Matuszewski, J.; Wnuk, M. Applying the radiated emission to the specific emitter identification. In Proceedings of the 15th International Conference on Microwaves, Radar and Wireless Communications, 2004, MIKON-2004, Warsaw, Poland, 17–19 May 2004; Volume 2, pp. 431–434.
17. Rehman, S.U.; Sowerby, K.W.; Coghill, C. Analysis of impersonation attacks on systems using {RF} fingerprinting and low-end receivers. *J. Comput. Syst. Sci.* **2014**, *80*, 591–601. [CrossRef]
18. Scanlon, P.; Kennedy, I.O.; Liu, Y. Feature extraction approaches to RF fingerprinting for device identification in femtocells. *Bell Labs Tech. J.* **2010**, *15*, 141–151. [CrossRef]
19. Baldini, G.; Steri, G.; Dimc, F.; Giuliani, R.; Kamnik, R. Experimental Identification of Smartphones Using Fingerprints of Built-In Micro-Electro Mechanical Systems (MEMS). *Sensors* **2016**, *16*, 818. [CrossRef] [PubMed]

20. Guo, Y.; Chen, X.; Wang, S.; Sun, R.; Zhao, Z. Wind Turbine Diagnosis under Variable Speed Conditions Using a Single Sensor Based on the Synchrosqueezing Transform Method. *Sensors* **2017**, *17*, 1149. [CrossRef]

21. Tary, J.B.; Herrera, R.H.; Han, J.; Baan, M. Spectral estimation—What is new? What is next? *Rev. Geophys.* **2014**, *52*, 723–749. [CrossRef]

22. Huang, N.E.; Shen, Z.; Long, S.R.; Wu, M.C.; Shih, H.H.; Zheng, Q.; Yen, N.C.; Tung, C.C.; Liu, H.H. The empirical mode decomposition and the Hilbert spectrum for nonlinear and non-stationary time series analysis. Royal Society of London A: Mathematical, Physical and Engineering Sciences. *R. Soc.* **1998**, *454*, 903–995. [CrossRef]

23. Auger, F.; Flandrin, P.; Lin, Y.T.; McLaughlin, S.; Meignen, S.; Oberlin, T.; Wu, H.T. Time-frequency reassignment and synchrosqueezing: An overview. *IEEE Signal Process. Mag.* **2013**, *30*, 32–41. [CrossRef]

24. Chen, H.; Lu, L.; Xu, D.; Kang, J.; Chen, X. The Synchrosqueezing Algorithm Based on Generalized S-transform for High-Precision Time-Frequency Analysis. *Appl. Sci.* **2017**, *7*, 769. [CrossRef]

25. Daubechies, I.; Lu, J.; Wu, H.T. Synchrosqueezed wavelet transforms: An empirical mode decomposition-like tool. *Appl. Comput. Harmonic Anal.* **2011**, *30*, 243–261. [CrossRef]

26. Baldini, G.; Steri, G.; Giuliani, R.; Gentile, C. Imaging time series for internet of things radio frequency fingerprinting. In Proceedings of the 2017 International Carnahan Conference on Security Technology (ICCST), Madrid, Spain, 23–26 October 2017; pp. 1–6.

27. Reising, D.R.; Temple, M.A.; Jackson, J.A. Authorized and Rogue Device Discrimination Using Dimensionally Reduced RF-DNA Fingerprints. *IEEE Trans. Inf. Forensics Secur.* **2015**, *10*, 1180–1192. [CrossRef]

28. Reising, D.R.; Temple, M.A.; Mendenhall, M.J. Improving Intra-Cellular Security Using Air Monitoring with RF Fingerprints. In Proceedings of the 2010 IEEE Wireless Communication and Networking Conference, Sydney, NSW, Australia, 18–21 April 2010; pp. 1–6. [CrossRef]

29. Cristianini, N.; Scholkopf, B. Support vector machines and kernel methods: the new generation of learning machines. *Ai Mag.* **2002**, *23*, 31.

![applied sciences logo] *applied sciences*

MDPI

Article

Interference Alignment in Multi-Hop Cognitive Radio Networks under Interference Leakage

Eylem Erdogan [1,*], Sultan Aldırmaz Çolak [2], Hakan Alakoca [3], Mustafa Namdar [4], Arif Basgumus [4] and Lutfiye Durak-Ata [3]

[1] Department of Electrical and Electronics Engineering, Istanbul Medeniyet University, Uskudar, Istanbul 34700, Turkey

[2] Department of Electronics and Communication Engineering, Kocaeli University, Izmit, Kocaeli 41380, Turkey; sultan.aldirmaz@kocaeli.edu.tr

[3] Informatics Institute, Istanbul Technical University, Maslak, Istanbul 34469, Turkey; alakoca@itu.edu.tr (H.A.); durakata@itu.edu.tr (L.D.-A.)

[4] Department of Electrical and Electronics Engineering, Kutahya Dumlupinar University, Kutahya 43100, Turkey; mustafa.namdar@dpu.edu.tr (M.N.); arif.basgumus@dpu.edu.tr (A.B.)

[*] Correspondence: eylem.erdogan@medeniyet.edu.tr; Tel.: +90-216-280-4021

Received: 6 November 2018; Accepted: 25 November 2018; Published: 4 December 2018

Abstract: In this work, we examine the interference alignment (IA) performance of a multi-input multi-output (MIMO) multi-hop cognitive radio (CR) network in the presence of multiple primary users. In the proposed architecture, it is assumed that linear IA is adopted at the secondary network to alleviate the interference between primary and secondary networks. By doing so, the secondary source can communicate with the secondary destination via multiple relays without causing any interference to the primary network. Even though linear IA can suppress the interference in CR networks considerably, interference leakages may occur due to a fast fading channel. To this end, we focus on the performance of the secondary network for two different cases: (i) the interference is perfectly aligned; (ii) the impact of interference leakages. For both cases, closed-form expressions of outage probability and ergodic capacity are derived. The results, which are validated by Monte Carlo simulations, show that interference leakages can deteriorate both system performance and the diversity gains considerably.

Keywords: cognitive radio; interference alignment; interference leakage; multi-hop relay network

---

1. Introduction

As mobile devices become widespread, wireless data traffic has been increasing significantly in the recent years. In order to meet the ever-increasing demand for high quality wireless communication standards, the Third-Generation Partnership Project (3GPP) and Long-Term Evolution (LTE) technologies have emerged [1,2]. Regarding the deployment of the next-generation wireless communication systems, the corresponding growth in the demand for wireless radio spectrum resources will appear. With a rapid increase in the number of connected devices and mobile users, improving spectrum utilization has now become an important concern in designing next-generation wireless communication networks [3]. Unfortunately, this situation will cause a severe shortage of spectrum resources. Thus, the solution methods for spectrum utilization have been attracting attention in recent years [4].

One of the candidates for solving the problem of spectrum shortage is the cognitive radio (CR) technology. CR has attracted considerable interest, as it can cope with the spectrum under-utilization phenomenon, as the efficient usage of the limited spectrum is important for mobile applications. CR can remedy this problem by allowing secondary (unlicensed) users (SUs) to share the same spectrum band with the primary (licensed) users (PUs). Thus, CR is a promising solution for providing quality of service and overcoming the problem of spectrum limitation in wireless networks. At the moment, spectrum usage is assigned for specific services with limited bandwidth based on the regulatory policy. This means that the unlicensed users will not be able to use the licensed frequency bands. However, the inefficient use of the licensed spectrum has been reported. CR allows the unlicensed users to exploit the unused frequency bands dynamically without causing harmful interference to the licensed users. For this reason, it has been proposed to improve spectral utilization and efficiency [5–8]. That is to say, CR is an inspiring approach for wireless communication systems that can alleviate the spectrum scarcity problem and utilize the existing spectrum resources efficiently. The CR network is composed of a primary network (PN) in which the licensed users of the spectrum are employed and a secondary network (SN), whose unlicensed users can access the spectrum opportunistically. SN users can access the licensed spectrum by three well-known techniques: underlay, overlay, and interweave [5–7]. In the underlay approach, the SU can simultaneously communicate with the PU using the PU's spectrum guaranteeing that the SU does not cause any harmful interference to the PU. In this scheme, the interference caused by the transmission of the SUs should not exceed an acceptable threshold. That means, the underlay method allows SU transmission as long as the interference remains under the predefined threshold value [9]. In the overlay approach, the SU has knowledge about the PU's transmitting information and how it is encoded. While the PU broadcasts its information periodically, the SU can obtain it by decoding the data sequence; thus, the interference can be partially or completely removed. The interweave paradigm is based on the concept of opportunistic communications. The idea was raised from the underutilized spectrum. Spectrum holes, which are not fully utilized most of the time, can be exploited by SUs to operate in the licensed bands. Thus, the spectrum utilization is enhanced by the opportunistic reuse of the spectrum holes. The interweave approach requires the detection of the PUs' activity in the licensed frequency band [10].

CR technology can be capable of utilizing the spectrum efficiently as long as the interference between PUs and SUs is perfectly aligned. To this end, interference alignment (IA) is an important approach for CR to recover the desired signal of the PU or SU by utilizing the precoding and suppression matrices of the channel matrix, which consolidates the interference beam or matrix into one subspace in order to eliminate them. This paper focuses on the interference alignment in CR networks considering multiple hops in the underlay scheme.

1.1. Related Works

There are various IA techniques that are trying to provide interference-free communications in CR networks. In the linear IA technique, the channel matrix is assumed to be perfectly known at the transmitter and receiver side of the PN [11–14]. In the literature, linear IA was adopted in CR interference channels in [15–18] and the references therein. In [15], the adaptive power allocation schemes were considered for linear IA-based CR networks where the outage probability and sum rate were derived. In [16], the adaptive power allocation was studied for linear IA-based CR using antenna selection at the receiver side, whereas [17] enhanced the security of CR networks by using a zero-forcing precoder. A similar work was proposed in [18] to improve the overall outage performance of the interference channel by using power allocation optimization. These studies show that interference management is an important issue for all multi-user wireless networks.

Most recently, multi-hop relaying, which is an effective way of enhancing reliability, connectivity, and coverage area, was introduced in CR networks [19–21]. In these papers, the authors studied the multi-hop cognitive relay networks under interference power constraints and provided a comprehensive performance study including the closed-form expressions for the outage probability, bit error rate (BER), and ergodic capacity. The paper [22] considered the performance of multi-hop CR networks with imperfect channel state information (CSI). Besides, the performance metrics of the secondary multi-hop networks covering outage probability, BER, and ergodic capacity were derived over Rayleigh fading channels. In [23,24], the authors considered the cognitive multi-hop system model and analyzed the performances over a generalized-*K* distribution. The outage probability analysis of a single-hop CR network was studied in [25] by considering multi-hop relaying in PN. Hussein et al. in [26] and the authors in [27] considered a multi-input multi-output (MIMO) multi-hop CR network and investigated its detailed performance. Finally, [28] demonstrated the effect of cluster-based relaying in the implementation complexity and provided the performance of a multi-hop cognitive relaying system in terms of outage probability, symbol error rate (SER), and ergodic capacity.

1.2. Motivation and Contributions

Even though cognitive multi-hop transmission offers numerous advantages to SUs, the primary-secondary interference is one of the most challenging issues to be solved in CR networks. To this end, IA, which can design coordinated signals to eliminate the interference in PU-SU, has become preferable [29]. Motivated by the advantages of multi-hop relaying and IA, herein, we investigate the interplay of the number of hops, relays, interference alignment, and interference leakage. Our main contributions are as follows:

- A decode-and-forward (DF) multi-hop SN is considered, and end-to-end SNRs are derived for two cases: (1) perfect interference alignment; (2) in the presence of interference leakages.
- Exact outage probability is derived for perfect IA and interference leakages.
- Approximate ergodic capacity expressions are derived for both cases.

1.3. Paper Organization

The rest of the paper is organized as follows: We introduce the signal and system model in Section 2. The outage probability analysis is given in Section 3. Section 4 presents the performance evaluations for the ergodic capacity. Numerical results are discussed in Section 5. Finally, Section 6 concludes the paper.

2. Signal and System Model

This paper considers a cognitive multi-hop relay-aided network with *L* PUs and two SUs in which the secondary source *S* wishes to communicate with the secondary destination *D* over $K - 1$ closely-located DF relays, as shown in Figure 1. We assume that all terminals are operating in a half-duplex fashion, and the direct path between *S* to *D* is not available due to heavy shadowing or large path loss effect. The uniformly-located relay terminals are clustered together and thus experiencing the same scale of fading even though the instantaneous SNR varies. In the SN, each node is equipped with *M* transmit/receive antennas applying maximal ratio transmission (MRT) and maximum ratio combining (MRC) techniques at the transmitter and receiver, respectively. In underlay CRNs, the transmit powers of the SUs are generally set to a predefined power level to meet the interference power constraints of the PUs [30]. However, in the proposed scheme shown in Figure 2, we adopt the linear IA method to mitigate the interference occurring at the SN without reducing the powers of the SUs. With the aid of precoding and linear suppression matrices, the single symbol detection at the *i*th hop (or $i + 1$th relay) can be expressed as [31]:

$$\mathbf{y}_{R_{i+1}} = \mathbf{U}_{R_{i+1}}^{\mathrm{H}} \mathbf{H}_{R_i \to R_{i+1}} \mathbf{V}_{R_i} \mathbf{x}_s + \sqrt{\alpha} \mathbf{U}_{R_{i+1}}^{\mathrm{H}} \sum_{j=1}^{L} \mathbf{H}_{P_j \to R_{i+1}} \mathbf{V}_{P_j} \mathbf{x}_j + \mathbf{U}_{R_{i+1}}^{\mathrm{H}} \mathbf{n}_{R_{i+1}}, \tag{1}$$

where $\mathbf{x}_s$ is the source signal, $\mathbf{x}_j$ is the information of the primary user, $\mathbf{H}_{R_i \to R_{i+1}}$ denotes the channel information at the *i*th hop of the SN, $\mathbf{H}_{P_j \to R_{i+1}}$ denotes the channel coefficients matrix between PN and SN, **V** and **U** denote the corresponding precoding and linear suppression matrices, α gives the interference leakage coefficient varying between 0 and 1 [32], $\mathbf{n}_{R_{i+1}}$ is the zero-mean unit-variance ($\sigma_{\mathbf{n}_{R_{i+1}}}^2 = \mathbf{I}$) circularly symmetric additive white Gaussian noise (AWGN) vector, and $(.)^{\mathrm{H}}$ stands for the Hermitian operation. Note that all signal and system model parameters are listed in Table 1.

The interference between PN and SN can be perfectly aligned if the following conditions are satisfied [31]:

$$\mathbf{U}_{R_{i+1}}^{H}\mathbf{H}_{P_j \rightarrow R_{i+1}}\mathbf{V}_{P_j} = 0$$

$$\text{Rank}\left(\mathbf{U}_{R_{i+1}}^{H}\mathbf{H}_{P_j \rightarrow R_{i+1}}\mathbf{V}_{P_j}\right) = d, \tag{2}$$

where d is the data stream transmitted by each user [33]. Using the ideal linear IA assumption, (1) can be expressed as:

$$\mathbf{y}_{R_{i+1}} = \widehat{\mathbf{H}}_{R_i \rightarrow R_{i+1}}\mathbf{x}_s + \widehat{\mathbf{n}}_{R_{i+1}}, \tag{3}$$

where $\widehat{\mathbf{H}}_{R_i \rightarrow R_{i+1}} \triangleq \mathbf{U}_{R_{i+1}}^{H}\mathbf{H}_{R_i \rightarrow R_{i+1}}\mathbf{V}_{R_i}$ and $\widehat{\mathbf{n}}_{R_{i+1}} \triangleq \mathbf{U}_{R_{i+1}}^{H}\mathbf{n}_{R_{i+1}}$.

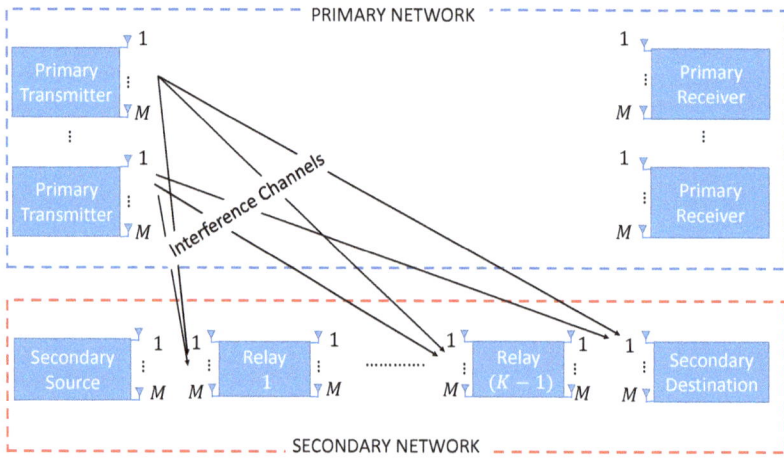

Figure 1. Multi-hop underlay cognitive radio network.

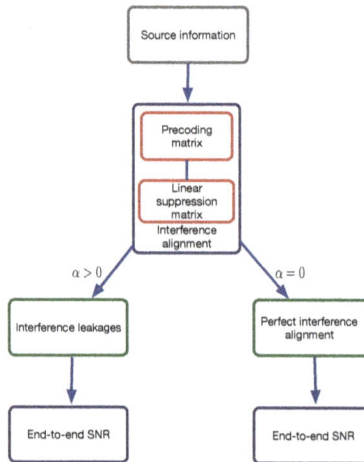

Figure 2. Block diagram of the system model.

Table 1. Parameters of the proposed system.

Parameter	Definition
M	The number of transmit and receive antennas
K	Number of hops
L	Number of primary users
$\mathbf{H}$	Channel coefficient matrix
$\mathbf{V}$	Precoding matrix
$\mathbf{U}$	Linear suppression matrix
$\mathbf{Q}$	QR decomposition matrix
$\mathbf{S}$	Singular-value decomposition matrix
α	The interference leakage coefficient

2.1. Interference Alignment Approach and End-to-End SNR Analysis

Throughout the paper, we apply the linear IA scheme to align the interference occurring between the PN and the SN. In order to suppress the interference signal, we use a minimum mean squared error (MMSE)-based decoder, which aims to maximize the capacity at the receiver part. The size of precoding matrix at each transmitter node $\mathbf{V}_c$, $(c \in P_j, R_i)$ and the suppression matrix of relay receiver at the ith hop $\mathbf{U}_{R_{i+1}}$ are $M \times \dfrac{M}{2}$ and $\dfrac{M}{2} \times M$, respectively, where M is a positive even number. We assume that $\mathbf{V}_c = \sqrt{\mathcal{P}_c}\mathbf{Q}_c\mathbf{X}_c$, where $\mathbf{Q}_c$ is an $M \times \dfrac{M}{2}$ matrix, whose columns form the orthonormal basis for $\mathbf{V}_c$ and $\mathbf{X}_c$. $\mathbf{X}_c$ is an $\dfrac{M}{2} \times \dfrac{M}{2}$ unitary matrix, which is obtained by using QR decomposition of the precoding matrix $\mathbf{V}_c$. Besides, $\mathcal{P}_c$ is the average power of each stream as $\mathcal{P}_c = 2P_c/M$ where $\mathcal{P}_c$, $(c \in P_j, R_i)$ is the total transmit power at the transmitter side.

The suppression matrix at the ith hop can be written as $\mathbf{U}_{R_{i+1}} = \tilde{\mathbf{U}}_{R_{i+1}}\bar{\mathbf{U}}_{R_{i+1}}$ [34]. When all CSI is known at each receiver node, the first term of the suppression matrix is obtained as:

$$\tilde{\mathbf{U}}_{R_{i+1}} = \mathbf{H}^{\mathrm{H}}_{ef,R_i \to R_{i+1}} \left(\sum_{j=1}^{L} \frac{\mathcal{P}_{P_j}}{\mathcal{P}_{R_i}} \mathbf{H}_{ef,P_j \to R_{i+1}} \mathbf{H}^{\mathrm{H}}_{ef,P_j \to R_{i+1}} + \frac{\sigma_n^2}{\mathcal{P}_{R_i}}\mathbf{I} \right)^{-1}, \tag{4}$$

where $\mathbf{H}_{ef,c \to R_{i+1}} = \mathbf{H}_{c \to R_{i+1}}\mathbf{Q}_c$, $(c \in P_j, R_i)$ denotes the effective channel matrix and σ_n^2 is the noise variance. Then, applying the Cholesky factorization as:

$$\bar{\mathbf{U}}_{R_{i+1}} \left(\sum_{j=1}^{L} \mathcal{P}_{P_j}\mathbf{H}_{ef,P_j \to R_{i+1}}\mathbf{H}^{\mathrm{H}}_{ef,P_j \to R_{i+1}} + \sigma_n^2\mathbf{I} \right) \bar{\mathbf{U}}^{\mathrm{H}}_{R_{i+1}} = \boldsymbol{\zeta}_{R_{i+1}}\boldsymbol{\zeta}^{\mathrm{H}}_{R_{i+1}}, \tag{5}$$

$\boldsymbol{\zeta}_{R_{i+1}}$ can be obtained. The second term of the suppression matrix is calculated as $\bar{\mathbf{U}}_{R_{i+1}} = \mathbf{S}^{\mathrm{H}}_{R_{i+1}}\boldsymbol{\zeta}^{-1}_{R_{i+1}}$, where $\mathbf{S}_{R_{i+1}}$ is obtained by using singular-value decomposition (SVD) as $\mathbf{H}_{R_{i+1}} = \mathbf{S}_{R_{i+1}}\boldsymbol{\Lambda}_{R_{i+1}}\mathbf{D}^{\mathrm{H}}_{R_{i+1}}$ and $\mathbf{H}_{R_{i+1}}$ denotes the $\dfrac{M}{2} \times \dfrac{M}{2}$ block channel matrix $\mathbf{H}_{R_{i+1}} = \boldsymbol{\zeta}^{-1}_{R_{i+1}}\tilde{\mathbf{U}}_{R_{i+1}}\mathbf{H}_{ef,R_i \to R_{i+1}}$. Finally, the suppression matrix can be written as a multiplication of these two terms:

$$\mathbf{U}_{R_{i+1}} = \mathbf{S}^{\mathrm{H}}_{R_{i+1}}\boldsymbol{\zeta}^{-1}_{R_{i+1}}\tilde{\mathbf{U}}_{R_{i+1}}. \tag{6}$$

Interested readers are referred to [34] and the references therein for a review of decoding matrix design. As $\mathbf{U}_{R_{i+1}}(\sum_{j=1}^{L}\mathcal{P}_{R_j}\mathbf{H}_{ef,R_i \to R_{i+1}}\mathbf{H}^{\mathrm{H}}_{ef,R_i \to R_{i+1}} + \sigma_n^2\mathbf{I})\mathbf{U}^{\mathrm{H}}_{R_{i+1}} = \mathbf{I}$, the interference can be aligned.

With the aid of the proposed IA approach, the interference can be perfectly aligned in the CR network. By doing so, the hop-by-hop transmission can be accomplished via the single-input and single-output (SISO) channel if one data stream is sent at each transmitter [15]. Thereby, the instantaneous SNRs between $S \to R_1$, $R_i \to R_{i+1}$, and $R_{K-1} \to D$ can be expressed as:

$$\Gamma_{S \to R_1} = \mathcal{P}_S \frac{|h_{S \to R_1}|^2}{\sigma_N^2},$$

$$\Gamma_{R_i \to R_{i+1}} = \mathcal{P}_{R_i} \frac{|h_{R_i \to R_{i+1}}|^2}{\sigma_N^2}, \text{ and}$$

$$\Gamma_{R_{K-1} \to D} = \mathcal{P}_{R_{K-1}} \frac{|h_{R_{K-1} \to D}|^2}{\sigma_N^2}, \tag{7}$$

where $\mathcal{P}_S$, $\mathcal{P}_{R_i}$, and $\mathcal{P}_{R_{K-1}}$ denote the transmit powers of S, R_i, and R_{K-1}. $h_{S \to R_1}$, $h_{R_i \to R_{i+1}}$, and $h_{R_{K-1} \to D}$ denote the channel fading coefficients between $S \to R_1$, $R_i \to R_{i+1}$, and $R_{K-1} \to D$ hops, respectively, which are modeled as zero mean and unit variance.

2.2. End-to-End SNRs in the Presence of Interference Leakage

In the presence of interference leakage, in other words, when the interference is not perfectly aligned, i.e., $\alpha \neq 0$, leakages occur, and the instantaneous SNRs can be expressed as:

$$\Gamma_{S \to R_1} = \frac{\mathcal{P}_S \frac{||\mathbf{H}_{S \to R_1}||^2}{\sigma_N^2}}{1 + \frac{\alpha \sum_{j=1}^{L} \mathcal{P}_j ||\mathbf{H}_{P_j \to R_1}||^2}{\sigma_N^2}},$$

$$\Gamma_{R_i \to R_{i+1}} = \frac{\mathcal{P}_{R_i} \frac{||\mathbf{H}_{R_i \to R_{i+1}}||^2}{\sigma_N^2}}{1 + \frac{\alpha \sum_{j=1}^{L} \mathcal{P}_j ||\mathbf{H}_{P_j \to R_{i+1}}||^2}{\sigma_N^2}} \text{ and}$$

$$\Gamma_{R_{K-1} \to D} = \frac{\mathcal{P}_{R_{K-1}} \frac{||\mathbf{H}_{R_{K-1} \to D}||^2}{\sigma_N^2}}{1 + \frac{\alpha \sum_{j=1}^{L} \mathcal{P}_j ||\mathbf{H}_{P_j \to D}||^2}{\sigma_N^2}}, \tag{8}$$

where $|| \cdot ||$ denotes the Frobenius norm.

3. Outage Probability Analysis

In this section, the exact outage probability expression is derived for two different cases: (i) the interference is perfectly aligned; (ii) the interference leakages occur due to imperfect IA. Outage probability can be defined as the outage probability of the overall system. In other words, the system is in outage if at least one of the hops is in outage. Mathematically, it can be expressed as:

$$P_{\text{out}} = 1 - \prod_{i=1}^{K-1} (1 - P_{\text{out}}^{(i)}), \tag{9}$$

where $P_{\text{out}}^{(i)}$ is the outage probability of the ith hop.

3.1. Outage Probability Performance of the Perfect IA Scheme

As described in the previous section, when the PN-SN interference is aligned, the system works in the SISO fashion if one data stream is sent at each transmitter [15]. With the aid of (7), $P_{\text{out},P}^{(i)}$ can be expressed as:

$$P_{\text{out},P}^{(i)} = \Pr[\Gamma_{R_i \to R_{i+1}}^{(i)} < \gamma_{th}], \tag{10}$$

where γ_{th} is the threshold value for acceptable communication quality. As we assume that all paths are modeled with independent and identically-distributed Rayleigh fading, the cumulative distribution function (cdf) of $\Gamma_{R_i \to R_{i+1}}^{(i)}$ can be expressed as:

$$P_{\text{out},P}^{(i)} = 1 - \exp\left(-\frac{\gamma_{th}}{\bar{\gamma}_{R_i \to R_{i+1}}}\right), \tag{11}$$

where $\bar{\gamma}_{R_i \to R_{i+1}}$ is the average SNR of the $R_i \to R_{i+1}$ hop describing the outage probability of the first $K - 2$ hops. For the last two hops, the outage probability can be expressed as [35]:

$$P_{\text{out},P}^{(K-1)} = \Pr[\min(\Gamma_{R_{K-2} \to R_{K-1}}^{(i)}, \Gamma_{R_{K-1} \to D}^{(i)}) < \gamma_{th}]$$
$$= 1 - \Pr[\Gamma_{R_{K-2} \to R_{K-1}} < \gamma_{th}] \Pr[\Gamma_{R_{K-1} \to D} < \gamma_{th}]. \tag{12}$$

Similar to (11), $P_{\text{out},P}^{(K-1)}$ can be expressed as:

$$P_{\text{out},P}^{(K-1)} = 1 - \exp\left(\frac{-\gamma_{th}}{\bar{\gamma}_{R_{K-2} \to R_{K-1}}}\right) \exp\left(\frac{-\gamma_{th}}{\bar{\gamma}_{R_{K-1} \to D}}\right). \tag{13}$$

By substituting (13) and (11) into (9), outage probability can be obtained.

3.2. Outage Probability in the Presence of Interference Leakages

To compute the outage probability of the first $K - 2$ hops in the presence of interference leakages, we express (8) as $\Gamma_{R_i \to R_{i+1}} = \frac{\gamma_{R_i \to R_{i+1}}}{1 + \gamma_{j \to R_{i+1}}^I}$, where $\gamma_{R_i \to R_{i+1}} = \frac{P_{R_i} \|H_{R_i \to R_{i+1}}\|^2}{\sigma_N^2}$ and $\gamma_{j \to R_{i+1}}^I = \frac{\sum_{j=1}^{L} P_j \|H_{P_j \to R_{i+1}}\|^2}{\sigma_N^2}$. Then, the probability density function (pdf) of $\gamma_{R_i \to R_{i+1}}$ can be expressed as [36]:

$$f_{\gamma_{R_i \to R_{i+1}}}(\gamma) = \frac{\gamma^{M^2-1} \exp\left(-\gamma / \bar{\gamma}_{R_i \to R_{i+1}}\right)}{\left(\bar{\gamma}_{R_i \to R_{i+1}}\right)^{M^2} (M^2 - 1)!}, \tag{14}$$

and the pdf of $\gamma_{j \to R_{i+1}}^I$ can be defined as:

$$f_{\gamma_{j \to R_{i+1}}^I}(\gamma) = \frac{\gamma^{LM^2-1} \exp\left(-\gamma / (\alpha \bar{\gamma}_{j \to R_{i+1}})\right)}{\left(\alpha \bar{\gamma}_{j \to R_{i+1}}\right)^{LM^2} (LM^2 - 1)!}. \tag{15}$$

Then, the cdf of $\Gamma_{R_i \to R_{i+1}}$ can be written as [27]:

$$F_{\Gamma_{R_i \to R_{i+1}}}(\gamma) = \int_0^\infty F_{\gamma_{R_i \to R_{i+1}}}((x+1)\gamma) f_{\gamma_{j \to R_{i+1}}^I}(x) dx. \tag{16}$$

We can find $F_{\gamma_{R_i \to R_{i+1}}}(\gamma)$ by taking the integral of $f_{\gamma_{R_i \to R_{i+1}}}(\gamma)$ with respect to γ. Then, by substituting $F_{\gamma_{R_i \to R_{i+1}}}(\gamma)$ and $f_{\gamma_{j \to R_{i+1}}^I}(\gamma)$ into (16), $F_{\Gamma_{R_i \to R_{i+1}}}(\gamma)$ can be obtained as:

$$F_{\Gamma_{R_i \to R_{i+1}}}(\gamma) = 1 - \exp\left(\frac{-\gamma}{\bar{\gamma}_{R_i \to R_{i+1}}}\right) \sum_{n=0}^{M^2-1} \left(\frac{\gamma}{\bar{\gamma}_{R_i \to R_{i+1}}}\right)^n \frac{1}{n!}$$
$$\times \frac{1}{\left(\alpha \bar{\gamma}_{j \to R_{i+1}}\right)^{LM^2}} \mathcal{U}\left(LM^2, LM^2 + 1, \frac{\gamma}{\bar{\gamma}_{R_i \to R_{i+1}}} + \frac{1}{\alpha \bar{\gamma}_{j \to R_{i+1}}}\right), \tag{17}$$

where $\mathcal{U}(\cdot,\cdot,\cdot)$ is Tricomi's confluent hypergeometric function [37]. As $P_{out,\mathcal{I}}^{(i)} = F_{\Gamma_{R_i \to R_{i+1}}}(\gamma_{th})$, the outage probability for the first $K-2$ hops can be derived. With the aid of (17), for the last two hops, $P_{out,\mathcal{I}}^{(K-1)}$ can be expressed as:

$$P_{out,\mathcal{I}}^{(K-1)} = F_{\Gamma_{R_{K-2} \to R_{K-1}}}(\gamma_{th}) + F_{\Gamma_{R_{K-1} \to D}}(\gamma_{th})$$
$$- F_{\Gamma_{R_{K-2} \to R_{K-1}}}(\gamma_{th}) F_{\Gamma_{R_{K-1} \to D}}(\gamma_{th}). \tag{18}$$

By substituting (17) into (18) and after replacing superscripts R_i with R_{K-2} and R_{K-1} and R_{i+1} with R_{K-1} and D, $P_{out,\mathcal{I}}^{(K-1)}$ can be obtained. With the aid of (9), outage probability can be derived.

4. Ergodic Capacity

Ergodic capacity can be defined as the maximum achievable mutual information from S to D, and it can be expressed as:

$$C_{erg} = \frac{1}{K}\mathbb{E}\left[\log_2(1+\gamma_{e2e})\right], \tag{19}$$

where $\mathbb{E}[\cdot]$ denotes the expectation operator. By substituting $\gamma_{e2e} = \min(\Gamma_{S \to R_1}, \Gamma_{R_1 \to R_2}, \dots, \Gamma_{R_{K-1} \to D})$ into (19), the ergodic capacity can be expressed as:

$$C_{erg} = \frac{1}{K}\mathbb{E}\left[\log_2(1+\min(\Gamma_{S \to R_1}, \Gamma_{R_1 \to R_2}, \dots, \Gamma_{R_{K-1} \to D}))\right]. \tag{20}$$

With the aid of Jensen's inequality, ergodic capacity can be upper bounded as:

$$C_{erg}^{up} \leq \frac{1}{K}\log_2\left(1+\min\left(\mathbb{E}[\Gamma_{S \to R_1}], \mathbb{E}[\Gamma_{R_1 \to R_2}], \dots, \mathbb{E}[\Gamma_{R_{K-1} \to D}]\right)\right), \tag{21}$$

and $\mathbb{E}[\Gamma_{R_i \to R_{i+1}}]$ can be obtained by using the following formula:

$$\mathbb{E}[\Gamma_{R_i \to R_{i+1}}] = \int_0^{\infty}\left(1 - F_{\Gamma_{R_i \to R_{i+1}}}(\gamma)\right)d\gamma. \tag{22}$$

4.1. Ergodic Capacity For Perfect IA

Using (11), after replacing γ_{th} with γ, the cdf of the ith hop can be expressed as:

$$F_{\Gamma_{R_i \to R_{i+1}}}(\gamma) = 1 - \exp\left(\frac{-\gamma}{\bar{\gamma}_{R_i \to R_{i+1}}}\right). \tag{23}$$

By substituting (23) into (22), $\mathbb{E}[\Gamma_{R_i \to R_{i+1}}]$ can be found as $\bar{\gamma}_{R_i \to R_{i+1}}$. Hence, C_{erg}^{up} can be expressed as:

$$C_{erg,P}^{up} \leq \frac{1}{K}\log_2\left(1+\min\left(\bar{\gamma}_{S \to R_1}, \bar{\gamma}_{R_1 \to R_2}, \dots, \bar{\gamma}_{R_{K-1} \to D}\right)\right). \tag{24}$$

Note that $\mathbb{E}[\Gamma_{S \to R_1}]$ and $\mathbb{E}[\Gamma_{R_{K-1} \to D}]$ can be found similarly.

4.2. Ergodic Capacity in the Presence of Interference Leakages

With the aid of Jensen's inequality and (20), ergodic capacity in the presence of interference leakages can be expressed as:

$$C_{erg}^{up}, \mathcal{I} = \frac{1}{K} \log_2 \left[\mathbb{E} \left(1 + \min \left(\frac{\mathcal{P}_S \frac{||H_{S \to R_1}||^2}{\sigma_N^2}}{1 + \frac{\alpha \sum_{j=1}^{L} \mathcal{P}_j ||H_{P_j \to R_1}||^2}{\sigma_N^2}}, \frac{\mathcal{P}_{R_1} \frac{||H_{R_1 \to R_2}||^2}{\sigma_N^2}}{1 + \frac{\alpha \sum_{j=1}^{L} \mathcal{P}_j ||H_{P_j \to R_2}||^2}{\sigma_N^2}}, \cdots, \frac{\mathcal{P}_{R_{K-1}} \frac{||H_{R_{K-1} \to D}||^2}{\sigma_N^2}}{1 + \frac{\alpha \sum_{j=1}^{L} \mathcal{P}_j ||H_{P_j \to D}||^2}{\sigma_N^2}} \right) \right) \right]. \quad (25)$$

5. Numerical Results

In this section, Monte Carlo simulations are carried out to verify the theoretical results. Without any loss of generality, we assume that the transmit powers at S and R_i are given as $\mathcal{P}_S = \mathcal{P}_{R_i} = P$. Moreover, the noise power is taken as N_0 for all hops, and $\gamma_{th} = 10$ dB, unless otherwise stated.

Figure 3 illustrates the outage probability performance of the SN for different numbers of hops when $M = L = 2$ and $\alpha = 0.005$. As can be seen from the figure, outage probability performance worsens as the number of hops increase. This is due to the fact that the number of interferers increase as the number of hops increase. As for example, almost 30 dB is needed to achieve $P_{out} = 10^{-2}$ at $K = 8$, while when $K = 2$, 18 dB is enough to achieve the same outage probability performance. Moreover, the theoretical curves verify the Monte Carlo simulations.

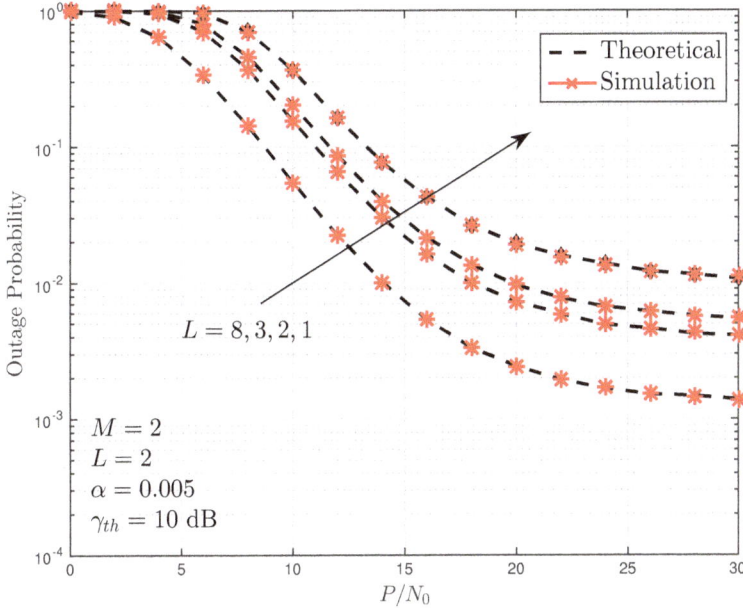

Figure 3. Outage probability of the secondary network versus P/N_0 for different numbers of hops, when $\alpha = 0.005$.

In Figure 4, the outage probability performance of the considered scheme is depicted for different numbers of interferers (PUs). As can be seen, as the number of interferers increase, the performance degrades. Moreover, the slopes of the curves verify that the diversity gain deteriorates as the number of interferers increase.

Figure 5 illustrates P_{out} with respect to P/N_0 of the considered scheme for three different interference leakage values, i.e., $\alpha = 0.01, 0.02, 0.05$. The other parameters are taken as $M = 2$,

$K = 2, L = 2$. As can be seen from Figure 5, the best P_{out} performance can be obtained when $\alpha = 0.01$, and the performance worsens as α increases.

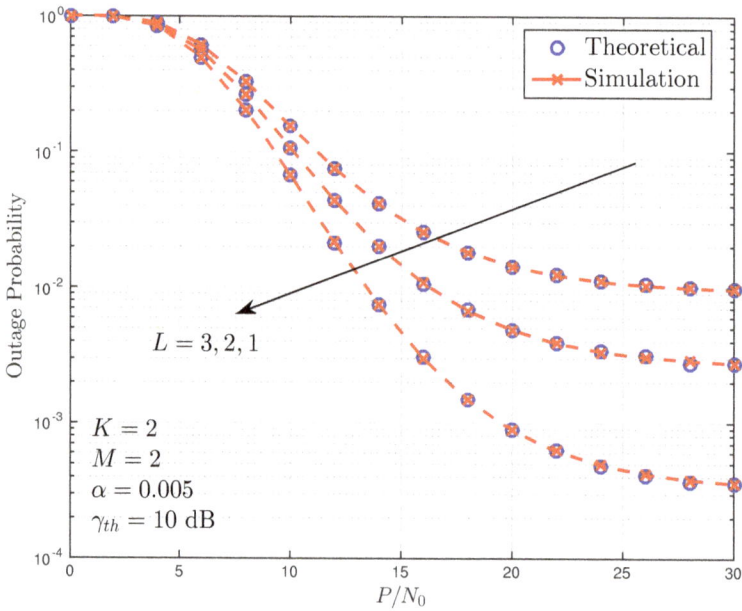

Figure 4. Outage probability of the secondary network versus P/N_0 for different numbers of primary users (interferers), when $\alpha = 0.005$.

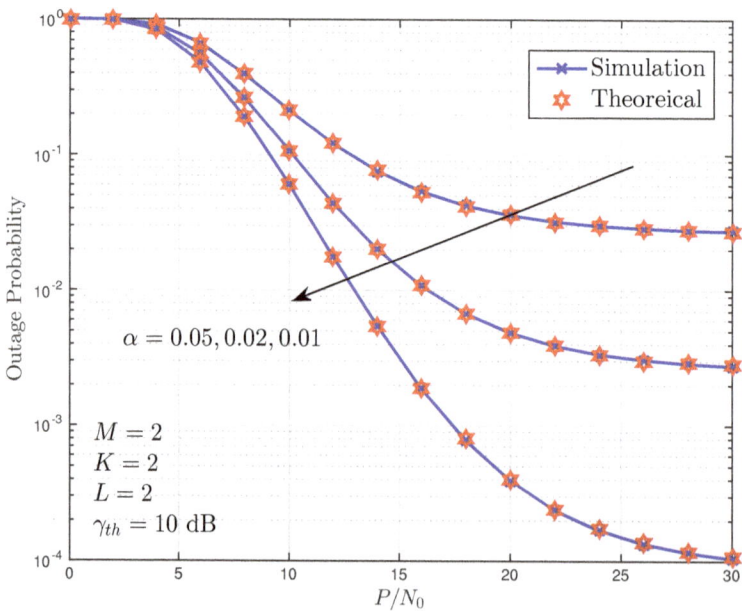

Figure 5. Outage probability of the secondary network versus P/N_0 for various interference leakage levels.

Figure 6 depicts the impact of various interference leakage levels on the performance of the multi-hop SN. As observed from the figure, ergodic capacity performance of the proposed scheme degrades as the impact of interference leakage enhances. On the contrary, when the interference is aligned, the capacity of the secondary network can achieve 10 bits/Hz at 50 dB. Comparing the derived approximate ergodic capacity with the simulation, it can be observed that the theoretical results match almost perfectly with the simulations.

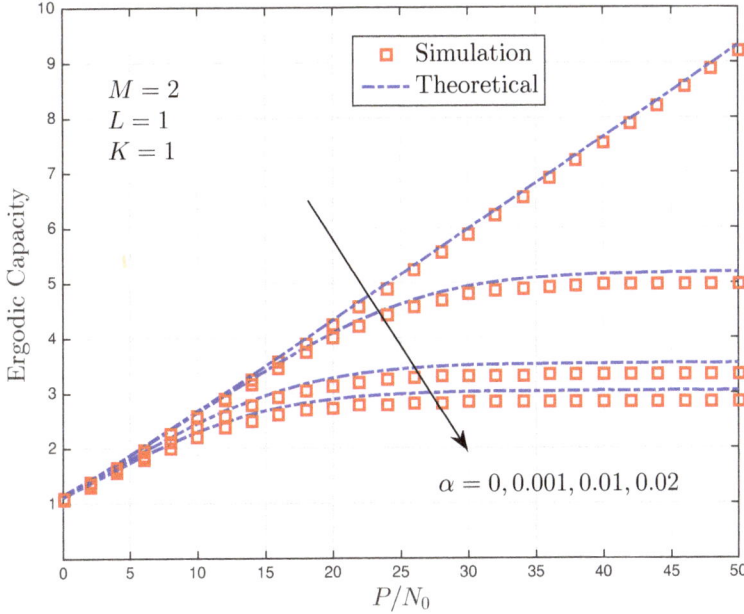

Figure 6. Ergodic capacity performance of the secondary network versus P/N_0 for various interference leakage levels.

6. Conclusions

In this work, we investigate the IA performance of the cognitive multi-hop network in the presence of multiple primary users. For the proposed scheme, we derived closed-form outage probability and ergodic capacity expressions for Rayleigh fading channel. The results, which were validated with the simulations, show that the system performance degraded as the number of interferers and/or leakage level increased.

This work can be extended to various practical scenarios where relays are affected both by primary-secondary interference and nodes mobility. Moreover, different clustering and/or relay selection approaches can be adopted, and the system performance of the overall multi-hop CR network can be presented.

Author Contributions: The authors E.E. and S.A.Ç. contributed to the methodology, validation, and writing; H.A., M.N., and A.B. contributed to the writing, investigation, and software; L.D.-A. contributed to the conceptualization, review, and editing.

Funding: This research received no external funding.

Acknowledgments: The content of this study has partially been submitted to the IEEE 41st International Conference on Telecommunications and Signal Processing (TSP 2018) [38].

Conflicts of Interest: The authors declare no conflict of interest.

References

1. Ericsson. Mobile Data Surpasses Voice, Stockholm, Sweden. 2010. Available online: http://www.ericsson.com/thecompany/press/releases/2010/03/1396928 (accessed on 29 October 2018).
2. Dahlman, E.; Parkvall, S.; Skold, J.; Beming, P. *3G Evolution: HSPA and LTE for Mobile Broadband*; Academic Press: New York, NY, USA, 2010.
3. Yucek, T.; Arslan, H. A survey of spectrum sensing algorithms for cognitive radio application. *IEEE Commun. Surv. Tutor.* **2009**, *11*, 116–130. [CrossRef]
4. Li, Q.; Hu, R.Q.; Qian, Y.; Wu, G. Cooperative communications for wireless networks: Techniques and applications in LTE-advanced systems. *IEEE Wirel. Commun.* **2012**, *19*, 22–29.
5. Haykin, S. Cognitive radio: Brain-empowered wireless communications. *IEEE J. Sel. Areas Commun.* **2005**, *23*, 201–220. [CrossRef]
6. Mitola, J.; Maguire, G.Q. Cognitive radio: Making software radios more personal. *IEEE Pers. Commun.* **1999**, *6*, 13–18. [CrossRef]
7. Cirik, A.C.; Biswas, S.; Taghizadeh, O.; Ratnarajah, T. Robust transceiver design in full-duplex MIMO cognitive radios. *IEEE Trans. Veh. Technol.* **2018**, *67*, 1313–1330. [CrossRef]
8. Sharma, S.K.; Bogale, T.E.; Chatzinotas, S.; Ottersten, B.; Le, L.B.; Wang, X. Cognitive radio techniques under practical imperfections: A survey. *IEEE Commun. Surv. Tutor.* **2015**, *17*, 1858–1884. [CrossRef]
9. Liang, W.; Ng, S.X.; Hanzo, L. Cooperative overlay spectrum access in cognitive radio networks. *IEEE Commun. Surv. Tutor.* **2017**, *19*, 1924–1944. [CrossRef]
10. Biglieri, E.; Goldsmith, A.J.; Greenstein, L.J.; Poor, H.V.; Mandayam, N.B. *Principles of Cognitive Radio*; Cambridge University Press: Cambridge, UK, 2013.
11. Razaviyayn, M.; Sanjabi, M.; Luo, Z.Q. Linear transceiver design for interference alignment: Complexity and computation. *IEEE Trans. Inf. Theory* **2012**, *58*, 2896–2910. [CrossRef]
12. Liu, T.; Yang, C. On the feasibility of linear interference alignment for MIMO interference broadcast channels with constant coefficients. *arXiv* **2012**, arXiv:1207.1517.
13. González, Ó.; Beltrán, C.; Santamaría, I. A Feasibility test for linear interference alignment in MIMO channels with constant coefficients. *IEEE Trans. Inf. Theory* **2014**, *60*, 1840–1856. [CrossRef]
14. Razaviyayn, M.; Lyubeznik, G.; Luo, Z.Q. On the degrees of freedom achievable through interference alignment in a MIMO interference channel. *IEEE Trans. Signal Process.* **2012**, *60*, 812–821. [CrossRef]
15. Zhao, N.; Yu, F.R.; Sun, H.; Li, M. Adaptive power allocation schemes for spectrum sharing in interference-alignment-based cognitive radio networks. *IEEE Trans. Veh. Technol.* **2016**, *65*, 3700–3714. [CrossRef]
16. Sam, R.P.; Govindaswamy, U.M. Antenna selection and adaptive power allocation for IA-based underlay CR. *IET Signal Process.* **2017**, *11*, 734–742.
17. Sultana, R.; Sarkar, M.; Hossain, M. Linear precoding techniques in enhancing security of cognitive radio networks. In Proceedings of the IEEE International Conference on Electrical, Computer & Telecommunication Engineering (ICECTE), Rajshahi, Bangladesh, 8–10 December 2016; pp. 1–4.
18. Zhao, N.; Yu, F.R.; Sun, H. Adaptive energy-efficient power allocation in green interference-alignment-based wireless networks. *IEEE Trans. Veh. Technol.* **2015**, *64*, 4268–4281. [CrossRef]
19. Bao, V.N.Q.; Thanh, T.T.; Nguyen, T.D.; Vu, T.D. Spectrum sharing-based multi-hop decode-and-forward relay networks under interference constraints: Performance analysis and relay position optimization. *J. Commun. Netw.* **2013**, *15*, 266–275. [CrossRef]
20. Hyadi, A.; Benjillali, M.; Alouini, M.S.; da Costa, D.B. Performance analysis of underlay cognitive multihop regenerative relaying systems with multiple primary receivers. *IEEE Trans. Wirel. Commun.* **2013**, *12*, 6418–6429. [CrossRef]
21. Phan, H.; Zepernick, H.J.; Tran, H. Impact of interference power constraint on multi-hop cognitive amplify-and-forward relay networks over Nakagami-*m* fading. *IET Commun.* **2013**, *7*, 860–866. [CrossRef]
22. Bao, V.N.Q.; Duong, T.Q.; Tellambura, C. On the performance of cognitive underlay multihop networks with imperfect channel state information. *IEEE Trans. Commun.* **2013**, *61*, 4864–4873. [CrossRef]
23. Kamga, G.N.; Fredj, K.B.; Aïssa, S. Multihop cognitive relaying over composite multipath/shadowing channels. *IEEE Trans. Veh. Technol.* **2015**, *64*, 3807–3812. [CrossRef]

24. Khoshafa, M.H.; Al-Ahmadi, S. On the capacity of underlay multihop cognitive relaying over generalized-K composite fading channels. *Wirel. Pers. Commun.* **2017**, *96*, 361–370. [CrossRef]
25. Park, J.; Jang, C.; Lee, J.H. Outage analysis of underlay cognitive radio networks with multihop primary transmission. *IEEE Commun. Lett.* **2016**, *20*, 800–803. [CrossRef]
26. Hussein, J.A.; Ikki, S.S.; Boussakta, S.; Tsimenidis, C.C.; Chambers, J. Performance analysis of a multi-hop UCRN with co-channel interference. *IEEE Trans. Commun.* **2016**, *64*, 4346–4364. [CrossRef]
27. Al-Qahtani, F.S.; Radaydeh, R.M.; Hessien, S.; Duong, T.Q.; Alnuweiri, H. Underlay cognitive multihop MIMO networks with and without receive interference cancellation. *IEEE Trans. Commun.* **2017**, *65*, 1477–1493. [CrossRef]
28. Boddapati, H.K.; Bhatnagar, M.R.; Prakriya, S. Performance analysis of cluster-based multi-hop underlay CRNs using max-link-selection protocol. *IEEE Trans. Cogn. Commun. Netw.* **2018**, *4*, 15–29. [CrossRef]
29. Shahjehan, W.; Shah, S.; Lloret, J.; Bosch, I. Joint Interference and Phase Alignment among Data Streams in Multicell MIMO Broadcasting. *Appl. Sci.* **2018**, *8*, 1237. [CrossRef]
30. Afana, A.; Asghari, V.; Ghrayeb, A.; Affes, S. Cooperative relaying in spectrum-sharing systems with beamforming and interference constraints. In Proceedings of the IEEE 13th International Workshop on Signal Processing Advances in Wireless Communications (SPAWC), Cesme, Turkey, 17–20 June 2012; pp. 429–433.
31. Basgumus, A.; Namdar, M.; Alakoca, H.; Erdogan, E.; Durak-Ata, L. Interference Alignment in Multi-Input Multi-Output Cognitive Radio-Based Network. In *Cognitive Radio in 4G/5G Wireless Communication Systems*; IntechOpen: Rijeka, Croita, 2018.
32. Ata, S.Ö.; Altunbaş, İ. Analog network coding over cascaded fast fading Rayleigh channels in the presence of self-interference. In Proceedings of the 2016 24th Signal Processing and Communication Application Conference (SIU), Zonguldak, Turkey, 16–19 May 2016; pp. 253–256.
33. Alakoca, H. *Linear Interference Alignment in Cognitive Radio Networks*; Istanbul Technical University, Informatics Institute: Istanbul, Turkey, 2018.
34. Sung, H.; Park, S.H.; Lee, K.J.; Lee, I. Linear precoder designs for K-user interference channels. *IEEE Trans. Wirel. Commun.* **2010**, *9*, 291–301. [CrossRef]
35. Ikki, S.; Ahmed, M.H. Performance analysis of cooperative diversity wireless networks over Nakagami-*m* fading channel. *IEEE Commun. Lett.* **2007**, *11*, 334–336. [CrossRef]
36. Lo, T.K. Maximum ratio transmission. In Proceedings of the IEEE International Conference on Communications (ICC'99), Vancouver, BC, Canada, 6–10 June 1999; Volume 2, pp. 1310–1314.
37. Spanier, J.; Oldham, K.B. *An Atlas of Functions*; Hemisphere: Washington, DC, USA, 1987.
38. Alakoca, H.; Ustunbas, S.; Namdar, M.; Basgumus, A.; Erdogan, E.; Durak-Ata, L. System performance of interference alignment in MIMO cognitive radio networks under interference leakage. In Proceedings of the IEEE 41st International Conference on Telecommunications and Signal Processing, Izmir, Turkey, 2–5 May 2018; Volume 2, pp. 569–572.

applied
sciences

MDPI

Article

Activation Process of ONU in EPON/GPON/XG-PON/NG-PON2 Networks

Tomas Horvath [1,*], Petr Munster [1], Vaclav Oujezsky [1] and Josef Vojtech [2]

[1]   Department of Telecommunication, Brno University of Technology, Technicka 12, 616 00 Brno,
     Czech Republic; munster@feec.vutbr.cz (P.M.); oujezsky@feec.vutbr.cz (V.O.)
[2]   Department of Optical Networks, CESNET a.l.e., Zikova 4, 160 00 Prague, Czech Republic; vojtech@cesnet.cz
*   Correspondence: horvath@feec.vutbr.cz; Tel.: +420-541-146-923

Received: 2 October 2018; Accepted: 13 October 2018; Published: 16 October 2018

✓ check for updates

Abstract: This article presents a numerical implementation of the activation process for gigabit and 10 gigabit next generation and Ethernet passive optical networks. The specifications are completely different because GPON, XG-PON and NG-PON2 were developed by the International Telecommunication Union, whereas Ethernet PON was developed by the Institute of Electrical and Electronics Engineers. The speed of an activation process is the most important in a blackout scenario because end optical units have a timer after expiration transmission parameters are discarded. Proper implementation of an activation process is crucial for eliminating inadvisable delay. An OLT chassis is dedicated to several GPON (or other standard) cards. Each card has up to eight or 16 GPON ports. Furthermore, one GPON port can operate with up to 64/128 ONUs. Our results indicate a shorter duration activation process (due to a shorter frame duration) in Ethernet-based PON, but the maximum split ratio is only 1:32 instead of up to 1:64/128 for gigabit PON and newer standards. An optimization improves the reduction time for the GPON activation process with current PLOAM messages and with no changes in the transmission convergence layer. We reduced the activation time from 215 ms to 145 ms for 64 ONUs.

Keywords: activation process; EPON; GPON; MATLAB; NG-PON2; timing; transmission convergence layer; XG-PON

---

1. Introduction

Passive optical networks (PONs) are the most promising solution for access networks. Because the European Union is seeking broadband access and continually rising bandwidth requirements for end users, some technologies are not sufficient, such as asymmetric digital subscriber line (ADSL) or wireless fidelity (WiFi). The current research goal is to not only use optical fibers for data transmission, but also for other special services, such as an accurate time, stable frequency and optical sensing for infrastructure defense [1,2].

The first specification of PON was approved in 1998 as the asynchronous transfer mode PON (APON) [3]. The basic topology, a cascade connection of splitters in the optical distribution network (ODN), corresponds to the first specification [4]. The Czech Republic is obliged to develop a broadband access Internet technology with a transmission speed of at least 30 Mbit/s (downstream direction) for current households and 100 Mbit/s for new customers by 2020. A proper technology for the main purpose is gigabit PON (GPON) technology and/or Ethernet PON (EPON). At present, EPON technology is not very popular around the world due to the decreasing price and better efficiency of GPON technology. Furthermore, the International Telecommunication Union (ITU) has more standards of PONs that are compatible with the previous standards. The ITU's last standard next generation PON Stage 2 (NG-PON2) can transfer 40 Gbit/s (by 4 λ, each with 10 Gbit/s), but the

Institute of Electrical and Electronics Engineers (IEEE) currently works on the first PON specification at 100 Gbit/s (by 4 λ, each with 25 Gbit/s). Note that ITU and IEEE are not compatible due to different encapsulation methods.

The rest of this paper is structured as follows. Section 2 introduces related works. Section 3 provides an overview of the GPON and EPON physical layer. Section 4 presents the activation processes for the G, XG, NG-PON(2) and EPONs. Section 5 presents the simulation model and a discussion of the results. Finally, Section 6 concludes the paper.

2. Related Works

In recent years, many publications on EPON have been published. Most of them involve the multi-point control protocol (MPCP) and energy saving. MPCP and energy saving issues belong to a transmission convergence layer of EPONs. The article [5] evaluates the exact mean packet delay for the MPCP. The authors proposed a reservation interval allocation method for the REPORTmessage. This message reports total queue size occupancy in ONU. The works [6,7] presented an EPON autodiscovery mechanism for fast network and service recovery and for scheduling traffic in the upstream direction. The mechanism improves the registration bandwidth efficiency from 0.13 to 0.6 with an arbitrary number of optical network units (ONUs), but it does not solve the activation process in an EPON. Other articles [8–10] considered the quality of service (QoS) and parameters in EPONs.

The authors in [11] proposed a Very High Speed Integrated Circuit Hardware Description Language (VHDL) implementation of the ONU autodiscovery process for EPONs. They moved the simplest autodiscovery process out of the ONU, which can be extended to data transfer in time slots via a VHDL implementation. The article [12] focused on dynamic bandwidth allocation (DBA) to provide QoS in an EPON and 10G-EPON coexistence scheme by distributed dynamic scheduling PON (DDSPON). The works [13–15] presented the latest phenomena in the EPON, which is energy saving for the ONU. On the one hand, ONU power consumption is not paid for by an Internet service provider (ISP), but by customers. On the other hand, Internet accessibility is not continuously needed by various customers throughout the day/night/weekends/holidays. Conversely, customers may want to be online at unexpected times when necessary. Each energy saving leads to a decrease in the power consumption bill. We evaluated the activation process in [16,17] for the GPON. The article [16] involved the activation process regarding the final verification by measurements. We have not considered the EPON standard for the comparison. Our simulation was designed only for the GPON, but the current model contains the EPON standard. Although the standards are different, both are found all over the world. In our paper [18], we provided the simulation of the transmission convergence layer in the next generation PONs. We implemented our own numerical model in MATLAB for the next generation PON (XG-PON). The work [19] evaluated the transmission convergence layer of GPON and the next generation PON (XG-PON). The results proved that the encapsulation method of ITU standards is highly efficient. For instance PON could be one of the solutions for the support of a cloud-radio access network (C-RAN), as described in [20,21]. The main contribution of this article is to demonstrate the importance of the activation process in passive optical (ITU and IEEE) networks and to reduce the total activation time of the ONU in GPON network with the current transmission convergence scheme and control messages format.

3. GPON and EPON Networks

The basic information on the transmission rate, used wavelengths, the maximum number of connected users and the actual range of the network is found in Table 1 (see the GPON column). The transfer medium is described in ITU's Telecommunication standardization sector ITU-T recommendation G.652, which describes a single-mode optical fiber. The optical signal is transmitted bidirectionally using the wavelength-division multiplexing (WDM)-based transfer medium via a single fiber or unidirectionally via two fibers. The non-return-to-zero (NRZ) code is used. The transmitter uses multi-longitudinal mode (MLM) and single-longitudinal mode (SLM) lasers. The attenuation classes of the GPON are defined in [22]:

- Class A: 5–20 dB
- Class B: 10–25 dB
- Class C: 15–30 dB

These attenuation specifications do assume the worst case scenario regarding losses on connectors, passive optical devices, fiber splices and optical fiber attenuation.

The GPON contains optical line termination (OLT), the ONU and the ODN. The OLT unit is the central unit and controls all communications on the network. ONU presents the end of the optical part of the ODN. In general, the ODN contains all transmissions between the OLT and ONU. The total minimum optical return loss (ORL) at the reference point R/S(before ONU) in the ODN must be below 32 dB. The maximum differential path losses (the difference between the largest and smallest loss in a single ODN) should be 15 dB [22,23].

Table 1. Basic comparison of GPON with EPON.

Parameters	GPON	EPON (Type 2)
Standard	ITU-T G.984	IEEE 802.3ah
Bitrate downstream	1.244/2.488 Gbit/s	1.25 Gbit/s
Bitrate upstream	1.244/2.488 Gbit/s	1.25 Gbit/s
λ for downstream	1480–1500 nm	1490 nm
λ for upstream	1260–1360 nm	1310 nm
Split ratio	64 (up to 128)	32
Network reach	20 km	20 km

The basic information on the transmission rate, used wavelengths, the maximum number of connected users and the range of the network can be found in Table 1 (see the EPON column where the point to multipoint (P2MP) topology is used). In the standard, the physical layer is divided into four sublayers: reconciliation sublayer (RS), physical medium attachment (PMA), physical medium dependent (PMD), medium dependent interface (MDI) and gigabit media independent interface (GMII). Descriptions of these sublayers and a further description of the associated interfaces are in [24]:

- MDI specifies the physical media and the mechanical and electrical interfaces between the transfer medium and the physical layer devices.
- PMD is responsible for linking to the transfer medium, and it is placed directly over the MDI. EPON uses WDM-based single-mode optical fibers. Two types of interfaces are supported, specifically: 1000BASE-PX10 with an overlapping distance of 10 km and 1000BASE-PX20 with an overlapping distance of 20 km.
- PMA provides functions for transmission, clock restart and phase alignment. This sublayer is primarily designed to specify the clock and data recovery (CDR) time interval.
- The physical coding sublayer (PCS) provides functions for link-coding of data (bits) that may be transmitted over the physical medium. EPON uses 8B/10B encoding [25].
- GMII specifies the interface between the media access control (MAC) layer and the physical layer.
- RS provides mapping of signals from GMII for the link layer.

4. xPON Activation Processes

The activation process describes the steps in which an inactive ONU connects or reconnects to a PON [26]. The activation process includes three phases, specifically: parameter learning, serial number acquisition and ranging. During the learning parameter phase, the ONU acquires the operational parameters that are needed in the upstream transmission. During the serial number acquisition phase, OLT discovers a new ONU (by serial number) and assigns an ONU identifier (ONU-ID) to it.

The ONU round trip delay (RTD) is the time interval between the downstream frame transmission and the corresponding upstream transmission burst from the given ONU. The RTD consists of a propagation delay that is directly proportional to the length of the fibers from the ONU and the response of the ONU. To ensure that transmission bursts from different ONUs are ordered at the interface of the same upstream GPON transmission convergence layer (GTC) frame, the delay time is assigned to each ONU to postpone the transfer of the upstream burst to the time not used for a common response time. This response time is called the equalization delay (EqD), and for each given ONU, the OLT is calculated based on the RTD measurement and consequently transmitted during the ranging state.

To avoid collisions with the upstream bursts transmitted during acquisition of the serial number and the range of the newly-connected ONU, the OLT must temporarily suppress the upstream transmission of the active ONU for the time that the arrival of upstream bursts from the new ONU is assumed. This time interval is referred to as the quiet window.

4.1. GPON Activation States

The following information is based on the recommendation [26].

State O1, Initialstate: In this state, the ONU switches on, waits for the downstream signal and synchronizes with it afterwards. Initially, a loss of signal (LoS) is set up to indicate the loss of a signal or a frame. It is also important for the synchronization machine of the ONU and OLT in the downstream direction to perform correct synchronization. In the synchronization state, ONU starts in the so-called Huntstate, in which it searches for the physical synchronization (PSync) field. When an error-free PSync array is found, the ONU moves to the next state, called the Pre-sync state, and sets the counter Nto one. The ONU then searches for the next PSync array that follows the previous one. For each error-free PSync array, the counter is incremented by one. If the ONU receives a corrupted PSync, it returns to the Hunt state. If the N counter in the Pre-sync state is equal to M1 (the recommended value for M1is two), the ONU moves to the Syncstate and begins processing the information from the physical control block downstream (PCBd) header. If the ONU in the Sync state receives M2 (the recommended value for M1 is five) consecutive frames with a corrupted PSync, it can declare the loss of the downstream signal and return to the Hunt state. The ONU then deletes all transmission convergence (TC) layer-based parameters known from the previous session such as: ONU-ID, default allocation identifier (Alloc-ID), delay compensation and Burst Headerparameters. Once the downstream transmission is received, the LoS and loss of frame (LoF) are cleared, and the ONU moves to the O2 state.

State O2, Standbystate: After State O1, the synchronization in the downstream direction is provided, yet the upstream direction synchronization is required and essential. Downstream transmission is received by the ONU and waits for global network parameters. Once the upstream overhead message is received, the ONU sets up the assigned parameters and moves to the O3 state.

State O3, Serialnumber state: In this state, the OLT requests broadcast ONUs to send their serial number. To prevent collisions with the normal traffic, the OLT creates as mentioned above a quiet window with a duration of 250 µs by sending a frame with an empty bandwidth map (BWmap) field. Subsequently, the previously mentioned SN request is sent (i.e., a request to send a serial number) with a random delay set between 0 and 48 µs. As a reply to the SN request, the ONU uses the serial number ONU message to enable the OLT to examine and detect the serial number. In addition, the OLT uses the AssignONU-ID message to assign the ONU-ID. Once the number is assigned, the ONU moves to the next state. The OLT can also send an Extended Burst Lengthmessage to all connected ONUs and hand over the extended overhead parameters. However, if the ONU receives this message before the request to send the serial number, it ignores such a message. In this state, the TO1 timer is used to cancel any unsuccessful activation attempt by setting the time during which the ONU can remain in this particular state. The recommended TO1 value is 10 s. After such a time, the ONU moves to the O2 state.

State O4, Rangingstate: Transmission in the upstream direction from different ONUs must be synchronized with the boundaries of the upstream GTC frame. To ensure the appearance of the ONUs, they are set at the same distance from the OLT, and the equalization delay for each ONU is required. The equalization delay is measured when the ONU is in this state. During this particular state, a quiet window with a duration of 202 μs is created. The OLT sends a ranging request, and the ONU replies with a Serial Numbermessage. Furthermore, the OLT sends the Ranging Timemessage, in which the allocated equalization delay is transmitted. Once this message is received by the ONU, it moves to its working state. In such a state, the TO1 timer is used.

State O5, Operationstate: In this state, the ONU can now send data, physical layer operations and administration and maintenance (PLOAM) messages according to the OLT instructions. Once the network is equalized and all the ONUs are working with the correct equalization delay, all upstream bursts will be synchronized among all ONUs.

State O6, POPUP state: An ONU enters this state when any of the LoS or LoF alarms (if the signal is lost or the frame is poorly assembled) is detected. Therefore, if this condition occurs, the ONUs immediately stop sending data. After the POPUP status occurs, the ONU first attempts to retrieve the optical signal, recover the synchronization of the GTC frame and remove the LoS/LoF alarm. The ONU goes either to the Operationstate or to the Rangingstate according to the particularly targeted POPUP messages. If the ONU receives the targeted POPUP message, it returns to the Ranging State. If the ONU cannot restore the optical signal or reset itself to recover the GTC frame synchronization, it does not receive a targeted POPUP message and is moved to the Initialstate. This is where the TO2 timer is used (the recommended time for the timer is 100 ms).

State O7, Emergencystop state: An ONU that receives a Disable Serial Numberwith the "deactivate" option goes to the emergency stop state and shuts off the laser. During this state, the ONU is not allowed to send any data. If a failure on the deactivated ONU is resolved, the OLT can activate the ONU to return it to its functional state. The activation is accomplished by sending the Disable Serial Number message with the "enable" option. Subsequently, the ONU returns to the Standbystate, and all parameters are discarded and retrieved.

4.2. XG-PON Activation Process

As mentioned in Section 4.1, the principles of the activation process for XG-PON are basically identical to those for GPON and are defined by the recommendations [26,27].

State O1, initial state: The ONU is in this state immediately upon switching on or after switching from other states when there is an error requiring a return to the initialization state. The transmission is switched off at this time, and all the previously set TC layer parameters (e.g., ONU-ID) are cleared. Synchronization in the downstream direction is provided by the synchronization machine. The ONU starts in the Hunt State, where it uses the downstream signal to search for the PSync pattern stored in the physical synchronization block downstream (PSBd). If it is found, the ONU verifies that a 64-bsuperframe counter (SFC) structure, which is also found in PSBd and secured by a self-repairing hybrid error correction (HEC), is valid. If the SFC is valid, the ONU stores its values and moves to the Pre-Syncstate. With the next successful validation (at this point, only 62 bits out of the total number of received 64 bits are sufficient), the ONU moves to the Sync state. However, if any of these validations fail, the ONU returns to the Hunt state. The unit remains in the Sync state (the unit has already been successfully synchronized) as long as the PSync and SFC authentication are successful. If the authentication fails, the ONU moves to the Re-Sync state. It moves to the Sync state only after successful validation. The recommended value for the M parameter is three. However, if M-1consecutive physical interface (PHY) frames validating the PSync or SFC fail, the ONU declares a loss of synchronization with the downstream frame, discards the saved SFC copy and returns to the Hunt state. This process is illustrated in Figure 1. Once synchronized with the downstream PHY frame, the ONU moves to the next state.

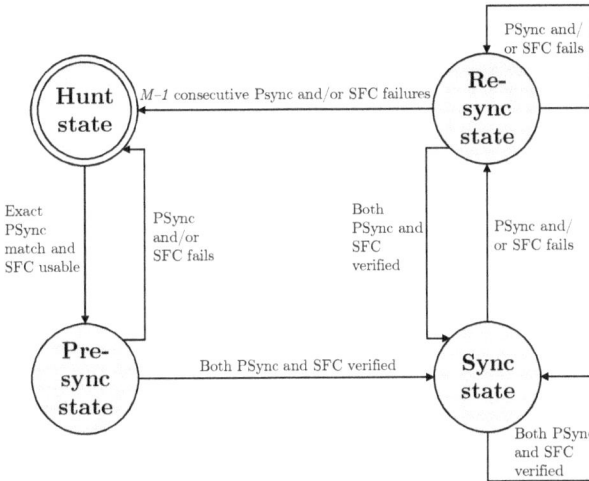

Figure 1. Synchronization state machine for XG-PON downstream [27].

State of O2-3, Serial Number State: In this state, the ONU activates its transmitter in a burst mode and waits for messages from OLT. The ONU analyzes the PLOAM section of the downstream XG-PON transmission convergence layer (XGTC) frame and begins to learn the burst profile specified in the Profilesection of the message. Upon receiving a serial number grant, it reports with the XGTC frame carrying the Serial Number ONU PLOAM message to send its serial number. As soon as it receives an AssignONU-ID PLOAM message with its serial number, it sets the allocated ONU-ID together with the other assigned parameters and moves to the next state. After receiving the Disable Serial Number PLOAM message (for its serial number or for all ONUs), it moves to the EmergencySTOPState. If OLT already knows the ONU that is returning to the network (e.g., during recovery, power failure, etc.), a problem with the Assign ONU-ID PLOAM message could occur. Therefore, the ONU can go directly to the state called the Ranging State when activated without responding to the serial number grant request.

State O4, Ranging state: In this state, the ONU receives a ranging grant with a known burst profile. Consequently, the XGTC frame containing the RegistrationPLOAM message is transmitted as a response. The ONU analyzes the PLOAM section of the downstream XGTC frame and responds only to the following messages: Profile, Ranging Time, Deactivate ONU-ID and Disable Serial Number. If the ONU receives the Ranging Time message with the absolute equalization delay, it moves to the next state. In this state, the TO1 timer is used to cancel unsuccessful attempts by limiting the time that the ONU can stay in that state. The recommended value for the TO1 timer is 10 s. If it expires, the unit discards the associated ONU-ID, as well as all other parameters, and returns to the Serial Number state.

State O5, Operationstate: The ONU already transmits data and PLOAM messages in the upstream direction as instructed by the OLT. At this point, the OLT can create additional connections with the ONU if they are required. Once the network is in operation and all ONUs are working with their assigned equalization delay, all upstream bursts are synchronized with all ONUs.

State O6, IntermittentLODS state: The ONU will move to this state from the Operation state when it does not synchronize with the downstream signal. Upon entering this state, the ONU will start the TO2 timer (the recommended value for this timer is 100 ms). After the timer expires, the ONU returns to the Initial state.

State O7, Emergency Stop State: The ONU moves to this state if it receives the Disable Serial Number message with the "Disable" option. In this state, it switches off the laser and rejects all TC settings (ONU-ID, equalization delay, burst profiles, etc.). The ONU keeps the downstream synchronization machine running and analyzes the XGTC frames in the downstream direction (at this point, however, it is forbidden to pass any downstream data or send any upstream data). If the problem is resolved, the OLT can re-enable the ONU and bring it back to normal operation by sending the Disable Serial Number with the "enable" option. As a result, the ONU returns to the Initial state.

4.3. NG-PON2 Activation Process

The activation process is provided by time and wavelength division multiplexing transmission convergence (TWDM-TC) and is defined by the recommendation [28]. In the NG-PON2 standard, there are two options for the PLOAM channel. The in-band option is a classical PLOAM message transmission, and the auxiliary management and control channel (AMCC) option is mandatory for ONUs that do not meet the specified calibration limits for a given upstream wavelength channel (see Figure 2).

State O1, Initial state: The ONU is in this state when it is turned on. At this point, scanning and downstream channel calibration occur. The unit can also move to this state when deactivated, or when the emergency stop is on. The transmitter is off and the previously set parameters, such as the ONU-ID, burst profiles and equalization delay, should be deleted. Next, the synchronization machine (see Figure 2) is started. The substate, O1.1, is called Off-Sync. In this state, the ONU searches for downstream synchronization attempts. As soon as the synchronization is finished, the ONU moves to the next substate, O1.2, known as the Profile Learning. When enough information has been gathered, the ONU evaluates the downstream wavelength of the channel. If the channel is suitable for activation, the ONU continues the process and moves to the next state. However, if it is not suitable, it searches for an alternative channel and returns to the O1.1 substate, retaining system and channel information, but discarding information about the burst profile.

State of O2-3, Serial number state: In this state, ONU activates its transmitter and tries to tune the upstream wavelength channel in line with the downstream wavelength channel. Once the ONU meets the minimum requirements for calibration accuracy for the required upstream wavelength channel, it receives a request known as an SN in-band grant to send the serial number. The message Serial Number ONU is sent as a response to this request. However, if the ONU does not meet the minimum calibration accuracy, it receives a request to send the AMCC type number. In this case, the AMCC Serial Number ONU PLOAM message is sent as a response to this request. Next, the ONU waits for an OLT response, which may be in the form of an Assign ONU-ID message, a Calibration Requestor an Adjust Tx Wave-length PLOAM message. Depending on the received message or request, the ONU either stays in this state and tunes the transmitter, returns to the initial state O1 so that another TWDM channel can be calibrated or moves to the next state and continues with the activation process. In this state, the ONU starts a discovery timer called TOZ. If this timer expires without the ONU receiving a response from the OLT, it returns to the O1 state. In this case, the unit discards all the accumulated system, channel and burst profile information.

State O4, Ranging state: In this state, the ONU responds to the ranging grant. If it receives a burst profile ranging grant from the previous Burst Profile PLOAM message, the FS burst carrying the Registration PLOAM message is transmitted. As soon as the ONU receives the Ranging Time message with the equalization delay, it moves to the next state. In this state, it starts the T01 timer with the recommended duration of 10 seconds. If the timer expires, the ONU deletes the allocated ONU-ID along with all the previously set parameters and returns to the O2-3 state, while retaining the collected profile information.

State O5, Operation state: In this state, the ONU is already processing frames in the downstream direction and transmits bursts in the upstream direction as instructed by the OLT. This particular state is divided into two substates. The entry point of this state is O5.1, which is called Associated. The ONU is associated with a specific TWDM channel, and the no Tuning ControlPLOAM message awaits processing. Another substate, O5.2, is called Pending. While the ONU completes upstream transmission of SDU units whose fragmentation already began in the previous subset, it performs further fragmentation if necessary and transfers any unfragmented SDU units.

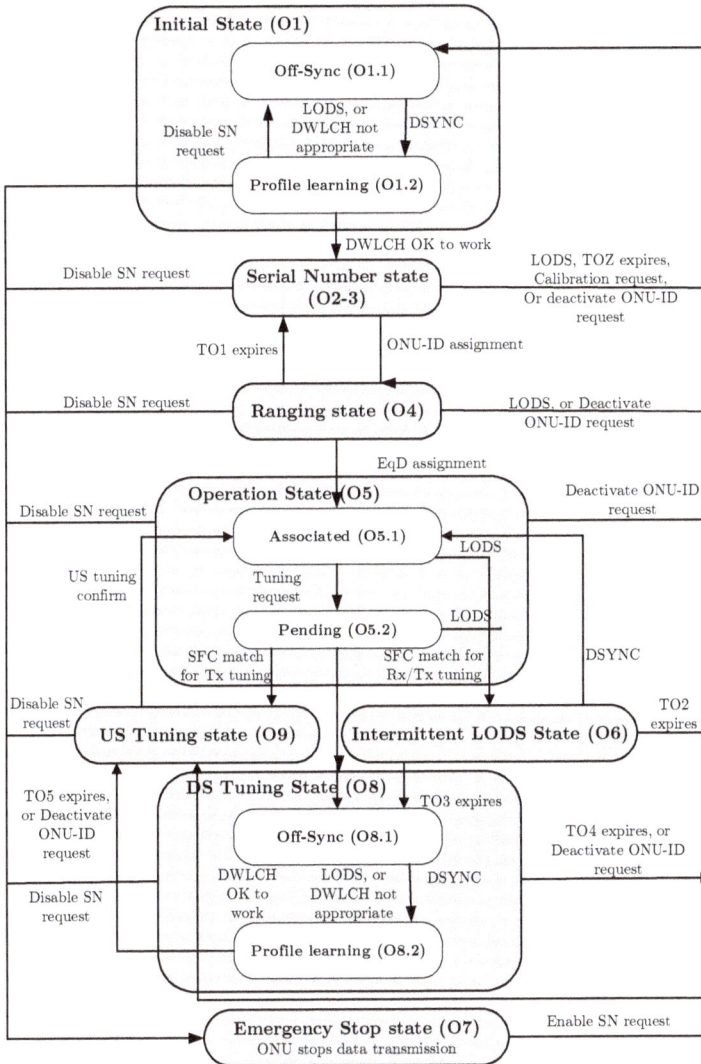

Figure 2. State diagram of ONU unit for NG-PON2 [28].

State O6, Intermittent LODS state: The ONU can reach this state from the O5 state in the case of downstream synchronization loss. Upon entering this state, the unit turns on the timer. When wavelength channel protection (WLCP) is enabled, the TO3 timer is turned on. If the WLCP is turned off, the TO2 timer is turned on. If the downstream signal is restored before any of the two timers expire, the ONU returns to the O5 state. However, once the TO2 timer expires, the ONU moves to the initial state O1. If, on the other hand, the TO3 timer expires, the ONU moves to the O8 state (to be described; see below).

State O7, Emergency Stop State: The ONU moves to this state if it receives the Disable Serial Number message with the "Disable" option on. In this case, it deactivates the laser. However, it keeps the downstream synchronization machine running and analyzes the PLOAM section of the downstream FS frames (however, at this point it is forbidden to pass any downstream data or send any upstream data). If the ONU receives the Disable Serial Number message with the "enable" option on, it returns to the O1 state.

State O8, Downstream tuning state: In this state, the ONU tries to restore the transmission using the new TWDM channel while maintaining the configuration of the TC layer except for its burst profiles. In this state, the TO4 timer is used. When it expires, the ONU returns to the initial O1 state and discards the TC layer configuration. In the O8.1 substate, also known as Off-Sync, the ONU tunes its receiver and tries to synchronize with the downstream signal. As soon as it is synchronized, it moves to the O8.2 substate, known as Profile Learning. In this state, it analyzes the downstream framing sublayer (FS) frame and starts collecting information about the system, channel and burst profile. When enough information has been gathered, the ONU will evaluate the downstream wavelength of the channel. If this channel is suitable for activation, the ONU continues the activation process and moves to the next state. However, if it is not suitable, it searches for an alternative channel and returns to the O8.1 substate, retaining the system and channel information, but discarding the burst profile information.

State O9, Upstream tuning state: As long as the ONU is in this state, it waits for a feedback from the OLT and performs a fine-tuning of its transmitter. Subsequently, it moves to the O5 state. In this state, the TO5 timer is started. If this timer expires, the ONU returns to the initial state.

4.4. EPON Activation Steps

The following subsection evaluates the EPON activation process according to [24]. The MPCP defines the autodiscovery mechanism used to detect the newly-connected ONUs, a circular delay and a MAC address [29]. This process is controlled by the OLT unit, which periodically creates an available discovery window, during which time it gives inactive units the ability to log in to the OLT. This periodicity is not specified by the standard and therefore depends on individual implementation. Autodiscovery uses the following four messages: GATE, REGISTER REQ, REGISTER and REGISTER ACK. These messages are transmitted in the MPCP frame. The autodiscovery process consists of four steps, which are illustrated in Figure 3.

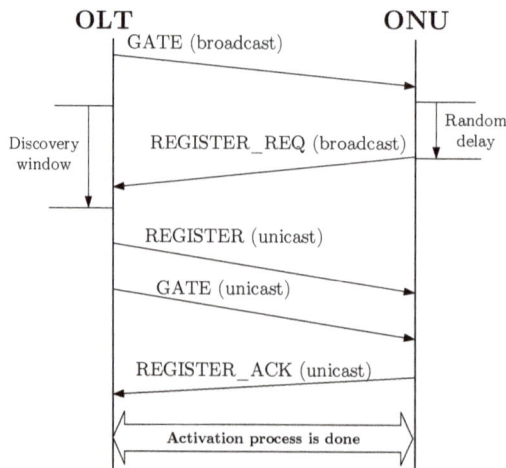

Figure 3. Activation process in EPON networks.

Step 1: The Discoveryagent decides to initiate the discovery process and assigns a discovery window (the time interval when none of the initialized ONUs can send data). The Discovery Agentinitiates the discovery process using the Discovery GATE message, which includes the starting time and the length of the slot. During the forwarding of the GATE discovery message, the MPCP stores the OLT time.

Step 2: Only previously uninitialized ONUs respond to the GATE message. After receiving such a message, the ONU sets the local time according to it. If the OLT clock reaches the starting time that is also included in the GATE discovery report, the ONU waits for a randomly selected time and then forwards the REGISTER REQ message. Accidental delays can lead to collisions when initiating multiple ONUs. The REGISTER REQ message contains the ONU source address and the time used to send the message from the ONU. When the OLT receives a REGISTER REQ message, it detects the MAC address and the circular delay.

Step 3: After analyzing and verifying the REGISTER REQ message, the OLT sends the message REGISTER directly to the given ONU using the MAC address obtained during the previous step. The REGISTER message contains a unique logical link identifier (LLID) that is assigned to all ONUs. Next, the OLT sends the GATE message to the same ONU.

Step 4: After the REGISTER and GATE ON messages are received, the REGISTER ACK confirms that an acceptance of the previous messages has been sent. The REGISTER ACK should be sent in the time interval granted by the GATE message.

5. Simulation Results

Simulations were performed in a MathWorks MATLAB™ environment for each standard separately according to their recommendations. A description of the triggering processes of the individual standards that constitute the simulations is in Section 4.

In the first instance, the user set some of the parameters influencing the simulation. At first, the standard to be simulated was chosen. The available choices were the following: GPON, XG-PON, NG-PON2, EPON and the optimized GPON. Furthermore, the number of ONUs to be connected to the OLT at the specified interval (the maximum number of ONUs that can connect was determined by the selection of the standard mentioned previously) and the length of the distribution network also need to be chosen at the specified interval needs. Then, the total activation time was calculated. The output was visualized as a graph that shows how long the activation of the ONUs that were connected to the OLT occurred.

5.1. Simulations for ITU Standards

The ITU activation processes are always described within the TC layer and are divided into several states. The simulation was based primarily on the ONU's activation description (see Section 4). Information was transmitted in PLOAM messages for 125 µs. The OLT operating status was the time granted to the ONU by the OLT to process the received message. It had a duration of 750 µs. The total ONU activation time consisted of the following: a synchronization time that was given to the synchronization state machine, the time of the transmission of the messages and their processing by the ONU, the generated window, the propagation delay that was added to each message, the calculated equalization delay and the activation time of all previously activated ONUs. All time constants were specified by the standards [24,26].

In Stage O1 of the GPON standard, synchronization with the downstream signal provided by the synchronization state machine was required. In the simulation, only the initial synchronization was solved. In all cases, two consecutive frames with the correct PSync were needed. Random and incorrect PSync were generated. If the PSync was correct, the N variable was incremented. Otherwise, it was set to zero. Synchronization ended when N was equal to M (M was set to two). Next, the OLT operation status followed. Because the simulation did not consider collisions, there were no standard timers to prevent the ONU from remaining in any part of the activation for an indefinite time.

Propagation delay was the time it took to move a message from the OLT to the ONU. It depended on the length of the distribution network to the unit and the transmission rate in the environment. In the simulation, the propagation delay was added to the downstream and upstream messages and was calculated for each unit separately according to the following formula:

$$T_{pd} = \frac{l_i}{\frac{c}{n}},$$ (1)

where l_i represents the distance between the ONU and OLT, c is the speed of light and n is the refractive index.

5.1.1. GPON

The simulation was based primarily on the description of the ONU activation process described in Section 4.1. The Activation passed through four different states until it reached the fifth state in which it was finally possible to transfer the data. The time flow and the transmitted messages are shown in Figure 4. As shown, all downstream PLOAM messages were sent three times. State O1 and State O2, described previously, were unified for each ONU. After creating a quiet window, the ONUs sent messages with their serial numbers. The unit whose message was received by the OLT first was activated, and the other units were forced to wait. Activation took place until all units were finally connected. In the O4 state, the equalization delay was measured and subsequently transmitted by the Ranging Time message. The ONU must use this delay in its message transfer. Because the GPON standard measures this delay, the computation performed during the simulation presented in this work must be considered as exploratory in nature. The computation was based on information proposed in [26].

$$EqD_i = T_{eqd} - RTD_i,$$ (2)

where T_{eqd} is the so-called zero-distance EqD (offset between the downstream frame and the obtained frame that was requested). It can be computed as follows [26]:

$$T_{eqd} = 2T_{pd} + RspTime + Prd + USframe,$$ (3)

where $RspTime$ is the response time of the ONU with its value set to 35 μs, Prd is the pre-assigned delay set to 202 μs and $USframe$ is the upstream frame duration.

RTD is the usual delay expressed as [26]:

$$RTD_i = 2T_{pd} + RspTime.$$ (4)

During the simulation, the program lists the current status, the transmitted messages and the total time of activation of the given ONU in milliseconds.

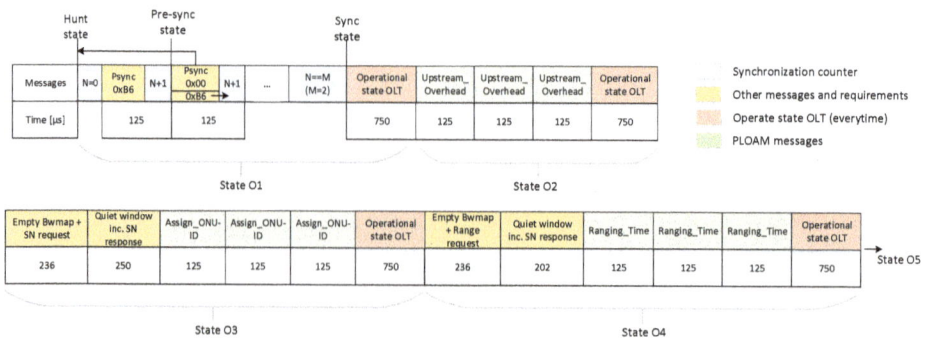

Figure 4. Visualization of the GPON activation.

5.1.2. XG-PON

In principle, activation for XG-PON is based on GPON. However, there have been several major changes such as: downstream PLOAM messages are only sent once; States O2 and O3 are merged together into a single state; some PLOAM messages are altered; and the duration of the quiet window depends on the fiber spacing distance, which is the value determined by the difference between the fiber length of the most remote and the nearest ONU from the OLT. In our simulation, the most remote unit was set by the user with the highest possible value of 40 km. At a 20-km difference, the quiet window duration was the same as for GPON, and at a distance of 40 km, the quiet window was increased by 200 µs. The nearest unit was set to 1 km. The entire activation process is described in Section 4.2, and the messages along with the transmission process are shown in Figure 5.

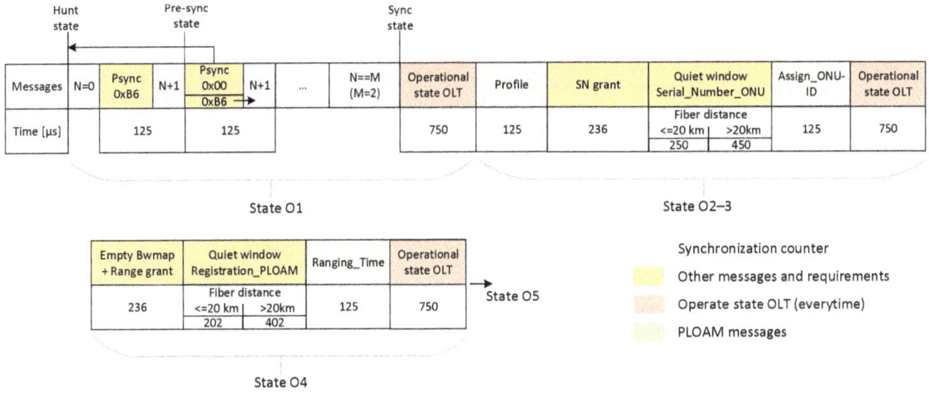

Figure 5. Visualization of the XGPON activation.

The basic equation for the computation of equalization delay was the same as in the case of GPON. However, the calculation of the individual components varied, as for XGPON EqD_i were computed as [27]:

$$T_{eqd} \geq RspTime_{max} + (L_{min} + D_{max}) \cdot \left(\frac{n_{1577} + n_{1270}}{c} \right), \tag{5}$$

where L_{min} represents the minimum distance, D_{max} is the maximum differential distance and the refractive index values for the wavelengths n_{1577} and n_{1270} are given as follows: $n_{1577} = 1.4686$ and $n_{1270} = 1.4677$. RTD is defined as [27]:

$$RTD_i = T_{pd} \frac{n1270 + 1577}{n1270} - RspTime_i. \tag{6}$$

5.1.3. NG-GPON2

For simulation of NG-PON2 (the activation process is described in Section 4.3), the PLOAM in-band message transmission option (i.e., a common transmission) was selected. In principle, the simulation is very similar to that of XG-PON, the only difference being the use of other types of messages. Time transmitted messages are shown in Figure 6.

For the NG-PON2 simulation, the same equations as the XG-PON simulation were used.

Hunt state — Pre-sync state — Sync state

Messages	N=0	Psync 0xB6	N+1	Psync 0x00 0xB6	N+1	...	N==M (M=2)	Operational state OLT
Time [µs]		125		125				750

- Synchronization counter
- Other messages and requirements
- Operate state OLT (everytime)
- PLOAM messages

State O1

Channel_Profile	SN grant	Quiet window inc. Serial_Number_ONU		Adjust_Tx_Wavelength	Assign_ONU-ID	Operational state OLT	Empty BWmap + Ranging grant	Quiet window inc. Serial_Number_ONU		Ranging_Time	Operational state OLT
		Diff. Fiber distance						Diff. Fiber distance			
		<=21 km	>21km					<=21 km	>21km		
125	236	250	450	125	125	750	236	202	402	125	750

State O2-3   State O4   State O5

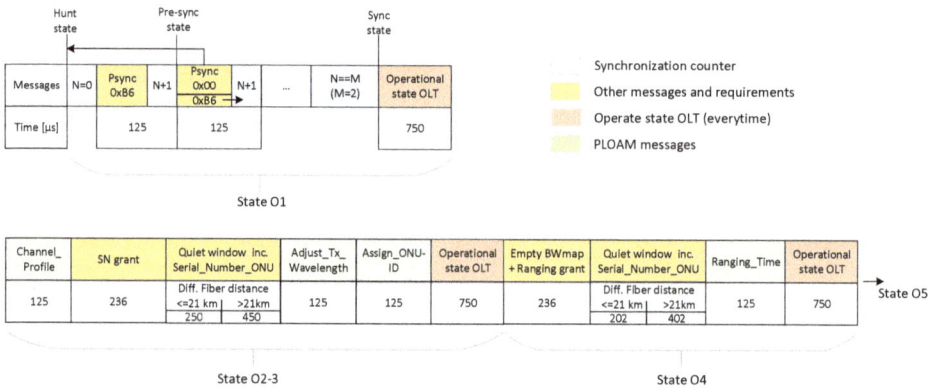

Figure 6. Visualization of the NG-PON2 activation.

5.2. EPON

In an IEEE-based standard, the connection of the units to the network is provided by the autodiscovery agent contained in the MPCP. Because of the activation process being controlled by the parent protocol, the activations for the EPON and 10G-EPON standards are identical, so only one simulation called EPON had been performed. In contrast with the ITU standard, IEEE does not directly define the accident time, discovery window, etc. Instead, these were calculated based on the number of connected units and the distance of the ONU from the OLT. However, the IEEE does not explicitly describe the procedures and mechanisms for these times in its recommendation, so the simulation was based not only on the information obtained from the recommendation, but also from [24,30].

The presented results were based on the graph comparing the EPON and GPON with 32/64 units and a 20-km ODN length. The graphs are shown in Figure 7. As shown, the activation was several times faster using EPON instead of GPON. This was primarily because the selected frame length in EPON was only 409.6 ns, whereas in GPON, it was 125 µs. The total activation time was much easier to evaluate for EPON than for GPON. Fewer messages were transmitted, and they were transmitted only once (GPON sends PLOAM messages three times); the minimum message processing time of the ONU was only 16.384 µs, whereas for GPON, it was 750 µs, and the equalization delay was not calculated during the simulation of EPON.

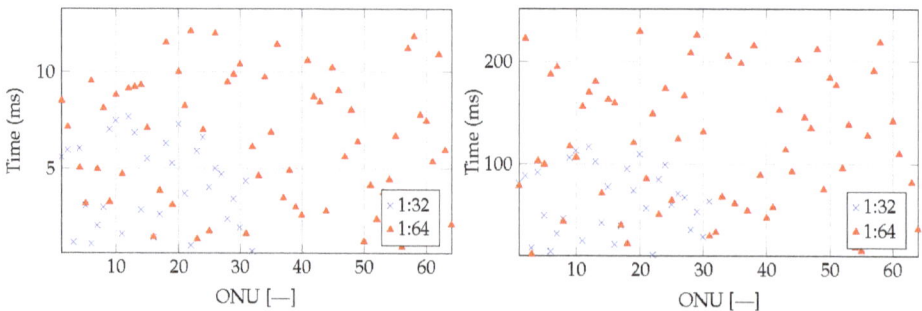

Figure 7. Graph of activation time for EPON (**left**) and GPON (**right**) 32/64 units and 20-km ODN.

Figure 8 compares the activation process in XG-PON and NG-PON2. As mentioned previously, the activation process for XG-PON and NG-PON was based on GPON. However, there were some new states of substates. In other words, the PLOAM messages were almost identical. We did not consider a tuning process for changing the wavelength in the downstream or upstream because this process lies at the OLT side.

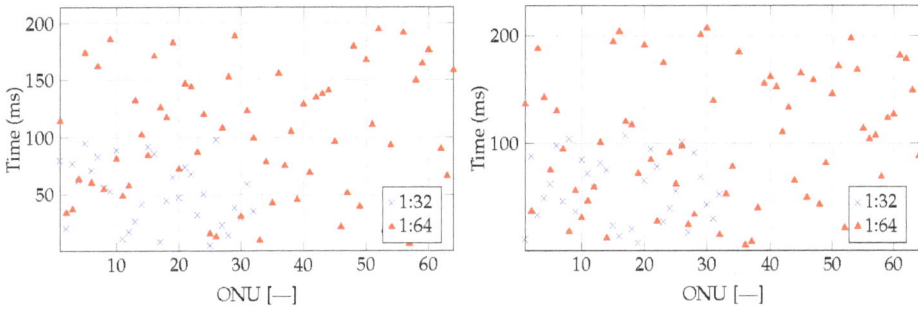

Figure 8. Graph of activation time for XG-PON (**left**) and NG-PON2 (**right**) 32/64 units and 20-km ODN.

On the left side, Figure 9 compares the XG-PON and NG-PON2 standards for 128 connected units and a 20-km ODN length. The maximum connection time was several ms higher for NG-PON2 than for XG-PON. The increase in connection time can be explained by the higher number of transmitted messages and the occurrence of other synchronization processes. A maximum connection time of up to 420 ms can be observed. Nevertheless, it was necessary to consider the limitations of the simulation, e.g., the simulation did not consider any collisions that can normally occur in the real network (the time necessary for the activation process would be increased in such cases), etc. If we consider the higher split ratio (1:256), which is the maximum split ratio in XG-PON and NG-PON2, we observe an almost two-fold higher value for the activation process for a 20-km ODN length.

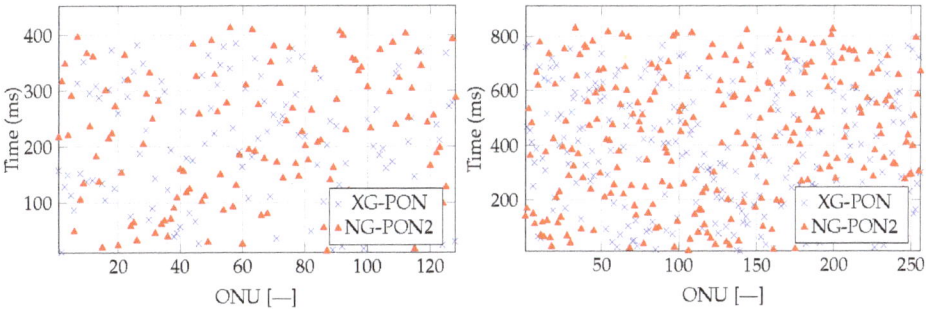

Figure 9. Graph of activation time for XG-PON and NG-PON2 with a 1:128 split ratio and 20-km length of ODN (**left**) and XG-PON and NG-PON2 with a 1:256 split ratio and a 20-km length of ODN (**right**).

The final aim of this study was an optimization of the GPON standard. To preserve the physical essence, the calculations and the length of the messages cannot be altered. Optimization, therefore, serves to reduce the number of messages and the length of the OLT operating state, which is the time that OLT grants the ONU to process incoming messages. In GPON, all messages are transmitted three times. Hence, in theory, ONU should require less processing time. This time was set to 350 µs. This is the sum of the duration of the sent message (125 µs), the maximum propagation delay at 20 km, which was approximately 100 µs, and the message processing time (125 µs). The entire timing of the transmitted messages is indicated in Figure 10. A real GPON does not contain only 32 or 64 ONUs, and the current OLT supports up to 16 GPON ports per card. The total amount of these cards depends on the OLT chassis. We considered 16 GPON ports with 32 ONUs per port. The total amount of ONUs was 512 ONUs, which must be activated. The current activation process can activate all ONUs in ≈112 s or 74 s with our optimization, respectively. Furthermore, our solution did not require a new PLOAM message or TC layer changes.

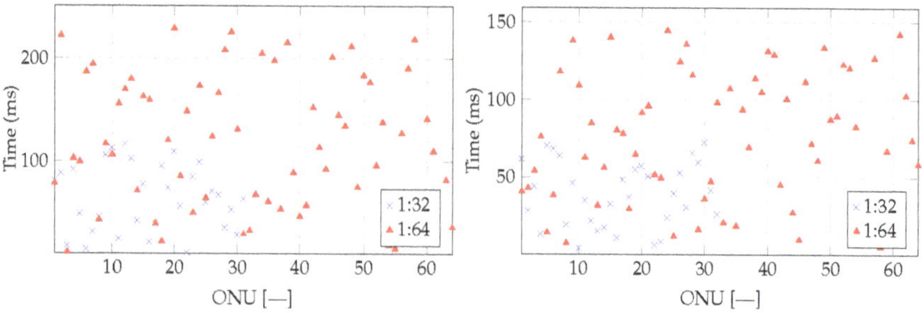

Figure 10. Graph of activation time for GPON (**left**) and optimized GPON (**right**) 32/64 units and 20-km ODN.

Our optimization was based on current PLOAM messages and transmission convergence layer. There were no changes required. Visualization of the optimized GPON activation process is shown in Figure 11.

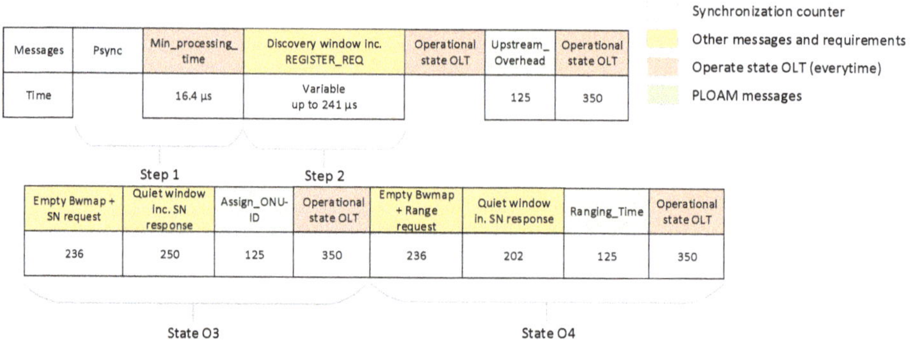

Figure 11. Visualization of the optimized GPON activation process.

6. Conclusions

In this paper, we introduced an activation process for IEEE and ITU PONs. Although both networks are standards for passive optical networks, they use completely different encapsulation methods, frame durations and frame structures. Our results demonstrate that EPONs have a faster activation process (with a maximum split ratio 1:32) and that GPONs operate with a split ratio of up to 1:128. We chose the most commonly-used split ratio, 1:32 or 1:64, with 20 km, because a higher distance does not follow either standard. However, GPON supports a higher split ratio, and there are some issues with the timing and managing of the time slots for the upstream direction. The main reason for the shorter activation time is the different frame durations of 16.384 µs and 750 µs for EPONs and GPONs, respectively.

XG-PON and NG-PON2 use approximately the same activation process and provide almost identical results as do GPONs. However, NG-PON2 supports wavelength tuning during transmission, but it has to be initialized by the OLT. Due to this fact, it is not a part of the activation process. An activation time of 256 end units requires approximately 820 ms, but the OLT can operate with eight or more NG-PON2 ports, which leads to a multiplication of this time. During a blackout scenario, the last end user will have to wait multiples of minutes.

Our optimization accelerates the activation process for 64 ONUs from 220 ms to 145 ms. In a real network with 512 ONUs, the operator can save approximately 40 ms. Note that our solution does not require a new PLOAM message or TC layer changes.

In future work, we would like to implement a collision scheme into our simulation models.

Author Contributions: Conceptualization, T.H. and P.M. Methodology, T.H. and P.M. Software, V.O. and J.V. Validation, T.H. and P.M. Formal analysis, J.V. Investigation, T.H. Resources, P.M. Writing, original draft preparation, T.H., P.M., V.O. and J.V. Writing, review and editing, T.H. Visualization, V.O. Project administration, P.M. Funding acquisition, P.M. and T.H.

Funding: The presented research has been supported by projects of the Ministry of the Interior under Grant No. VI20172019072 registration, the National Sustainability Program under Grant No. LO1401, and E-infrastructure CESNET – modernization, registration number CZ.02.1.01/0.0/0.0/16_013/0001797.

Conflicts of Interest: The authors declare no conflict of interest.

Abbreviations

The following abbreviations are used in this manuscript:

ADSL	Asymmetric digital subscriber line
Alloc-ID	Allocation identifier
AMCC	Auxiliary management and control channel
APON	Asynchronous transfer mode passive optical network
BWmap	Bandwidth map
C-RAN	Cloud-radio access network
CDR	Clock and data recovery
DBA	Dynamic bandwidth allocation
DDSPON	Distributed dynamic scheduling passive optical network
EPON	Ethernet passive optical network
EqD	Equalization delay
FS	Framing sublayer
GMII	Gigabit media independent interface
GPON	Gigabit passive optical network
GTC	GPON transmission convergence layer
HEC	Hybrid error correction
IEEE	Institute of Electrical and Electronics Engineers
ISP	Internet services provider
ITU	International Telecommunication Union
LLID	Logical link identifier
MAC	Medium access control
MDI	Medium-dependent interface
MLM	Multi-longitudinal mode
MPCD	Multi-point control protocol
NG-PON2	Next Generation Passive Optical Network Stage 2
NRZ	Non-return-to-zero
LoF	Loss of frame
LoS	Loss of signal
ODN	Optical distribution network
OLT	Optical line termination
ONU	Optical network unit
ONU-ID	ONU identifier
ORL	Optical return loss
P2MP	Point to multipoint
PCBd	Physical control block downstream
PCS	Physical coding sublayer
PHY	Physical interface
PLOAM	Physical layer operations, administration and maintenance
PMA	Physical medium attachment
PMD	Physical medium dependent
PON	Passive optical network
PSBd	Physical synchronization block downstream

PSync   Physical synchronization
QoS   Quality of service
RS   Reconciliation sublayer
RTD   Round trip delay
SFC   Superframe counter
SLM   Single-longitudinal mode
TC   Transmission convergence
TWDM-TC   Time and wavelength division multiplexing transmission convergence
VHDL   Very High Speed Integrated Circuit Hardware Description Language
WDM   Wavelength-division multiplexing
WiFi   Wireless fidelity
XG-PON   Next generation passive optical network
XGTC   XG-PON transmission convergence layer
WLCP   Wavelength channel protection

References

1. Horvath, T.; Munster, P.; Vojtech, J.; Velc, R.; Oujezsky, V. Simultaneous transmission of accurate time, stable frequency, data, and sensor system over one fiber with ITU 100 GHz grid. *Opt. Fiber Technol.* **2018**, *40*, 139–143. [CrossRef]
2. Vojtech, J.; Slapak, M.; Skoda, P.; Radil, J.; Havlis, O.; Altmann, M.; Munster, P.; Velc, R.; Kundrat, J.; Altmannova, L.; et al. Joint accurate time and stable frequency distribution infrastructure sharing fiber footprint with research network. *Opt. Eng.* **2017**, *56*, 027101. [CrossRef]
3. Ansari, N.; Zhang, J. *Media Access Control and Resource Allocation*; Springer: New York, NY, USA, 2013; p. 103, ISBN 978-1-4614-3939-4.
4. Ballance, J.W.; Rogers, P.H.; Halls, M.F. ATM Access through a Passive Optical Network. *Electron. Lett.* **1990**, *26*, 558–560. [CrossRef]
5. Miyata, S.; Baba, K.I.; Yamaoka, K. Exact Mean Packet Delay for Delayed Report Messages Multipoint Control Protocol in EPON. *J. Opt. Commun. Netw.* **2018**, *10*, 209–219. [CrossRef]
6. Chen, W.; Montuno, D.Y.; Felske, K. Enhanced EPON auto-discovery for fast network and service recovery. In Proceedings of the Canadian Conference on Electrical and Computer Engineering 2004 (IEEE Cat. No. 04CH37513), Niagara Falls, ON, Canada, 2–5 May 2004; pp. 1683–1687.
7. Hajduczenia, M.; de Silva, H.J.A.; Monteiro, P.P. Extended GATE/REPORT MPCP DUs for EPONs. In Proceedings of the 11th IEEE Symposium on Computers and Communications (ISCC'06), Cagliari, Italy, 26–29 June 2006; pp. 1–6.
8. Liem, A.T.; Hwang, I.-S.; Nikoukar, A.; Yang, C.-Z.; Ab-Rahman, M.S.; Lu, C.-H. P2P Live-Streaming Application-Aware Architecture for QoS Enhancement in the EPON. *IEEE Syst. J.* **2016**, *12*, 648–658. [CrossRef]
9. Petridou, S.; Basagiannis, S.; Mamatas, L. Energy-Efficiency Analysis under QoS Constraints Using Formal Methods: A Study on EPONs. In Proceedings of the 2017 IEEE International Conference on Communications (ICC), Paris, France, 21–25 May 2017; pp. 1–6.
10. Li, L.; Xin, S.T.; Duan, D.G. Research of DBA Schemes and QoS in PON System. In Proceedings of the 2016 2nd IEEE International Conference on Computer and Communications (ICCC), Chengdu, China, 14–17 October 2016; pp. 2148–2153.
11. Mady, A.E.D.; Tonini, A. A VHDL Implementation of ONU Auto-discovery Process for EPON. In Proceedings of the ICNM: 2009 International Conference on Networking & Media Convergence, Cairo, Egypt, 24–25 March 2009; pp. 36–41.
12. Garfias, P.; Gutierrez, L.; Sallent, S. Enhanced DBA to Provide QoS to Coexistent EPON and 10G-EPON Networks. *J. Opt. Commun. Netw.* **2012**, *4*, 978–988. [CrossRef]
13. Pajcin, B.; Matavulj, P.; Radivojevic, M. MPCP adjusting for improving QoS in Green WDM EPON. In Proceedings of the 2016 International Workshop on Fiber Optics in Access Network (Foan), Lisbon, Portugal, 18–19 October 2016; pp. 1–6.

14. Zhang, M.; Liu, X.F.; Song, C.; Zhang, H. Grouped LR-EPON with WDM-OFDM-TDM DBA Algorithm and Sleep-Based Energy-Efficient Mechanism. In Proceedings of the 2016 15th International Conference on Optical Communications and Networks (ICOCN), Hangzhou, China, 24–27 September 2016; pp. 1–3.
15. Liu, C.P.; Wu, H.T.; Ke, K.W. Asymmetric Loading on the Energy Saving Scheme for EPON Networks. In Proceedings of the 2017 IEEE International Conference on Consumer Electronics—Taiwan (ICCE-TW), Taipei, Taiwan, 12–14 June 2017.
16. Horvath, T.; Munster, P.; Jurcik, M.; Filka, M. Novel Algorithm in Activation Process of GPON Networks. *J. Commun. Softw. Syst.* **2015**, *11*, 204–209. [CrossRef]
17. Horvath, T.; Munster, P.; Jurcik, M.; Koci, L.; Filka, M. Timing measurement and simulation of the activation process in gigabit passive optical networks. *Opt. Appl.* **2015**, *45*, 459–471. [CrossRef]
18. Koci, L.; Horvath, T.; Munster, P.; Jurcik, M.; Filka, M. Transmission Convergence Layer in XG-PON. In Proceedings of the 2015 38th International Conference on Telecommunications and Signal Processing (TSP), Prague, Czech Republic, 9–11 July 2015; pp. 104–108.
19. Luo, Y.Q.; Effenberger, F.; Gao, B. Transmission Convergence Layer Framing in XG-PON1. In Proceedings of the 2009 IEEE Sarnoff Symposium, Conference Proceedings, Princeton, NJ, USA, 30 March–1 April 2009; pp. 298–302.
20. Eramo, V.; Listanti, M.; Lavacca, F.G.; Iovanna, P.; Bottari, G.; Ponzini, F. Trade-Off Between Power and Bandwidth Consumption in a Reconfigurable Xhaul Network Architecture. *IEEE Access* **2016**, *4*, 9053–9065. [CrossRef]
21. Eramo, V.; Listanti, M.; Lavacca, F.; Iovanna, P. Dimensioning Models of Optical WDM Rings in Xhaul Access Architectures for the Transport of Ethernet/CPRI Traffic. *Appl. Sci.* **2018**, *8*, 612. [CrossRef]
22. International Telecommunication Union. G.984.1: Gigabit-Capable Passive Optical Networks (GPON): General Characteristics. Available online: https://www.itu.int/rec/T-REC-G.984.1-200803-I (accessed on 18 September 2018).
23. International Telecommunication Union. G.984.2: Gigabit-Capable Passive Optical Networks (G-PON): Physical Media Dependent (PMD) Layer Specification. Available online: https://www.itu.int/rec/T-REC-G.984.2-200303-I/en (accessed on 18 September 2018).
24. *IEEE Standard for Information Technology—Local and Metropolitan Area Networks—Specific Requirements—Part 3: CSMA/CD Access Method and Physical Layer Specifications Amendment 1: Physical Layer Specifications and Management Parameters for 10 Gb/s Passive Optical Networks*; IEEE Std 802.3av-2009 (Amendment to IEEE Std 802.3-2008); IEEE: New York, NY, USA, 2009; pp. 1–227. [CrossRef]
25. Widmer, A.X.; Franaszek, P.A. Transmission Code for High-Speed Fibre-Optic Data-Networks. *Electron. Lett.* **1983**, *19*, 202–203. [CrossRef]
26. International Telecommunication Union. G.984.3: Gigabit-Capable Passive Optical Networks (G-PON): Transmission Convergence Layer Specification. Available online: https://www.itu.int/rec/T-REC-G.984.3-201401-I/en (accessed on 18 September 2018).
27. International Telecommunication Union. G.987.3: 10-Gigabit-Capable Passive Optical Networks (XG-PON): Transmission Convergence (TC) Layer Specification. Available online: https://www.itu.int/rec/T-REC-G.987.3 (accessed on 18 September 2018).
28. International Telecommunication Union. G.989.3: 40-Gigabit-Capable Passive Optical Networks (NG-PON2): Transmission Convergence Layer Specification. Available online: https://www.itu.int/rec/T-REC-G.989.3 (accessed on 18 September 2018).
29. Kramer, G. *Ethernet Passive Optical Networks (Professional Engineering)*, 1st ed.; McGraw-Hill Education: New York, NY, USA, 2005; 307p, ISBN 978-0071445627.
30. Cui, Q.P.; Ye, T.; Lee, T.T.; Guo, W.; Hu, W.S. Stability and Delay Analysis of EPON Registration Protocol. *IEEE Trans. Commun.* **2014**, *62*, 2478–2493. [CrossRef]

applied
sciences

MDPI

Article

Validation of Fractional-Order Lowpass Elliptic Responses of $(1 + \alpha)$-Order Analog Filters [†]

David Kubanek [1], Todd J. Freeborn [2,*], Jaroslav Koton [1] and Jan Dvorak [1]

[1] Department of Telecommunications, Brno University of Technology, Technicka 12, 616 00 Brno, Czech Republic; kubanek@feec.vutbr.cz (D.K.); koton@feec.vutbr.cz (J.K.); dvorakjan@phd.feec.vutbr.cz (J.D.)

[2] Department of Electrical and Computer Engineering, The University of Alabama, Tuscaloosa, AL 35487, USA

[*] Correspondence: tjfreeborn1@eng.ua.edu; Tel.: +1-205-348-6634

[†] This paper is an extended version of paper published in 2018 41st International Conference on Telecommunications and Signal Processing (TSP), Athens, Greece, 4–6 July, 2018.

Received: 15 November 2018; Accepted: 10 December 2018; Published: 13 December 2018

Abstract: In this paper, fractional-order transfer functions to approximate the passband and stopband ripple characteristics of a second-order elliptic lowpass filter are designed and validated. The necessary coefficients for these transfer functions are determined through the application of a least squares fitting process. These fittings are applied to symmetrical and asymmetrical frequency ranges to evaluate how the selected approximated frequency band impacts the determined coefficients using this process and the transfer function magnitude characteristics. MATLAB simulations of $(1 + \alpha)$ order lowpass magnitude responses are given as examples with fractional steps from $\alpha = 0.1$ to $\alpha = 0.9$ and compared to the second-order elliptic response. Further, MATLAB simulations of the $(1 + \alpha) = 1.25$ and 1.75 using all sets of coefficients are given as examples to highlight their differences. Finally, the fractional-order filter responses were validated using both SPICE simulations and experimental results using two operational amplifier topologies realized with approximated fractional-order capacitors for $(1 + \alpha) = 1.2$ and 1.8 order filters.

Keywords: fractional-order filters; fractional calculus; Chebyshev filters; low-pass filters; magnitude responses

1. Introduction

Fractional-order filter circuits are a class of electronic circuits that use concepts from fractional-calculus [1–3], which refers to the branch of mathematics concerning non-integer order differentiation and integration, to realize magnitude and phase characteristics that are not easily achievable using traditional integer-order design techniques. For example, while traditional lowpass filters typically yield $-20n$ dB/decade stopband attenuations (where n is the filter integer order), fractional-order filters provide attenuations of $-20(n + \alpha)$ dB/decade, where $0 < \alpha < 1$ is the fractional-component of the order. Fractional-order filters are being actively investigated with recent works exploring the circuit theory [4–8], implementation [9–13], and applications [14,15] of this class of circuits.

To date, there have been two approaches to realize fractional-order filter circuits: (1) using approximations of s^α in a fractional-order transfer function to realize an integer-order filter that implements the fractional response [10,11,14]; and (2) using fractional-order capacitors ($Z = 1/s^\alpha C$ where $0 < \alpha < 1$ and C is a pseudocapacitance with units F s$^{\alpha-1}$) as elements in traditional filter topologies [4,5,9]. It is important to note that using a fractional-order capacitor implies a fractional derivative of order $0 < \alpha < 1$. Therefore, the current–voltage relationship for this component is defined as:

$$i(t) = C\frac{d^\alpha v(t)}{dt^\alpha} \tag{1}$$

where $i(t)$ and $v(t)$ are the time-dependent current and voltage, respectively. One definition of a fractional derivative of order α is given by the Grünwald–Letnikov definition

$$_aD_t^\alpha f(t) = \lim_{h\to 0} \frac{1}{h^\alpha} \sum_{m=0}^{\left[\frac{t-a}{h}\right]} (-1)^m \frac{\Gamma(\alpha+1)}{m!\,\Gamma(\alpha-m+1)} f(t-mh) \tag{2}$$

where $\Gamma(\cdot)$ is the gamma function, $n-1 \leq \alpha \leq n$, and a and t are the terminals of fractional differentiations [2]. The Grünwald–Letnikov definition is presented here because this definition leads to a correct generalization of the current linear systems theory [3], but it is important to note that other definitions such as the Riemann–Liouville and Caputo definitions are also available for describing fractional-derivatives. Applying the Laplace transform to the fractional derivative of Equation (2) with zero initial conditions with lower terminal $a = 0$ yields

$$\mathscr{L}\left\{_0D_t^\alpha f(t)\right\} = s^\alpha F(s) \tag{3}$$

The transfer-function representations of fractional-order differential equations are widely used during the design of fractional-order analog filters. Using these representations does not require the computation of their time-domain fractional-order differential equation representations, which reduces their design complexity. The numerical complexity for simulations of fractional-order differential equations stems from the number of computations required to capture the significant number of addends [2]. Consider Equation (2), which is a series that requires a greater number of operations for greater values of t to capture the entire history of the function $f(t)$, especially for $t \gg a$ [2]. Podlubny noted, however, that for large t the history of the function at the lower terminal ($t = a$) can be neglected under certain assumptions. This reduces the numerical complexity required to simulate a fractional-order differential equation by applying the "short memory" principle. The "short memory" principle approximates the lower limit a with a moving lower limit ($t - L$) in cases where the behavior of the function is driven by the memory of the "recent past" [2], that is:

$$_aD_t^\alpha f(t) \approx _{(t-L)}D_t^\alpha f(t) \tag{4}$$

where $t > a + L$. This approximation does reduce the accuracy of the simulations, although the "memory length" (L) can be determined to meet a required accuracy (ϵ) with details given in [2]. Further, Podlubny noted that using the short-memory principle leads to reductions in accumulated rounding errors during simulations as a result of the fewer addends [2]. The "short memory" principle has recently been applied in electronics for FPGA implementations of fractional-order systems [16,17]. These works highlight the impact of different window sizes on the accuracy and necessary hardware to realize Grünwald–Letnikov implementations. While the complexities of simulating fractional-order differential equations are discussed here, this work employs transfer-function representations of fractional-order systems and does not implement the time-domain simulations of the underlying fractional-order differential equations.

Recent studies have presented methods to approximate the passband and stopband characteristics of traditional filter responses with fractional-order attenuations using fractional-order transfer functions. This approach requires appropriate selection of the transfer function coefficients to achieve the desired responses. To date, this method has been applied to realize lowpass Butterworth [18], Chebyshev [19], Inverse Chebyshev [5] and elliptic [20] fractional-order filter responses. While the coefficients of a $(1 + \alpha)$ fractional-order transfer function to approximate the passband and stopband ripple characteristics of a second-order elliptic lowpass filter are presented in [20], this work expands on those results to: (i) explore how the selected bandwidth in the least squares coefficient selection impacts the coefficients and resulting magnitude response; and (ii) validate the elliptic responses using circuit simulations and experimental measurements from topologies realized with approximated fractional-order capacitors. The least-squares fitting process is applied in this work to multiple frequency ranges to evaluate the impact of the selected frequency band on the coefficients and resulting transfer function magnitude characteristics. MATLAB simulations of the magnitude responses of $(1 + \alpha)$ order lowpass filters with fractional steps from $\alpha = 0.1$ to $\alpha = 0.9$ using the determined coefficients are presented to highlight the fractional-step compared to the second-order elliptic response. Further, SPICE simulations and experimental measurements of $(1 + \alpha) = 1.2$ and 1.8 order filters implemented with operational amplifier topologies using coefficients from the fitting process are given to validate the magnitude characteristics and stability of the proposed circuits.

2. Approximated Lowpass Elliptic Response

The elliptic or Cauer filter approximation is characterized by having ripples in both passband and stopband of the magnitude response. This results in a magnitude response that has a faster attenuation increase through that transition from passband to stopband than comparable order Butterworth, Chebyshev, or Inverse-Chebyshev responses [21]. A second-order elliptic lowpass filter can be realized using lowpass notch circuits described by the transfer function

$$H(s) = k\frac{s^2 + \omega_z^2}{s^2 + s\frac{\omega_0}{Q} + \omega_0^2} \tag{5}$$

where k is a gain factor, ω_0 is the pole frequency in rad/s, ω_z is the zero frequency in rad/s, and Q is the quality factor. A second-order elliptic filter designed with a minimum attenuation of 50 dB in the stopband and 5 dB passband ripple is:

$$E_2(s) = 0.0031622\frac{s^2 + 108.0248}{s^2 + 0.4562s + 0.607502} \tag{6}$$

The magnitude response of Equation (6) is shown in Figure 1 as a solid line. Note that this response has the expected DC gain of -5 dB and high frequency gain (HFG) of -50 dB. To realize this magnitude response requires both poles and zeros to realize the ripples in stopband and passband. This differentiates the elliptic response from the Butterworth and Chebyshev responses which only require poles in their realization. To approximate this ripple behavior in stopband and passband using a fractional-order transfer function requires a form that can also realize fractional-order poles and zeros. A lowpass notch filter with a $(1 + \alpha)$ order transfer function that achieves this is given by

$$H_{1+\alpha}(s) = a_4\frac{a_1 s^{1+\alpha} + 1}{a_2 s^{1+\alpha} + a_3 s^{\alpha} + 1} \tag{7}$$

This transfer function will have a notch shaped magnitude response with low and high frequency gains of a_4 and $a_4 a_1/a_2$, respectively, and an attenuation from passband to stopband dependent on α. The magnitude expression for Equation (7) is given by

$$|H_{1+\alpha}(j\omega)| = a_4\frac{\sqrt{a_1^2\omega^{2+2\alpha} + 2a_1\omega^{1+\alpha}\cos\left(\frac{(1+\alpha)\pi}{2}\right) + 1}}{\sqrt{a_2^2\omega^{2+2\alpha} + a_3^2\omega^{2\alpha} + 2a_2\omega^{1+\alpha}\cos\left(\frac{(1+\alpha)\pi}{2}\right) + 2a_3\omega^{\alpha}\cos\left(\frac{\alpha\pi}{2}\right) + 1}} \tag{8}$$

where ω is the frequency in rad/s.

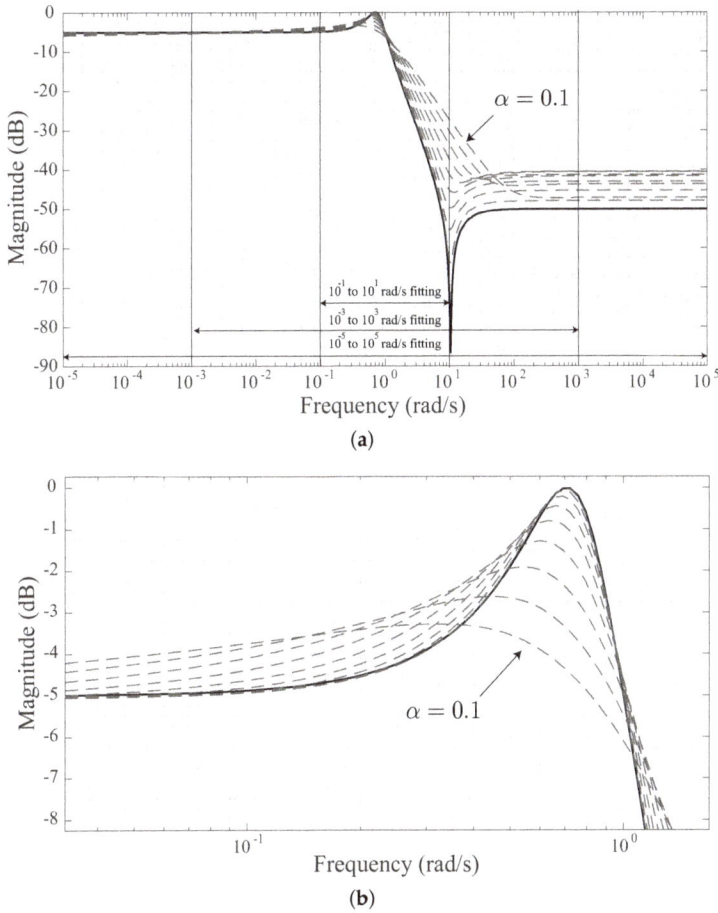

Figure 1. (a) Simulated magnitude responses of $(1 + \alpha)$ lowpass fractional order transfer function for $\alpha = 0.1$ to 0.9 in steps of 0.1 with coefficients selected to approximate elliptic response; and **(b)** details of passband ripple.

2.1. Coefficient Determination

In the design of traditional integer-order analog filters, the acceptable passband and stopband attenuations are used to calculate the necessary filter order based on the desired approximation (Butterworth, Chebyshev, Inverse Chebyshev, and elliptic) [21]. This calculated order is often a real-number that is rounded-up to the nearest integer value which is necessary to be realizable using integer-order filters. From this order, the necessary coefficients can be determined through provided design equations or from tables of available coefficients, which are further used in the implementation and realization of the necessary circuits. Additionally, optimization methods have also been explored to design these types of circuits [22,23].

Currently, fractional-order filter design does not yet have a comprehensive set of design procedures and tables of coefficients to support their realization. However, studies are ongoing to develop methods to design these filters [5,6,19], which also provides the motivation behind this study. Realizing an approximated elliptic response using Equation (7) that has both passband and stopband ripple characteristics requires the appropriate selection of the transfer function coefficients $[a_1, a_2, a_3, a_4]$. One method to determine these coefficients is through the application of optimization routines, and have been previously utilized for other filter designs [5,18,19]. In this work, a non-linear least squares optimization routine is applied to search for the coefficients of the fractional notch transfer function given by Equation (7) that yields the least error when compared to the second-order elliptic response given by Equation (6). This optimization problem is described by:

$$\min_{x} \sum_{i=1}^{k} \left(|H_{1+\alpha}(x, \omega_i)| - |E_2(\omega_i)| \right)^2 \tag{9}$$

where x is the vector of filter coefficients, $|H_{1+\alpha}(x, \omega_i)|$ is the magnitude response using Equation (7) with x at frequency ω_i (rad/s), $|E_2(\omega_i)|$ is the second order elliptic magnitude response given by Equation (6) at frequency ω_i (rad/s), and k is the total number of data points in the magnitude responses. The number of data points used in the fitting procedure ($k = 9001$) was selected to ensure that the ripple characteristics of the elliptic response were represented with sufficient frequency resolution within the dataset. The routine was implemented in MATLAB (v.2015b, 8.6.0.267246) using the *fminsearch* function with the default termination tolerances. The *fminsearch* function uses the simplex search method [24].

2.2. Symmetrical Fittings

It was noted previously in [20] that the frequency band used in the fitting procedure could impact the coefficients. However, this impact has not yet been formally investigated, providing the motivation for this exploration. This work quantifies the differences (in terms of coefficients and overall magnitude characteristics) that result from using different frequency bands in the optimization procedure. The least squares fitting procedure described by Equation (9) was applied to three symmetrical test cases. The three frequency bands were $\omega_{B1} \in [10^{-5}, 10^5]$ rad/s, $\omega_{B2} \in [10^{-3}, 10^3]$ rad/s, and $\omega_{B3} \in [10^{-1}, 10^1]$ rad/s. These frequency bands, labeled in Figure 1a, were selected because they each capture the passband ripple and transition to the stopband (which occurs between approximately 10^{-1} rad/s and 10^1 rad/s). The wider ranges (ω_{B1}, ω_{B2}) capture more data in the flat regions of the stopband and passband.

The coefficients determined for the fitting with each frequency band for $\alpha = 0.1$ to $\alpha = 0.9$ in 0.01 steps are given in Figure 2. The solid lines in Figure 2 represent the coefficients using ω_{B1}, dashed lines those using ω_{B2}, and dash-dotted those using ω_{B3}. The coefficients (a_4, a_3, a_2, a_1) are very similar for $\alpha > 0.7$ for all fitting cases. The greatest differences between coefficients determined using the different frequency bands are observed at lower α values. Therefore, the difference in fitted frequency bands do not have a significant effect on the determined coefficients for $\alpha > 0.7$ in these specific cases. Coefficients (a_2, a_3, a_4) all display a general trend of decreasing values for the cases using smaller frequency fitting bands (i.e., the fitting for ω_{B3} shows the lowest coefficient values for a_2, a_3, and a_4).

MATLAB simulations of Equation (8) using the ω_{B1} coefficients for $\alpha = 0.1$ to $\alpha = 0.9$ in steps of 0.1 are given in Figure 1a as dashed lines. From these simulations, it is clear that this approximation method does realize the fractional-step attenuations (with the slope in the transition band strictly dependent on α) but that there are further differences compared to the elliptic response. The most significant difference is that the HFG of each fractional-order case is higher than the -50 dB magnitude of the elliptic response (detailed in Figure 1a). Further, the passband ripple is smaller for each fractional-order case (detailed in Figure 1b) with the smallest passband ripple (i.e., the lowest passband gain) occurring for the $\alpha = 0.1$ case. This case, given in Figure 1b, reaches approximately -3.6 dB compared to 0 dB of the second-order elliptic response. This is understandable, as α approaches zero the filter order is approaching 1, with first-order filters not able to provide any ripple or higher Q.

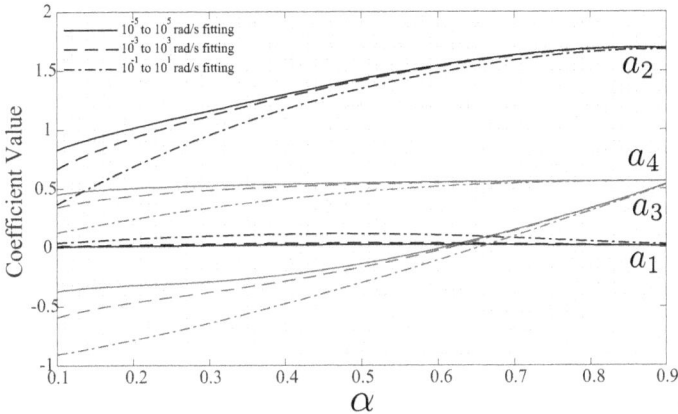

Figure 2. Coefficients of fractional-order transfer function in Equation (7) to approximate second-order elliptic characteristics applying least squares fitting to three symmetrical frequency ranges.

The differences in DC gain and HFG for each set of coefficients are further detailed in Figure 3 for $\alpha = 0.1$ to $\alpha = 0.9$ in steps of 0.01. The solid, dashed, and dash-dotted lines correspond to those cases using the coefficients from the ω_{B1}, ω_{B2}, and ω_{B3} frequency bands, respectively. Observing the DC gains, the magnitude for each fitting is very close to the theoretical value of -5 dB for $\alpha > 0.7$, but it decreases at lower values of α. These deviations are most significant for the ω_{B3} fitting, which has a DC gain of approximately -19 dB for the $\alpha = 0.1$ case. A similar trend is observed in the HFG, with the greatest deviations observed for the ω_{B3} fitting; though the most significant differences occur near $\alpha = 0.5$ for the HFG. These low and high frequency gain differences highlight the impact that the selection of frequency range has on the determined coefficients using the optimization procedure. The ω_{B1} range contains the widest range of frequencies, which places a greater weighting on the optimization procedure to return coefficients that best fit these regions. This results in the DC gain and HFG that is closest to the elliptic case compared to coefficients determined using the ω_{B2} and ω_{B3} frequency bands.

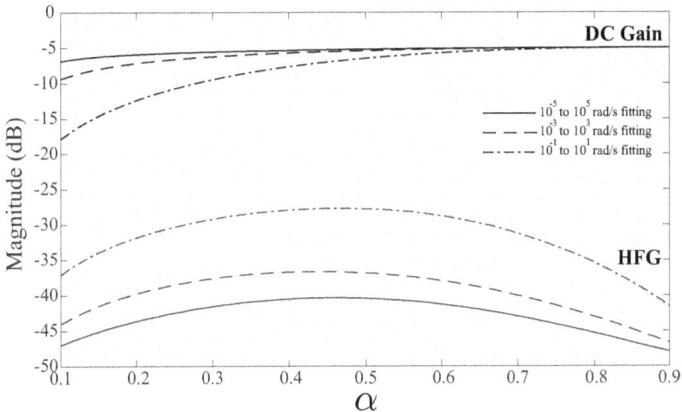

Figure 3. DC and high-frequency gain of fractional-order transfer function using coefficients to approximate elliptic characteristics.

2.3. Asymmetrical Fittings

In the previous fittings, each of the selected frequency bands were symmetrical in terms of the number of frequency decades included above and below 1 rad/s. However, it is possible to choose an asymmetrical frequency band to manipulate the distribution of data in the passband and stopband within the fitting routine. To evaluate the differences this causes in the returned coefficients, the fitting routine was applied to five different frequency bands: (i) $\omega_{B1} \in [10^{-5}, 10^5]$ rad/s; (ii) $\omega_{A2} \in [10^{-5}, 10^3]$ rad/s; (iii) $\omega_{A3} \in [10^{-5}, 10^1]$ rad/s; (iv) $\omega_{A4} \in [10^{-3}, 10^5]$ rad/s; and (v) $\omega_{A5} \in [10^{-1}, 10^5]$ rad/s. The frequency bands $\omega_{A2,A3}$ were selected to introduce a greater passband frequency range into the fitting procedure and $\omega_{A4,A5}$ were selected to introduce a greater stopband frequency range. These ranges are all within the frequency bands explored for the symmetrical fittings. Applying the previously described optimization fitting routine, the coefficients extracted from each frequency band for $\alpha = 0.1$ to $\alpha = 0.9$ in 0.01 steps are given in Figure 4a,b. In Figure 4a, the solid lines represent the coefficients using ω_{B1}, dashed lines those using ω_{A2}, and dash-dotted those using ω_{A3}. Note that in these cases the set of returned coefficients shows very few differences based on the different frequency bands. This suggests that the frequency range of the stopband has little effect on the coefficients when a large sampling of the passband is used. In Figure 4b, the solid lines represent the coefficients using ω_{B1}, dashed lines those using ω_{A4}, and dash-dotted those using ω_{A5}. Similar to the symmetric results, the coefficients (a_4, a_3, a_2, a_1) are very similar for $\alpha > 0.7$ for all fitting cases. That is, the greatest differences between the coefficients extracted using the different frequency bands are observed at lower α values, with a decreasing trend for coefficients (a_2, a_3, a_4) observed at lower α values.

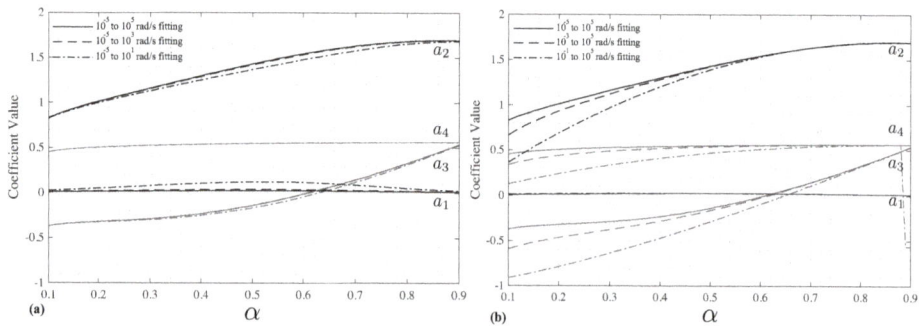

Figure 4. Coefficients of fractional-order transfer function in Equation (7) to approximate second-order elliptic characteristics applying least squares fitting to asymmetrical frequency ranges with: (a) lower limits of $\omega = 10^{-5}$ rad/s; and (b) upper limits of $\omega = 10^5$.

To further evaluate the differences of the asymmetric fitting frequency bands, the DC gain and HFG for $\alpha = 0.1$ to $\alpha = 0.9$ in steps of 0.01 are detailed in Figure 5. Figure 5a presents the gains for ω_{B1} (solid), ω_{A2} (dashed), and ω_{A3} (dash-dotted). Observing the DC gains, the magnitude for each fitting is very close to the theoretical value of -5 dB for $\alpha > 0.2$. This is not unexpected, since the DC gain is related to coefficient a_4, which is noted in Figure 4 to show very little variation for this set of asymmetrical fittings. Similar to the symmetric fittings, the HFG shows variation based on the frequency band, with the frequency band that includes the greatest number of data in the stopband (ω_{A3}) showing the closest agreement with the elliptic case. Figure 5b presents the gains for ω_{B1} (solid), ω_{A4} (dashed), and ω_{A5} (dash-dotted). In these cases, the HFG show similar values for all frequency fittings and the DC gains showing the greatest deviations. This confirms the previous expectations that the distribution of data in the passband and stopbands impacts the fitting process. With those frequency bands that have a larger representation of data yielding filter characteristics closer to the elliptic characteristics.

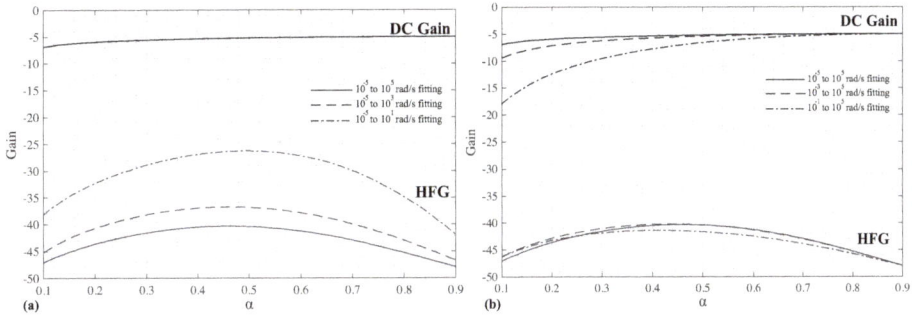

Figure 5. DC and high-frequency gain of fractional-order transfer function using coefficients from asymmetrical fittings using frequency ranges with: (**a**) lower limits of $\omega = 10^{-5}$ rad/s; and (**b**) upper limits of $\omega = 10^5$ to approximate elliptic characteristics.

2.4. Stability

Prior to designing hardware realizations of the fractional-order transfer functions, it is necessary to analyze them with the determined coefficients to ensure that they realize stable responses. Analyzing the stability of fractional-order systems is often accomplished by converting the s-domain transfer function to the W-plane defined in [25]. This transforms the transfer function from fractional order to integer order. This reduces the complexity of analyzing the stability by allowing traditional analysis methods to be employed. Applying this process to the denominator of Equation (7) yields the characteristic equation in the W-plane given by:

$$a_2 W^{m+k} + a_3 W^k + 1 \;\;=\;\; 0 \tag{10}$$

From this characteristic equation, the roots of Equation (10) for $\alpha = 0.1$ to 0.9 were calculated with $k = 10$ to 90, respectively, when $m = 100$ using all sets of coefficients from both the symmetrical and asymmetrical fittings. The minimum root angles, $|\theta_W|_{min}$, for each case of the symmetrical and asymmetrical fittings are given in Figures 6 and 7. The minimum root angles are all greater than the minimum required angle for stability, $|\theta_W| > \frac{\pi}{2m} = 0.9°$. This confirms that the fractional-order filters using the determined coefficients are stable for all fitted frequency ranges. It is interesting to note that the minimum phase angle for all fitted frequency ranges approaches a similar value (1.07°) as α approaches 0.9 and the ω_{B1} case has the largest stability margin compared to the other cases.

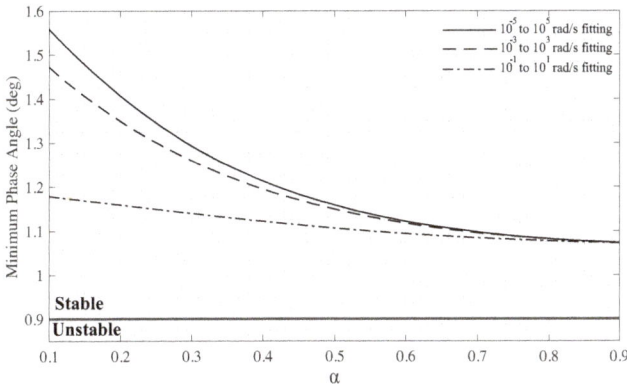

Figure 6. Minimum phase angle of roots of Equation (10) using coefficients from symmetrical frequency fittings.

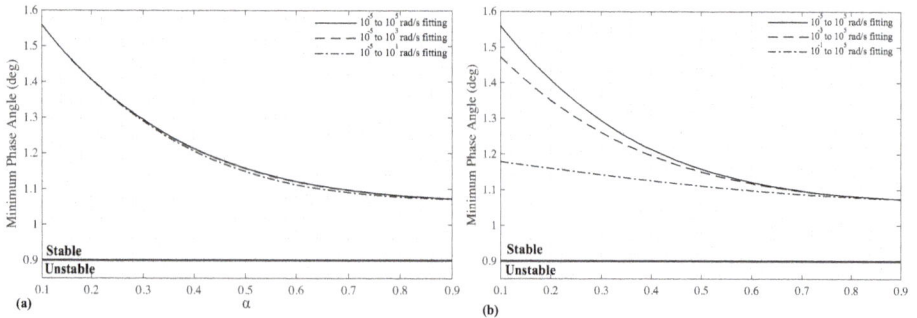

Figure 7. Minimum phase angle of roots of Equation (10) using coefficients from asymmetrical frequency fittings with: (**a**) lower limits of $\omega = 10^{-5}$ rad/s; and (**b**) upper limits of $\omega = 10^5$.

3. Fitted Frequency Range Comparison

To quantify the differences using coefficients from the different fitted frequency ranges, MATLAB simulations of the magnitude responses of Equation (7) for $\alpha = 0.75$ and $\alpha = 0.25$ are given in Figure 8. The solid, dashed, and dash-dotted lines represent the simulated responses using the coefficients from the optimization fittings applied to the ω_{B1}, ω_{B2}, and ω_{B3} frequency ranges, respectively. Each of these simulations used the coefficient values detailed in Figure 2. For the $\alpha = 0.75$ case, the most significant differences are observed above 5 rad/s. Above this frequency, the stopband ripple occurs at the lowest frequency with the least attenuation when the coefficients from the smallest frequency band, ω_{B3} are used. These simulations support the results in Figure 3, with the ω_{B1} case yielding the closest high frequency gain to the elliptic case.

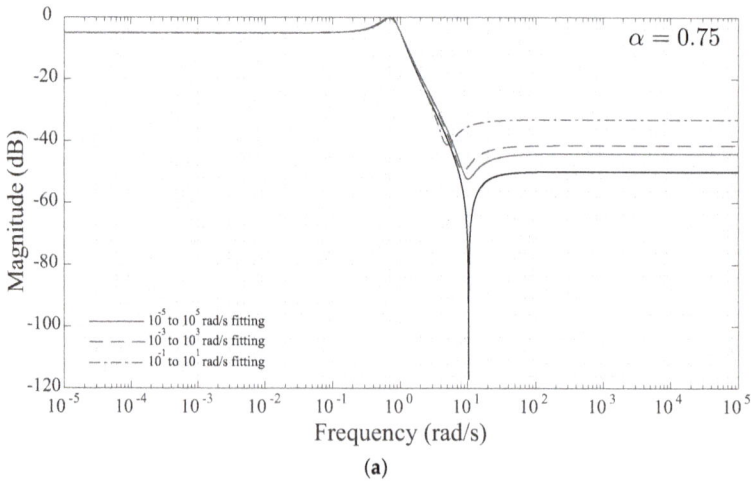

Figure 8. *Cont.*

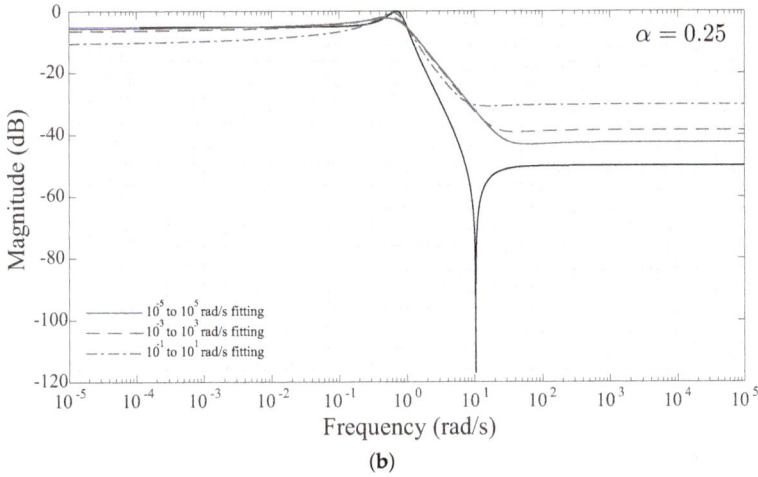

Figure 8. Simulated magnitude responses of (**a**) $(1 + 0.75)$ and (**b**) $(1 + 0.25)$ lowpass fractional order transfer function using coefficients from different symmetrical frequency fitting ranges.

The magnitude characteristics of Equation (7) using the extracted coefficients from the asymmetrical fittings in Figure 4 for $\alpha = 0.25$ and $\alpha = 0.75$ are given in Figures 9 and 10, respectively. These magnitude responses highlight the previous DC and HFG differences in Figure 5. For the coefficients derived using frequency bands with greater passband data ($w_{A2,A3}$), the magnitude responses show very good agreement for both $\alpha = 0.25$ and $\alpha = 0.75$ up to 1 rad/s. At higher frequencies, the responses with fewer stopband data show poorer agreement with the elliptic response. This behavior is reversed in the magnitude responses that use the coefficients derived from the $w_{A4,A5}$. That is, the magnitude characteristics above 1 rad/s show very good agreement for all sets of coefficients for both $\alpha = 0.25$ and $\alpha = 0.75$ cases, resulting from the stopband containing the greatest number of data in the fittings, which results in the best fit of these region.

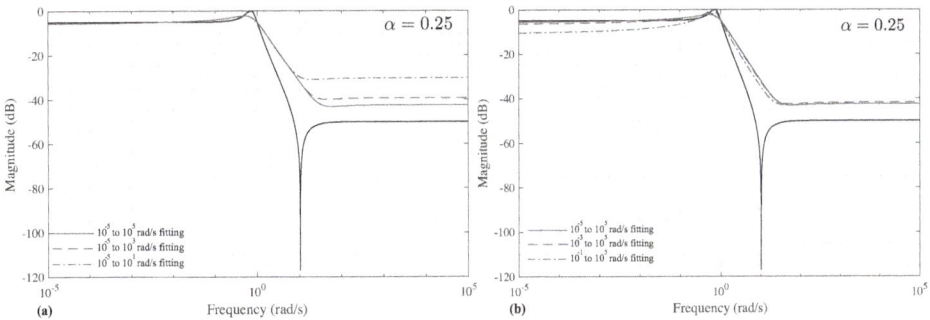

Figure 9. Simulated magnitude responses of $(1 + 0.25)$ lowpass fractional order transfer function using coefficients from different asymmetrical frequency fitting ranges with: (**a**) lower limits of $w = 10^{-5}$ rad/s; and (**b**) upper limits of $w = 10^5$.

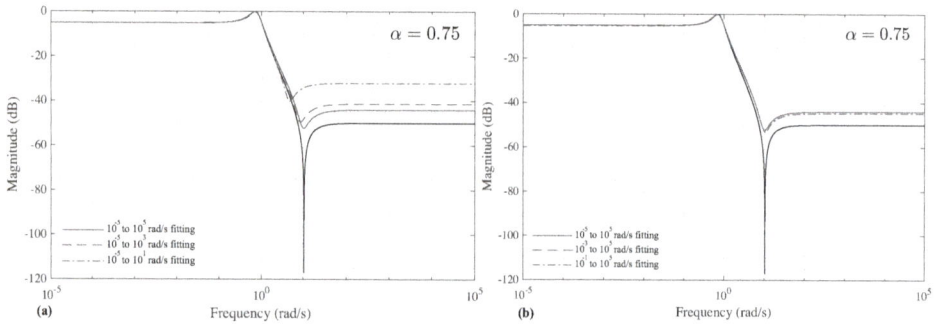

Figure 10. Simulated magnitude responses of $(1 + 0.75)$ lowpass fractional order transfer function using coefficients from different asymmetrical frequency fitting ranges with: (**a**) lower limits of $\omega = 10^{-5}$ rad/s; and (**b**) upper limits of $\omega = 10^5$.

Comparing the $\alpha = 0.25$ cases of the symmetrical fittings in Figure 8, the coefficients have a significant impact on the HFG, DC gain and passband ripples of the magnitude characteristics. While the ω_{B3} case yields the most significant low- and high-frequency differences compared to the elliptic response, it has the closest approximation of the passband ripple. The elliptic response reaches a maximum of 0 dB, which is 5 dB above the DC gain, while the ω_{B3} case reaches -0.40 dB. For comparison, the ω_{B2} and ω_{B1} cases reach -1.79 dB and -2.26 dB, respectively. This is expected to be an impact of the frequency range, with the ω_{B2} and ω_{B1} fittings having a larger weighting (based on the distribution of data) to fit the high and low frequency bands which reduces the weighting of the interior band. As a result, the ω_{B3} case provides the coefficients that best approximate the magnitude characteristics in this band. This is also observed in the asymmetrical fittings of Figure 9, with the $\omega_{A4,A5}$ cases showing better agreement with the passband ripple than the $\omega_{A2,A3}$ cases. Again, this is a result of the $\omega_{A4,A5}$ cases having fewer data in the flat passband region to influence the fitting process.

It is clear the selection of frequency band for the optimization procedure does impact the coefficients. Therefore, the frequency band should be considered when employing optimization fitting processes to design a fractional-order filter. Designers must weigh the trade-offs that result between selecting a wider frequency band (necessary to capture the DC or high frequency gain) or a smaller specific band (necessary to capture the ripple characteristics); and decide based on which design features are most important for their specific application. Additionally, further studies should investigate the sensitivity of the fractional-order transfer functions magnitude characteristics to the coefficients, similar to that presented in [7]. This is important to understand during the physical implementation.

4. Circuit Simulations

To validate the proposed fractional-order filters with approximated elliptic characteristics, two cases were simulated and experimentally verified against the theoretical expectations. For these validations, two circuit topologies were selected and are given in Figure 11 where the component C_1 is a fractional-order capacitor with impedance $Z_{C_1} = 1/s^\alpha C_1$ and C_2 is a traditional integer-order capacitor. The topologies in Figure 11a,b realize Equation (7) when the coefficient a_3 is positive and negative, respectively. It is necessary to present two topologies to realize the complete range of responses with the identified coefficients in this work, which take both positive and negative values depending on the value of α.

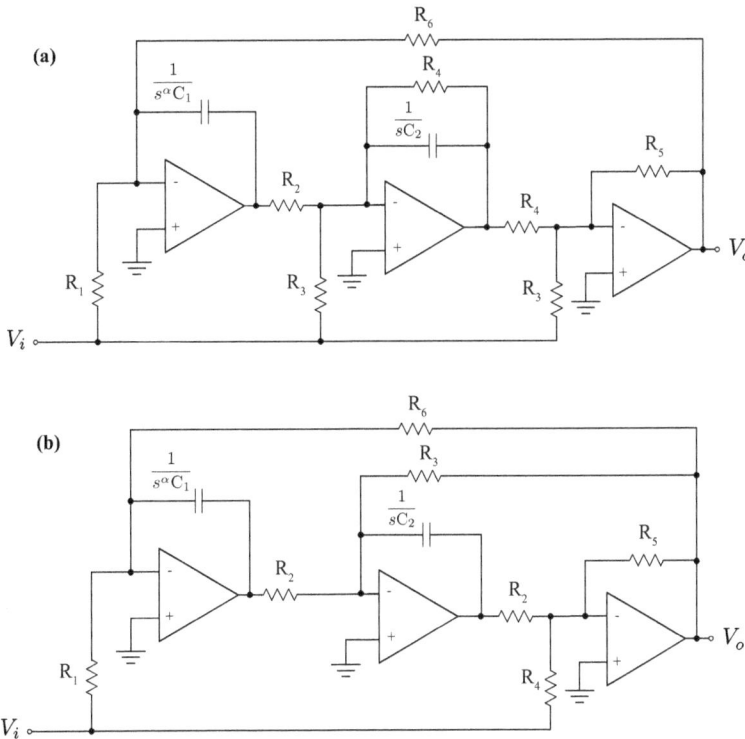

Figure 11. Circuit topologies to realize fractional-order low-pass notch filter response given by Equation (7) when a_3 is: (**a**) positive; and (**b**) negative. Note that in both topologies C_1 is a fractional-order capacitor with impedance $Z_{C_1} = 1/s^\alpha C_1$.

The topology in Figure 11a realizes the transfer function given by:

$$T_1^{1+\alpha}(s) = -\frac{R_6}{R_1} \frac{s^{1+\alpha} \frac{C_1 C_2 R_1 R_2 R_4}{R_3} + 1}{s^{1+\alpha} \frac{C_1 C_2 R_2 R_4 R_6}{R_5} + s^\alpha \frac{C_1 R_2 R_6}{R_5} + 1} = -a_4 \frac{\frac{s^{1+\alpha}}{\omega_0^{1+\alpha}} a_1 + 1}{\frac{s^{1+\alpha}}{\omega_0^{1+\alpha}} a_2 + \frac{s^\alpha}{\omega_0^\alpha} a_3 + 1} \tag{11}$$

and the topology in Figure 11b realizes the transfer function given by:

$$T_2^{1+\alpha}(s) = -\frac{R_6}{R_1} \frac{s^{1+\alpha} \frac{C_1 C_2 R_1 R_2^2}{R_4} + 1}{s^{1+\alpha} \frac{C_1 C_2 R_2^2 R_6}{R_5} - s^\alpha \frac{C_1 R_2 R_6}{R_3} + 1} = -a_4 \frac{\frac{s^{1+\alpha}}{\omega_0^{1+\alpha}} a_1 + 1}{\frac{s^{1+\alpha}}{\omega_0^{1+\alpha}} a_2 + \frac{s^\alpha}{\omega_0^\alpha} a_3 + 1} \tag{12}$$

Note that in both cases the term ω_0 is the frequency scaling factor to shift the response from the normalized frequency of 1 rad/s.

The topology in Figure 11a with transfer function given by Equation (11) was used to realize a $(1+\alpha) = 1.8$ order filter. For this case, the utilized coefficients were $a_1 = 0.01626$, $a_2 = 1.6844$, $a_3 = 0.3317$, and $a_4 = 0.5622$, selected from coefficients returned using the 10^{-5} rad/s to 10^5 fitting bandwidth given in Figure 2. Using these coefficients, the necessary resistances and capacitances for the topology were calculated using the system of equations built by equating the coefficients in Equation (11) (i.e., $C_1 C_2 R_1 R_2 R_4/R_3 = a_1/\omega_0^{1+\alpha}$; $C_1 C_2 R_2 R_4 R_6/R_5 = a_2/\omega_0^{1+\alpha}$; $C_1 R_2 R_6/R_5 = a_3/\omega_0^\alpha$; $R_6/R_1 = a_4$). The specific resistors and capacitors are given in Table 1 when a frequency of $\omega_0 = (2\pi)10$ krad/s is used. At this time, fractional-order capacitors are not available for either simulations or physical realizations and require the use of approximations; however, it is important to note the recent progress in realizing devices with fractional-order impedances [26–28]. For the simulations and experimental circuits in this work, the fractional-capacitors were approximated using the 5th order Foster-I topology given in Figure 12. The circuit components required to realize the $\alpha = 0.8$ device were calculated using the process detailed in [29], with these calculated values given in Table 2. The experimental results were collected using an Omicron Bode 100 network analyzer from a circuit realized using discrete components on a breadboard. The approximated fractional-order capacitor was realized on a custom printed-circuit board (PCB) interfaced to the breadboard setup. This experimental test setup is shown in Figure 13a with the circuit implementation detailed in Figure 13b. The fractional-order capacitor PCB implementation is outlined in Figure 13b using the dashed box. The magnitude responses collected from both SPICE simulations and experimental implementations, using LT1361 operational amplifiers, are given in Figure 14a as dashed and dashed-dotted lines, respectively. For comparison to the simulation and experimental results, the theoretical magnitude response given by Equation (8) is a solid line in Figure 14a. Note that both simulations and experimental results show very good agreement with the theoretical response. The simulations show less than 0.21 dB difference compared to the theoretical for frequencies below 70 kHz with a maximum deviation of 1.73 dB occurring at 100 kHz.

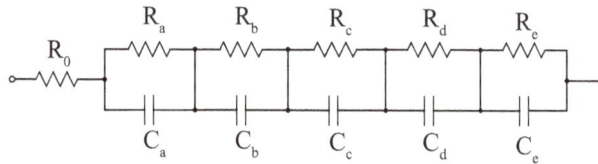

Figure 12. Foster-I circuit topology to realize a fifth order approximation of a fractional-order capacitor.

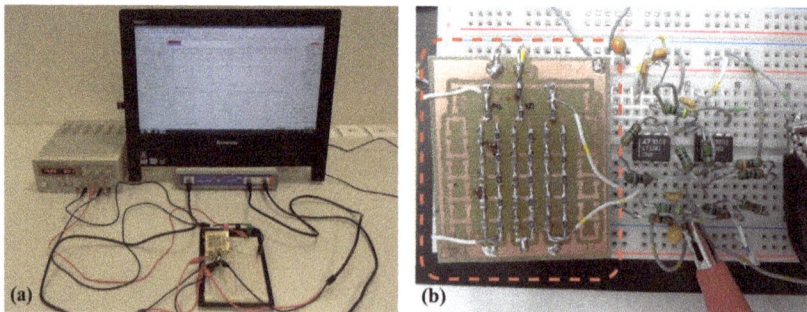

Figure 13. (a) Experimental setup to measure magnitude responses of approximated $(1 + \alpha)$ order filters; and (b) breadboard implementation of topology using fifth order approximation of a fractional-order capacitor (outlined using the dashed box).

Table 1. Component values to realize Equation (7) for the Figure 2 coefficients when $\alpha = 0.8$ and 0.2 using the topologies in Figure 11a,b, respectively.

α	C_1 (F s$^{\alpha-1}$)	C_2 (nF)	R_1 (Ω)	R_2 (Ω)	R_3 (Ω)	R_4 (Ω)	R_5 (Ω)	R_6 (Ω)
0.8	62n	10	1k	680	90.8k	8.08k	492	562
0.2	46.9μ	6.8	4.7k	1k	3.17k	65k	428	2.38k

Table 2. Component values to realize fractional-order capacitors with $\alpha = 0.8$ and 0.2 using fifth order Foster I topology centered at 10 kHz.

α	R_0 (Ω)	R_a (Ω)	R_b (Ω)	R_c (Ω)	R_d (Ω)	R_e (Ω)	C_a (nF)	C_b (nF)	C_c (nF)	C_d (nF)	C_e (nF)
0.8	58.8	65.1	326.4	1.47k	7.06k	84.23k	12.8	16.2	22.6	29.7	15.7
0.2	931.5	374.9	573.6	837.2	1.23k	1.93k	1.28	5.29	22.85	98.5	393.8

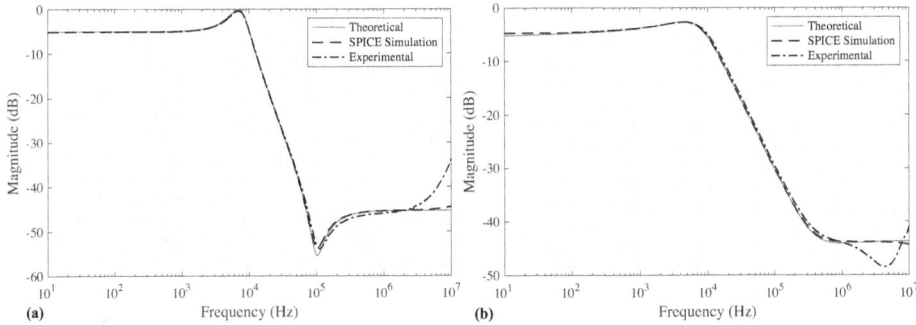

Figure 14. Theoretical (solid), SPICE simulated (dashed), and experimental (dashed-dotted) magnitude responses of (a) $(1+\alpha) = 1.8$ and (b) 1.2 order approximated elliptic filter responses.

Further, the topology in Figure 11b with transfer function given by Equation (12) was used to realize a $(1+\alpha) = 1.2$ order filter. For this case, the utilized coefficients were $a_1 = 0.01320$, $a_2 = 1.1037$, $a_3 = -0.3208$, and $a_4 = 0.5055$, selected from coefficients returned using the 10^{-5} rad/s to 10^5 fitting bandwidth given in Figure 2. Using these coefficients, the necessary resistances and capacitances for the topology were calculated using the system of equations built by equating coefficients in Equation (12) (i.e., $C_1 C_2 R_1 R_2^2 / R_4 = a_1 / \omega_0^{1+\alpha}$; $C_1 C_2 R_2^2 R_6 / R_5 = a_2 / \omega_0^{1+\alpha}$; $-C_1 R_2 R_6 / R_3 = a_3 / \omega_0^\alpha$; $R_6 / R_1 = a_4$). The specific resistors and capacitors are given in Table 1 when a frequency of $\omega_0 = (2\pi)10$ krad/s is used and the C_1, C_2, R_1, R_2 values are initially chosen, and the other values are computed. The circuit components required to realize the $\alpha = 0.2$ device are also given in Table 2. The magnitude responses collected from both SPICE simulations and experimental implementations are given in Figure 14b as dashed and dashed-dotted lines, respectively, compared to the theoretical response given as a solid line. Again, both simulation and experimental results show very good agreement with the theoretical response. The simulations show less than 0.1 dB difference compared to the theoretical for 100 Hz $< f <$ 100 kHz with a maximum deviation of 0.49 dB occurring at 10 Hz. From the experimental results of both filters in Figure 14, the deviation at frequencies above 1 MHz are likely a result of parasitics in the breadboard implementation of these circuits.

To validate the stability of each constructed filter circuit, the transient responses were collected for both $(1+\alpha) = 1.2$ and 1.8 order filters when applying a 800 Hz square wave input signal. Both the input (solid) and output (dashed) waveforms during this transient test are given in Figure 15a,b for the 1.2 and 1.8 order filters, respectively. In both cases, the output waveforms confirm that the circuits are stable and validate the previous stability analyses. The oscillations in the transient response of the 1.8 order filter in Figure 15b confirm the results in the stability analysis, that is, that the filters with higher α have less stability margin. Both simulation and experimental results serve to validate the proposed $(1+\alpha)$ fractional-order elliptic filter responses and that both topologies are appropriate in realizing these designs.

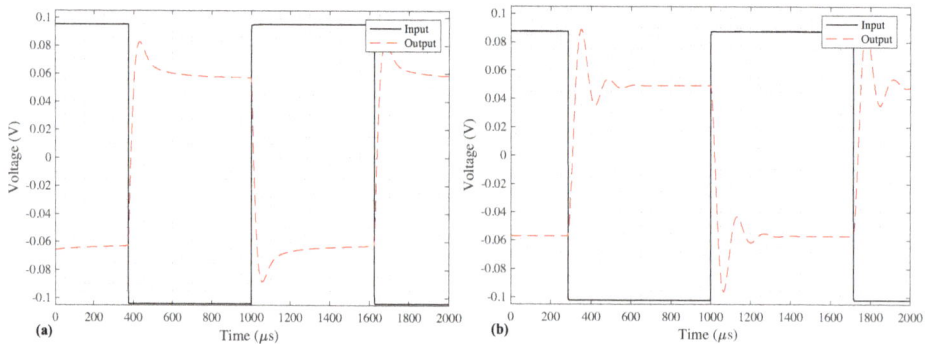

Figure 15. Transient responses (dashed) of experimentally realized (**a**) $(1 + \alpha) = 1.2$ and (**b**) 1.8 order approximated elliptic filter responses when applying a 800 Hz square wave input (solid).

5. Summary of Optimization Procedure

While the coefficients presented in Figures 2 and 4 can be used to implement the approximated fractional-order elliptic responses in this work, to realize responses with different ripple characteristics requires calculation of different sets of coefficients. To support the adoption of the process used in this work to calculate further sets of coefficients for different characteristics, the steps are summarized below:

1. Select desired second-order elliptic magnitude characteristics to approximate with fractional-order filter:

   (a) Passband ripple (5 dB in this work).
   (b) Stopband attenuation (−50 dB in this work).

2. Select target frequency band to use for procedure fitting.
3. Implement objective function given by Equation (9) using desired elliptic response and fractional-order transfer function given by Equation (7).
4. Apply optimization solver to objective function to evaluate coefficients for target filter order $(1 + \alpha)$.
5. Evaluate stability of fractional-order filter with solved coefficients using W-plane transformations.
6. Select appropriate circuit topology to realize fractional-order transfer function.
7. Calculate necessary component values to realize target coefficients for desired center frequency.

6. Conclusions

An optimization fitting procedure was applied in this work to design low-pass fractional-order filters with passband and stopband ripples to approximate the traditional elliptic characteristics. The fractional-order filter coefficients were determined using symmetrical and asymmetrical frequency bands, with all coefficients yielding stable responses. While this method does realize fractional-step attenuations in the transition from passband to stopband, it also shows limits in the presented cases to approximate the low- and high-frequency behavior. Of significant note is that the choice of frequency range used with the optimization procedure significantly impacts the coefficients and magnitude response of the approximated transfer function. For the cases explored in this work, the wider frequency ranges yielded better DC gains and HFG, but at the cost of poorer approximations of the ripple characteristics. The simulated and experimental magnitude responses from filter circuits with order $(1 + \alpha) = 1.2$ and 1.8 realized using approximated fractional-order capacitors validate the proposed circuits and their fractional-order characteristics, and confirm their stability. The simulation and experimental validation of the filter structure to realize negative a_3 coefficients in this work expands the range of available topologies for designers exploring fractional-order filter implementations.

Author Contributions: Formal analysis, D.K. and T.F.; Methodology, D.K., T.F. and J.K.; Validation, D.K. and T.F.; Writing—original draft, D.K. and T.F.; and Writing—review and editing, J.K. and J.D.

Funding: Research described in this paper was financed by the Czech Science Foundation under grant No. 16-06175S, by the Czech Ministry of Education within the frame of the National Sustainability Program under grant LO1401 and within the Inter-Cost Program under grant LD15034. For the research, infrastructure of the SIX Center was used.

Acknowledgments: This article was based on work from COST Action CA15225, a network supported by COST (European Cooperation in Science and Technology). The authors also acknowledgement the financial support of this work detailed in the funding section.

Conflicts of Interest: The authors declare no conflict of interest.

References

1. Oldham, K.B.; Spanier, J. *The Fractional Calculus: Theory and Applications of Differentiation and Integration to Arbitrary Order*; Academic Press: New York, NY, USA, 1974.
2. Podlubny, I. *Fractional Differential Equations*; Academic Press: London, UK, 1999.
3. Ortigueira, M.D. *Fractional Calculus for Scientists and Engineers*; Springer: Heidelberg, Germany, 2011.
4. Soltan, A.; Radwan, A.G.; Soliman, A.M. Fractional order Sallen-Key and KHN filters: Stability and poles allocation. *Circ. Syst. Signal Process.* **2015**, *34*, 1461–1480. [CrossRef]
5. Freeborn, T.J.; Elwakil, A.S.; Maundy, B. Approximated fractional-order Inverse Chebyshev lowpass filters. *Circ. Syst. Signal Process.* **2016**, *35*, 1973–1982. [CrossRef]
6. Kubanek, D.; Freeborn, T.J. $(1 + \alpha)$ Fractional-order transfer functions to approximate low-pass magnitude responses with arbitrary quality factor. *AEU Int. J. Electron. Commun.* **2018**, *83*, 570–578. [CrossRef]
7. Kubanek, D.; Freeborn, T.J.; Koton, J.; Herencsar, N. Evaluation of $(1 + \alpha)$ fractional-order approximated Butterworth high-pass and band-pass filter transfer functions. *Elektronika ir Elektrotechnika* **2018**, *24*, 37–41. [CrossRef]
8. Baranowski, J.; Pauluk, M.; Tutaj, A. Analog realization of fractional filters: Laguerre approximation approach. *AEU-Int. J. Electron. Commun.* **2018**, *81*, 1–11. [CrossRef]
9. Ahmadi, P.; Maundy, B.; Elwakil, A.S.; Belostotski, L. High-quality factor asymmetric-slope band-pass filters: A fractional-order capacitor approach. *IET Circ. Devices Syst.* **2012**, *6*, 187–198. [CrossRef]
10. Bertsias, P.; Khateb, F.; Kubanek, D.; Khanday, F.A.; Psychalinos, C. Capacitorless Digitally Programmable Fractional-Order Filters. *AEU Int. J. Electron. Commun.* **2017**, *78*, 228–237. [CrossRef]
11. Jerabek, J.; Sotner, R.; Dvorak, J.; Polak, J.; Kubanek, D.; Herencsar, N.; Koton, J. Reconfigurable fractional-order filter with electronically controllable slope of attenuation, pole frequency and type of approximation. *J. Circ. Syst. Comput.* **2017**, *26*, 1750157. [CrossRef]
12. AbdelAty, A.M.; Soltan, A.; Ahmed, W.A.; Radwan, A.G. On the analysis and design of fractional-order Chebyshev complex filter. *Circ. Syst. Signal Process.* **2018**, *37*, 915–938. [CrossRef]
13. Hamed, E.M.; AbdelAty, A.M.; Said, L.A.; Radwan, A.G. Effect of different approximation techniques on fractional-order KHN filter design. *Circ. Syst. Signal Process.* **2018**, 1–31. [CrossRef]
14. Tsirimokou, G.; Laoudias, C.; Psychalinos, C. 0.5-V fractional-order companding filters. *Int. J. Circ. Theory Appl.* **2015**, *43*, 1105–1126. [CrossRef]
15. Baranowski, B.; Piatek, P. Fractional band-pass filters: Design, implementation and application to EEG signal processing. *J. Circ. Syst. Comput.* **2017**, *26*, 1750170. [CrossRef]
16. Tolba, M.F.; AbdelAty, A.M.; Soliman, N.S.; Said, L.A.; Madian, A.H.; Azar, A.T.; Radwan, A.G. FPGA implementation of two fractional order chaotic systems. *AEU Int. J. Electron. Commun.* **2017**, *78*, 162–172. [CrossRef]
17. Tolba, M.F.; AboAlNaga, B.M.; Said, L.A.; Madian, A.H.; Radwan, A.G. Fractional order integrator/differentiator: FPGA implementation and FOPID controller application. *AEU Int. J. Electron. Commun.* **2019**, *98*, 220–229. [CrossRef]
18. Mahata, S.; Saha, S.K.; Kar, R.; Mandal, D. Optimal design of fractional-order low pass Butterworth filter with accurate magnitude response. *Digit. Signal Process.* **2018**, *72*, 96–114. [CrossRef]
19. Freeborn, T.J.; Maundy, B.; Elwakil, A.S. Approximated fractional order Chebyshev lowpass filters. *Math. Probl. Eng.* **2014.**, 2014. [CrossRef]

20. Freeborn, T.J.; Kubanek, D.; Koton, J.; Dvorak, J. Fractional-order lowpass elliptic responses of $(1 + \alpha)$-order transfer functions. In Proceedings of the 2018 IEEE International Conference on Digital Signal Processing, Athens, Greece, 4–6 July 2018; pp. 346–349. [CrossRef]

21. Schaumann, R.; Xiao, H.; Van Valkenburg, M.E. *Design of Analog Filters*; Oxford University Press: Oxford, UK, 2010.

22. Le, T.H.; Van Barel, M. A convex optimization method to solve a filter design problem. *J. Comput. Appl. Math.* **2014**, *255*, 183–192. [CrossRef]

23. Le, T.H.; Van Barel, M. A convex optimization model for finding non-negative polynomials. *J. Comput. Appl. Math.* **2016**, *301*, 121–134. [CrossRef]

24. Lagarias, J.C.; Reeds, J.A.; Wright, M.H.; Wright, P.E. Convergence properties of the Nelder-Mead simplex method in low dimensions. *SIAM J. Optim.* **1998**, *9*, 112–147. [CrossRef]

25. Radwan, A.; Soliman, A.; Elwakil, A.; Sedeek, A. On the stability of linear systems with fractional-order elements. *Chaos Solit. Fract.* **2009**, *40*, 2317–2328. [CrossRef]

26. Agambayev, A.; Patole, S.; Bagci, H.; Salama, K.N. Tunable fractional-order capacitor using layered ferroelectric polymers. *AIP Adv.* **2017**, *7*, 095202. [CrossRef]

27. Agambayev, A.; Rajab, K.H.; Hassan, A.H.; Farhat, M.; Bagci, H.; Salama, K.N. Towards fractional-order capacitors with broad tunable constant phase angles: Multi-walled carbon nanotube-polymer composite as a case study. *J. Phys. D Appl. Phys.* **2018**, *51*, 065602. [CrossRef]

28. Agambayev, A.; Farhat, M.; Patole, S.P.; Hassan, A.H.; Bagci, H.; Salama, K.N. An ultra-broadband single-component fractional-order capacitor using MoS_2-ferroelectric polymer composite. *Appl. Phys. Lett.* **2018**, *113*, 093505. [CrossRef]

29. Tsirimokou, G. A systematic procedure for deriving RC networks of fractional-order elements emulators using MATLAB. *AEU Int. J. Electron. Commun.* **2017**, *78*, 7–14. [CrossRef]

applied
sciences

MDPI

Article

Identification and Monitoring of Parkinson's Disease Dysgraphia Based on Fractional-Order Derivatives of Online Handwriting [†]

Jan Mucha [1], Jiri Mekyska [1], Zoltan Galaz [1,2], Marcos Faundez-Zanuy [3],
Karmele Lopez-de-Ipina [4], Vojtech Zvoncak [1], Tomas Kiska [1], Zdenek Smekal [1],
Lubos Brabenec [2] and Irena Rektorova [2,5,*]

[1] Department of Telecommunications and SIX Research Centre, Brno University of Technology, Technicka 10, 61600 Brno, Czech Republic; muchajano@phd.feec.vutbr.cz (J.M.); mekyska@feec.vutbr.cz (J.M.) z.galaz@phd.feec.vutbr.cz (Z.G.); vojtech.zvoncak@phd.feec.vutbr.cz (V.Z.); kiskatomas@phd.feec.vutbr.cz (T.K.); smekal@feec.vutbr.cz (Z.S.)

[2] Applied Neuroscience Research Group, Central European Institute of Technology, Masaryk University, Kamenice 5, 62500 Brno, Czech Republic; lubos.brabenec@ceitec.muni.cz

[3] Escola Superior Politecnica, Tecnocampus Avda. Ernest Lluch 32, 08302 Mataro, Barcelona, Spain; faundez@tecnocampus.cat

[4] Department of Systems Engineering and Automation, University of the Basque Country UPV/EHU, Av de Tolosa 54, 20018 Donostia, Spain; karmele.ipina@ehu.eus

[5] First Department of Neurology, Masaryk University and St. Anne's University Hospital, Pekarska 53, 65691 Brno, Czech Republic

[*] Correspondence: irena.rektorova@fnusa.cz; Tel.: +420-543-182-639

[†] This paper is an extended version of our paper published in Mucha, J.; Zvoncak, V.; Galaz, Z.; Faundez-Zanuy, M.; Mekyska, J.; Kiska, T.; Smekal, Z.; Brabenec, L.; Rektorova, I.; Ipina, K.L. Fractional Derivatives of Online Handwriting: a New Approach of Parkinsonic Dysgraphia Analysis. In Proceedings of the 2018 41st International Conference on Telecommunications and Signal Processing (TSP), Athens, Greece, 4–6 July 2018; pp. 1–4.

Received: 30 October 2018; Accepted: 6 December 2018; Published: 11 December 2018

Abstract: Parkinson's disease dysgraphia affects the majority of Parkinson's disease (PD) patients and is the result of handwriting abnormalities mainly caused by motor dysfunctions. Several effective approaches to quantitative PD dysgraphia analysis, such as online handwriting processing, have been utilized. In this study, we aim to deeply explore the impact of advanced online handwriting parameterization based on fractional-order derivatives (FD) on the PD dysgraphia diagnosis and its monitoring. For this purpose, we used 33 PD patients and 36 healthy controls from the PaHaW (PD handwriting database). Partial correlation analysis (Spearman's and Pearson's) was performed to investigate the relationship between the newly designed features and patients' clinical data. Next, the discrimination power of the FD features was evaluated by a binary classification analysis. Finally, regression models were trained to explore the new features' ability to assess the progress and severity of PD. These results were compared to a baseline, which is based on conventional online handwriting features. In comparison with the conventional parameters, the FD handwriting features correlated more significantly with the patients' clinical characteristics and provided a more accurate assessment of PD severity (error around 12%). On the other hand, the highest classification accuracy (ACC = 97.14%) was obtained by the conventional parameters. The results of this study suggest that utilization of FD in combination with properly selected tasks (continuous and/or repetitive, such as the Archimedean spiral) could improve computerized PD severity assessment.

Keywords: Parkinson's disease dysgraphia; micrographia; online handwriting; kinematic analysis; fractional-order derivative; fractional calculus

1. Introduction

As a second most common neurodegenerative disorder, Parkinson's disease (PD) is expected to impose an increasing social and economic burden on societies as populations age [1]. Its prevalence rate is estimated to approximately 1.5% for people aged over 65 years [2]. The risk of being affected by PD strongly increases with age, and, in the next 15 years, the incidence of PD is expected to be doubled [3,4]. The rapid degeneration of dopaminergic cells in the substantia nigra pars compacta [5] arose as the most significant biological finding associated with the disease, but the exact pathophysiological cause of PD has not yet been discovered. PD cardinal motor symptoms involve bradykinesia (slowness of movement), tremor at rest, rigidity, gait impairment, and postural instability [6–8]. A variety of non-motor symptoms may emerge as well—for instance, cognitive impairment, dementia, depression, sleep disorders, or anxiety [6,9,10].

Handwriting requires cognitive, perceptual, and fine motor abilities. In conjunction with motor dysfunctions in people suffering from PD, it has been proven that disrupted handwriting may be used as a significant biomarker for PD diagnosis [11,12]. Micrographia, which is associated with the progressive decrease in letters' amplitude, is the most commonly observed handwriting abnormality in patients with PD [13,14]. Moreover, according to McLennan et al. [14], in approximately 5% of PD patients, micrographia may be observed even before the onset of the cardinal motor symptoms.

The recent advantage of new technologies coming hand-in-hand with Health 4.0 systems enables the acquisition of online handwriting signals, where temporal information is added to the x and y position. Therefore, by using a digitizing tablet, the analysis is not limited to spatial features which mainly quantify PD micrographia. In addition, we are able to quantify temporal, kinematic, and dynamic manifestations of PD dysgraphia, such as hesitations, pauses, and slow movement [7], which cannot be studied objectively using a classical paper-and-pen method. Due to this complexity, Letanneux et al. [15] started to refer to these manifestations using the generalized term PD dysgraphia.

Several research teams have explored the impact of quantitative PD dysgraphia analysis utilizing simple handwriting/drawing tasks (e.g., separate characters, a combination of two or three characters, repetitive loops, circles), as well as more complex ones (e.g., words, sentences, figures, 3D objects, and the Archimedean spiral) [8,16–20]. An overview of recent related works (2015–present) can be seen in Table 1. Most of them confirm the irreplaceability of kinematic features in PD dysgraphia analysis. Additionally, the researchers usually employ temporal, spatial, and dynamic features. Some more advanced parameters are reported too. For instance, Drotar et al. [8,16,17] demonstrated a combination of kinematic, pressure, energy, or empirical mode decomposition (EMD)-based features that resulted in a classification accuracy of up to 89% using several handwriting tasks. Kotsavasilogloua et al. [21] achieved an average prediction accuracy of 91% using simple horizontal lines and features describing the variability in the pen tip's velocity, a deviation from the horizontal plane, and the trajectory's entropy. Other works report even higher classification accuracies (approximately 97%), e.g., Loconsole et al. [18], who used computer vision and electromyography signal processing techniques, or Taleb et al. [22], who used a combination of features related to the correlation between kinematic and pressure characteristics (but, in this case, applied to a very small dataset). Another promising approach was published by Moetesum et al. [23], who reached an 83% classification accuracy by employing convolutional neural networks (CNN) that were used to extract discriminating visual features from handwriting data transformed into the offline mode. In 2018, Impedovo et al. reported the results of a study focused only on the early stages of PD; the best accuracy was 74.76% for a combination of three handwriting tasks. Finally, in our previous work [20], we proposed a new approach of advanced kinematic feature extraction that utilizes fractional-order derivatives (FD). This approach increased the classification accuracy by 10% (72.39%) for Archimedean spiral tasks in comparison with the baseline [20].

Table 1. Overview of related works focused on computerized analysis of Parkinson's disease (PD) dysgraphia.

First Author	Year	PD/HC	Handwriting Task	Analysis	Features	Conclusions
Drotar * [17]	2015	37/38	letters, words, sentences	differential analysis (SVM)	kinematic, temporal, spatial, entropy, EMD, signal energy	The highest classification accuracy after feature selection approach was 88.13%.
Drotar * [16]	2015	37/38	letters, words, sentences	differential analysis (SVM)	kinematic, temporal, spatial, entropy, EMD, pressure	Classification performance was at its peak with on-surface features equal to AUC = 89.09%.
Heremans [24]	2015	34/10	up/down strokes at varying amplitudes	ANOVA	spatial and kinematic	Significant difference between groups was in spatial ($F(2.41) = 3.97$; $p = 0.03$).
Pereira [25]	2015	37/18	Archimedean spiral	differential an. (SVM, NB, OPF)	mean relative tremor and spatial parameters	The best results were obtained by NB classifier that provided around 79% classification accuracy.
Drotar * [8]	2016	37/38	letters, words, Archimedean spiral, sentences	differential an. (SVM, K-NN, ADA)	kinematic, temporal, spatial, entropy, EMD, pressure	Combining all exercises, SVM proved to be the best classifier with 82.5% accuracy.
Heremans [26]	2016	30/15	repetitive cursive loops	ANOVA, correlation an.	writing amplitude and velocity	PD dysgraphia is more severe in patients with freezing of gait.
Pereira [27]	2016	14/21	Archimedean spiral, meander	differential an. (CNN, OPF)	pen-based features	The best result was obtained by CNN with 87.14% classification accuracy using meander task.
Kotsavasil [21]	2017	24/20	horizontal lines	differential analysis (NB)	kinematic	Average classification accuracy was 91%.
Loconsole [18]	2017	4/7	sentence, repetitive loops	differential analysis (ANN)	temporal, kinematic, spatial	Highest classification accuracy (96.81%) was achieved using all the extracted features.
Taleb [22]	2017	16/16	letters, waves, words	differential analysis (SVM)	kinematic, stroke, pressure, entropy, energy, EMD	The highest classification accuracy 96.88% for 12 kinematic and pressure features.
Moetesum * [23]	2018	37/38	Archimedean spiral, letters, words, sentence, loops	differential analysis (SVM)	CNN-based features	Extraction of features using CNN applied on raw handwriting data resulted in 83% classification accuracy.
Mucha * [20]	2018	30/36	Archimedean spiral	differential analysis (RF, SVM)	fractional derivatives based kinematic features	Improvement of classification accuracy by 10% (72.38%) in comparison to the baseline.
Impedovo * [28]	2018	37/38	Archimedean spiral letters, words, sentence	differential an. (RF, SVM, K-NN, NB, LDA, ADA)	kinematic, temporal, spatial, entropy, EMD, pressure	Analysis focused on PD diagnosis at earlier stages resulted in 74.76% classification accuracy.

SVM—support vector machine; EMD—empirical mode decomposition; K-NN—K-nearest neighbors; ANOVA—analysis of variance; NB—naïve Bayes classifier; OPF—optimum path forest; ANN—artificial neural network; CNN—convolutional neural network; RF—random forests; LDA—Linear Discriminant Analysis; ADA—AdaBoost; AUC—area under the receiver operating characteristics (ROC) curve; articles are sorted by the year of release and then alphabetically; * analyzes performed on the same database (Parkinson's disease handwriting database (PaHaW) [8]).

Although the authors of the previously mentioned studies reported high classification accuracies, further signal processing and machine learning pipeline improvements are expected to make the differential analysis even more accurate. One possible approach could involve an advanced feature extraction methodology based on fractional calculus (FC) [29,30], which enables the use of an arbitrary order of derivatives and/or integrals. Generally, FC has many applications in different fields of science [31–33]. For instance, it has been advantageously used during the modeling of different diseases, such as human immunodeficiency virus (HIV) [34] and malaria [35]. In addition, FC-based analytical tools have outperformed classical techniques in geology [36,37], economics and finance [38,39], etc. Moreover, in our recent paper [20], we identified a high potential for the use of FC in the kinematic analysis of PD drawings. Based on these preliminary results, we assume that FD-based handwriting features may bring improvements to PD diagnosis and assessment. In the frame of this article, we would like to go further and deeply explore the impact of FD on the PD dysgraphia diagnosis and its monitoring. More specifically, we aim to:

- investigate the relationship between newly designed FD handwriting features and a patient's clinical data and compare these results with a baseline (i.e., results based on conventional parameters),
- evaluate the discrimination power of the FD features in terms of binary classification accuracy and compare the results to the baseline,
- use the newly designed features to establish regression models that will estimate the severity of PD and compare its performance to that of a baseline.

The rest of this paper is organized as follows: Section 2 describes the cohort of patients and the methodology, and Section 3 includes the results. A discussion is presented in Section 4, and, finally, conclusions are drawn in Section 5.

2. Materials and Methods

2.1. Dataset

For the purpose of this work, the Parkinson's disease handwriting database (PaHaW) [8], which consists of multiple handwriting/drawing samples from 37 PD patients and 38 age- and gender-matched healthy controls (HC), was used. Since the Archimedean spiral drawing task is missing for some participants, we reduced the analyzed cohort to 33 PD patients and 36 HC. Demographic and clinical data of the participants can be found in Table 2. The participants were enrolled at the First Department of Neurology, St. Anne's University Hospital in Brno, Czech Republic. All participants reported the Czech language as their native language and were right-handed. The patients completed their tasks approximately 1 h after their regular dopaminergic medication (L-dopa). All participants signed an informed consent form approved by the local ethics committee. Unified Parkinson's disease rating scale, part V (UPDRS V): Modified Hoehn and Yahr staging score [40], was used to assess clinical symptoms of PD. In the frame of this work, the duration of the disease was considered as well. Descriptive visualization (histograms, regression, and residual plots) of the clinical data for the subjects participating in this study can be seen in Figure 1.

Table 2. Demographic and clinical data of the enrolled participants.

Gender	N	Age [years]	PD dur [years]	UPDRS V	LED [mg/day]
Parkinson's disease patients					
Females	17	71.76 ± 10.93	9.88 ± 5.27	2.18 ± 0.86	1146.03 ± 543.89
Males	16	66.50 ± 13.44	7.44 ± 4.04	2.31 ± 0.75	1673.38 ± 616.66
All	33	69.21 ± 11.10	8.70 ± 4.82	2.24 ± 0.80	1401.72 ± 630.71
Healthy controls					
Females	17	61.59 ± 10.17	-	-	-
Males	19	63.32 ± 13.14	-	-	-
All	36	62.50 ± 11.70	-	-	-

PD—Parkinson's disease; N—number of subjects; PD dur—PD duration; UPDRS V—Unified Parkinson's disease rating scale, part V: Modified Hoehn and Yahr staging score [40]; LED—L-dopa equivalent daily dose [41].

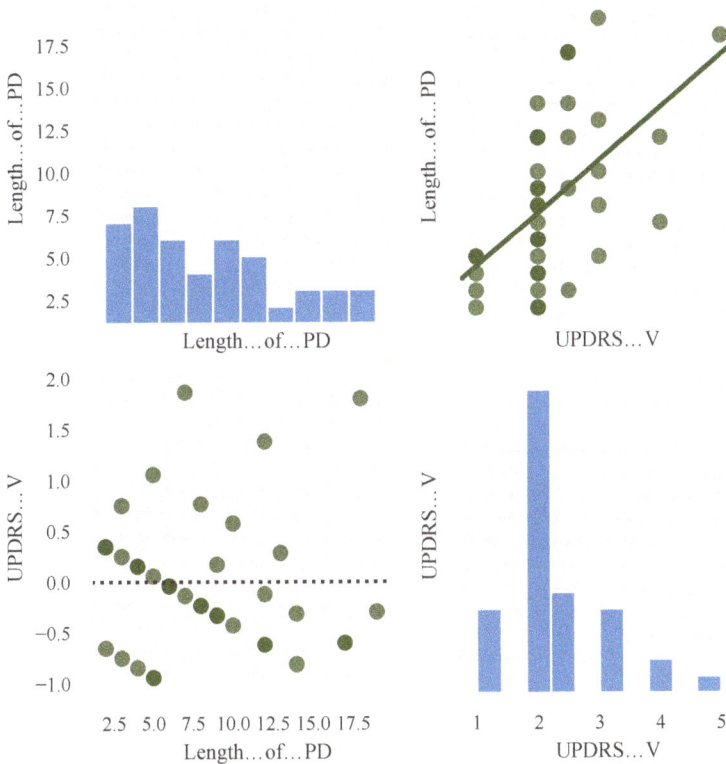

Figure 1. Descriptive graphs of patients' clinical characteristics: Unified Parkinson's disease rating scale (UPDRS V) and Parkinson's disease (PD) duration (in years). Histograms are visualized on the diagonal. A scatterplot with a line fitted using linear regression is visualized in the top-right corner. Residuals of the trained linear model are visualized in the bottom-left corner.

2.2. Data Acquisition

The PaHaW database [8] includes nine different handwriting tasks written in the Czech language. Their description and translation to English can be found in Table 3. During all handwriting tasks, the participants were rested and seated in a comfortable position with the possibility to look at the prefilled template (see Figure 2). A digitizing tablet (Wacom Intuos 4M, Wacom, Kazo, Saitama, Japan) was overlaid with an empty paper template and participants were asked to perform all tasks using a special Wacom inking pen that gave the patients immediate visual feedback. Online handwriting signals were recorded with a sampling frequency of $f_s = 150$ Hz. The following time sequences were acquired: x and y coordinates ($x[t]$, $y[t]$); time-stamp (t); in-air/on-surface (on-surface movement is a movement of a pen when its tip is touching the surface, e.g., paper (i.e., it provides the information about the pen writing/drawing on the paper); vice versa, in-air movement is a movement of a pen when its tip is up to 1.5 cm above the surface [42,43]) status ($b[t]$); pressure ($p[t]$); azimuth ($az[t]$); and altitude ($al[t]$).

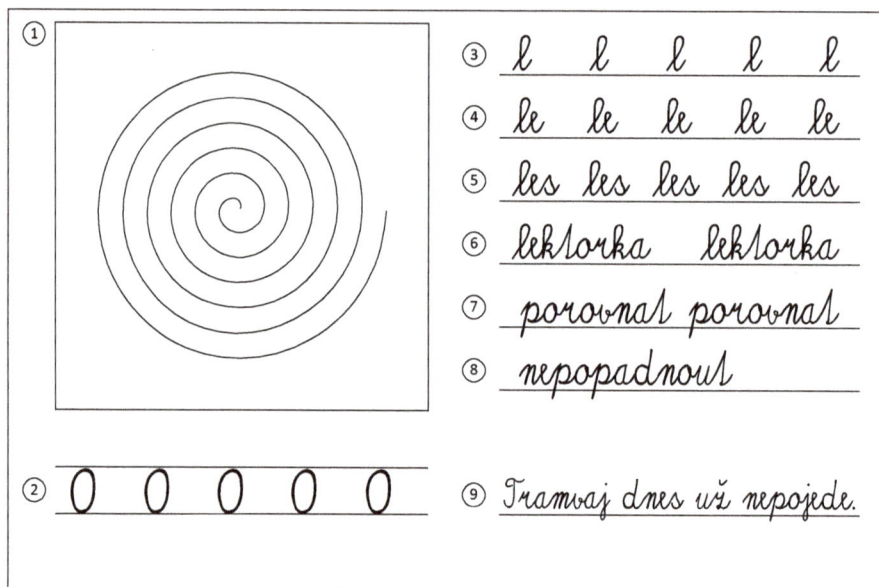

Figure 2. Filled template of the PaHaW database.

Table 3. Description of the PaHaW handwriting tasks.

N	Task	Czech (Original)	English (Translation)
1	Archimedean spiral	-	-
2	repetitive loops	-	-
3	letter	l	l
4	syllable	le	le
5	word	les	forest
6	word	lektorka	lecturer
7	word	porovnat	compare
8	word	nepopadnout	not grasped
9	sentence	Tramvaj dnes už nepojede.	The tram will no longer go today.

2.3. Feature Extraction

The main goal of this work is to compare a set of commonly used kinematic features with newly proposed FD-based features in terms of quantitative PD dysgraphia analysis. All of the handwriting features were computed using both on-surface as well as in-air movements. The two movements were quantified separately using *velocity* (rate at which the position of the pen changes with time [mm/s]), *acceleration* (rate at which the velocity of the pen changes with time [mm/s^2]), *jerk* (rate at which the acceleration of the pen changes with time [mm/s^3]), and their horizontal and vertical variants [8,44,45]. FD-based features were extracted for different values of α. In the frame of this work, α ranging from 0.1

to 1.0 with a step of 0.1 was used. Subsequently, the statistical properties of the computed handwriting features were described using the mean, median, standard deviation (std), and maximum (max). Finally, all of the extracted features were divided into nine different feature sets according to the type of the movement (on-surface, in-air, and combined) and the calculation approach, i.e., the type of feature (FD-based, conventional, and combined). For more information, see Table 4.

Table 4. Feature sets matrix.

Movement	FD (Count)	Conventional (Count)	Together (Count)
on-surface	4536	618	5154
in-air	2916	404	3320
together	7452	1022	8474

Fractional-Order Derivatives

Utilization of the FD as a substitution for the conventional differential derivative during calculation of the basic kinematic features provides a new advanced approach. The advantage of FDs is in their wide range of settings and many different approaches to approximation, e.g., Riemann–Liouville, Caputo, or Grünwald–Letnikov formulations [31,46,47]. For the purpose of this work, Jonathan Hadida's FD Matlab implementation was used following the Grünwald–Letnikov approximation [31,48]. A direct definition of the FD $D^\alpha y(t)$ is based on the finite differences of an equidistant grid in $[0, \tau]$, assuming that the function $y(\tau)$ satisfies certain smoothness conditions in every finite interval $(0, t), t \leq T$. Choosing the grid [31],

$$0 = \tau_0 < \tau_1 < \dots < \tau_{n+1} = t = (n+1)h \tag{1}$$

with

$$\tau_{k+1} - \tau_k = h \tag{2}$$

and using the notation of finite differences

$$\frac{1}{h^\alpha} \Delta_h^\alpha y(t) = \frac{1}{h^\alpha} \left(y(\tau_{n+1}) - \sum_{v=1}^{n+1} c_v^\alpha y(\tau_{n+1-v}) \right), \tag{3}$$

where

$$c_v^\alpha = (-1)^{v-1} \binom{\alpha}{v}. \tag{4}$$

The Grünwald–Letnikov implementation is defined as

$$D^\alpha y(t) = \lim_{h \to 0} \frac{1}{h^\alpha} \Delta_h^\alpha y(t), \tag{5}$$

where $D^\alpha y(t)$ denotes a derivative with order α of function $y(t)$, and h represents a sampling lattice.

2.4. Statistical Analysis

Prior to providing a description of the analytical setup, it is important to note that the effect of well-known confounding factors, also known as covariates, was controlled for in all of the analytical steps described below. In the frame of this work, we controlled for the effect of participants' age, gender, and L-dopa [41] (dopaminergic medication).

To assess the strength of the relationship between the computed handwriting features and patient's clinical data (UPDRS V and PD duration), we computed the partial Pearson's correlation coefficient (assessment of a linear relationship), as well as the partial Spearman's correlation coefficient (assessment of a monotonic relationship). With this approach, we aimed to identify the handwriting features that are significantly correlated with the clinical measures under focus and also to compare the FD features with conventional ones. A significance level of correlation (*p*) of 0.05 was selected for both of the correlation types. Only the results with a *p*-value below the significance level in both correlation coefficients were considered statistically significant.

Next, to evaluate and compare the power of the handwriting features to discriminate PD patients and HC, multivariate binary classification analysis was performed. For this purpose, state-of-the-art gradient boosted trees were employed. Specifically, we used the famous XGBoost algorithm [49]. The XGBoost algorithm was chosen for its ability to achieve a good performance, even for small datasets; its inherent robustness to outliers; its ability to model complex interdependencies in the data; and also its recent successes in the field of machine learning (e.g., the winning algorithm in many www.kaggle.com competitions). To train and evaluate the models, we used the following supervised learning setup: stratified 10-fold cross-validation with 20 repetitions. The performance of the trained classification models was evaluated by Matthew's correlation coefficient (MCC) [50], classification accuracy (ACC), sensitivity (SEN), and specificity (SPE), which are defined as follows:

$$MCC = \frac{TP \times TN - FP \times FN}{\sqrt{(TP+FP)(TP+FN)(TN+FP)(TN+FN)}}, \tag{6}$$

$$ACC = \frac{TP+TN}{TP+TN+FP+FN} \cdot 100\,[\%], \tag{7}$$

$$SEN = \frac{TP}{TP+FN} \cdot 100\,[\%], \tag{8}$$

$$SPE = \frac{TN}{TN+FP} \cdot 100\,[\%], \tag{9}$$

where TP is the number of true positives, TN is the number of true negatives, FP is the number of false positives, and FN is the number false negatives.

Finally, to evaluate and compare the power of the handwriting features' ability to predict the values of the selected clinical characteristics (UPDRS V and PD duration), multivariate regression analysis was performed. For this purpose, the same boosting tree algorithm (XGBoost) and the supervised learning setup were used. The performance of the trained regression models was evaluated by the mean absolute error (MAE), root mean square error (RMSE), and estimated error rate (EER), which are defined as follows:

$$MAE = \frac{1}{n}\sum_{i=1}^{n}|y_i - \hat{y}_i|, \tag{10}$$

$$RMSE = \sqrt{\frac{1}{n}\sum_{i=1}^{n}(y_i - \hat{y}_i)^2}, \tag{11}$$

$$EER = \frac{1}{n \cdot r}\sum_{i=1}^{n}|y_i - \hat{y}_i| \cdot 100\,[\%], \tag{12}$$

where y_i represents the true label of the *i*th observation, $\hat{y}_i$ denotes the predicted label of the *i*th observation, *n* is the number of observations, and *r* is the range of the values of the predicted clinical characteristic (not the range that can be theoretically reached, but the actual range of the values in the dataset). Therefore, the EER describes a percentage of error predictions in regard to the statistical properties of the data.

3. Results

In Table 5, the results of partial correlation analysis between the handwriting features (FD-based features, conventional features) and patients' clinical characteristics (UPDRS V, PD duration) are summarized. The table shows the five best features according to Spearman's correlation coefficient for each movement (on-surface, in-air).

In the case of UPDRS V (on-surface movement), the following FD-based features achieved a statistical significance of correlation: the median of jerk ($\alpha = 0.3$, $\alpha = 0.4$) and horizontal velocity ($\alpha = 0.1$) for the repetitive letter *l*, the mean of vertical acceleration ($\alpha = 0.7$) for repetitive loops, and the standard deviation of the vertical velocity ($\alpha = 0.3$) for the sentence. The following conventional features achieved a statistical significance of correlation (*p*-value of only one of the coefficients was below the threshold): the maximum of horizontal jerk and velocity for the repetitive letters *le*, the maximum of horizontal jerk and horizontal velocity for the repetitive letter *l*, and the maximum of horizontal velocity for the letter *l*. Regarding UPDRS V (in-air movement), the following FD-based features achieved a statistical significance of correlation: the median of vertical velocity ($\alpha = 0.9$, $\alpha = 0.8$, $\alpha = 0.7$) for the sentence and the median of horizontal velocity ($\alpha = 0.5$) and vertical jerk ($\alpha = 0.3$) for the repetitive letters *le*. The following conventional features achieved a statistical significance of correlation (*p*-value of only one of the coefficients was below the threshold): the mean of acceleration for the repetitive word *lektorka*, the maximum of horizontal jerk for the word *porovnat*, the median of the vertical velocity for the repetitive letter *l*, and the median of the horizontal velocity of the repetitive letters *le*.

Table 5. Results of partial correlation analysis between handwriting features and clinical data.

UPDRS V								
FD on-suface						Conventional on-surface		
feature name	α	task	r_p	r_s	r_s	r_p	task	feature name
jerk (median)	0.3	r. letters l	0.37 *	0.48 **	−0.45 *	−0.24	r. letters le	h. jerk (max)
jerk (median)	0.4	r. letters l	0.43 *	0.46 *	−0.43 *	−0.2	r. letters le	velocity (max)
h. velocity (std)	0.1	r. letters l	−0.42 *	−0.41 *	−0.42 *	0.25	r. letters l	h. jerk (max)
v. acceleration (mean)	0.7	r. loops	0.48 **	0.40 *	−0.42 *	−0.16	r. letters l	h. velocity (max)
v. velocity (std)	0.3	sentence	0.40 *	0.40 *	−0.41 *	−0.15	letter l	h. velocity (max)
FD in-air						Conventional in-air		
feature name	α	task	r_p	r_s	r_s	r_p	task	feature name
v. velocity (median)	0.9	sentence	0.44 *	0.53 **	0.43 *	0.28	r. word lektorka	acceleration (mean)
v. velocity (median)	0.8	sentence	0.40 *	0.52 **	−0.37 *	−0.31	word porovnat	h. jerk (max)
h. velocity (median)	0.5	r. letters le	−0.38 *	−0.49 **	0.36 *	0.25	r. letters l	v. velocity (median)
v. jerk (median)	0.3	r. letters le	−0.43 *	−0.49 **	0.35	0.41 *	r. letters le	h. velocity (median)
v. velocity (median)	0.7	sentence	0.37 *	0.48 **	0.35	0.19	r. word lektorka	acceleration (median)
PD Duration								
FD on-surface						Conventional on-surface		
feature name	α	task	r_p	r_s	r_s	r_p	task	feature name
velocity (max)	0.1	spiral	0.54 **	0.55 **	−0.46 *	−0.40 *	r. letters l	h. velocity (max)
acceleration (max)	0.8	spiral	0.54 **	0.54 **	−0.40 *	−0.37 *	r. letters l	h. jerk (max)
acceleration (max)	0.6	spiral	0.54 **	0.54 **	−0.38 *	−0.37 *	r. letters l	velocity (max)
acceleration (max)	0.2	spiral	0.54 **	0.54 **	0.46 **	0.34	spiral	v. velocity (mean)
acceleration (max)	0.7	spiral	0.54 **	0.53 **	0.40 *	0.14	r. loops	h. acceleration (mean)
FD in-air						Conventional in-air		
feature name	α	task	r_p	r_s	r_s	r_p	task	feature name
jerk (median)	0.4	sentence	−0.37 *	−0.49 **	−0.44 *	−0.38 *	word lektorka	h. jerk (median)
jerk (max)	0.1	r. word les	0.57 **	0.46 *	0.38 *	0.40 *	word nepopad.	velocity (max)
jerk (max)	0.3	r. word les	0.57 **	0.45 *	0.37 *	0.42 *	word lektorka	h. n. jerk (mean)
velocity (max)	0.1	r. word les	0.57 **	0.45 *	−0.47 **	−0.13	r. word lektorka	h. velocity (mean)
jerk (max)	0.2	r. word les	0.57 **	0.45 *	−0.42 *	−0.13	word nepopad.	h. velocity (mean)

α—order of FD; r_p—Pearson's correlation coefficient; r_s—Spearman's correlation coefficient; v.—vertical; h.—horizontal; r.—repetitive task; *—$p < 0.05$; **—$p < 0.01$; rows are ordered by the absolute value of Spearman's correlation coefficient.

For PD duration (on-surface movement), the following FD-based features achieved a statistical significance of correlation (of note: all of these features satisfied the stronger threshold for statistical significance of correlation $p < 0.01$): the maximum of the velocity ($\alpha = 0.1$) and acceleration ($\alpha = 0.8$, $\alpha = 0.7$, $\alpha = 0.6$, $\alpha = 0.2$) for the Archimedean spiral. The following conventional features achieved a statistical significance of correlation (*p*-value of only one of the coefficients was below the threshold):

the maximum of horizontal velocity, horizontal jerk, and velocity for the repetitive letter *l*; the mean of the vertical velocity for the Archimedean spiral; and the mean of horizontal acceleration for repetitive loops. For PD duration (in-air movement), the following FD-based features achieved a statistical significance of correlation: the median of jerk ($\alpha = 0.4$) for sentence, the maximum of jerk ($\alpha = 0.1$, $\alpha = 0.2$, $\alpha = 0.3$) and velocity ($\alpha = 0.1$) for repetitive word *les*. The following conventional features achieved a statistical significance of correlation (*p*-value of only one of the coefficients was below the threshold): the median and mean of horizontal jerk for the word *lektorka*, the maximum of the velocity for the word *nepopadnout*, and the mean of horizontal velocity for the repetitive word *lektorka* and the word *nepopadnout*.

The results of the multivariate binary classification analysis are summarized in Table 6. In total, we built and evaluated nine different classification models. These models were selected according to the following criteria: movement type (on-surface, in-air, all), feature type (FD features, conventional features, all). We built models based on the combinations of these criteria as well. For more information, see Table 4.

Table 6. Results of multivariate binary classification analysis (PD/HC).

Feature Set	MCC	ACC [%]	SEN [%]	SPE [%]	Feat
conventional on-surface	0.83 ± 0.18	91.19 ± 9.65	93.00 ± 15.52	70.00 ± 0.46	1
conventional in-air	0.95 ± 0.10	**97.14 ± 5.71**	95.50 ± 9.07	100.00 ± 0.00	1
conventional together	0.95 ± 0.11	**97.14 ± 5.71**	95.50 ± 9.07	100.00 ± 0.00	1
FD on-surface	0.95 ± 0.12	87.14 ± 13.48	82.00 ± 21.24	90.00 ± 30.00	1
FD in-air	0.95 ± 0.13	81.43 ± 12.86	71.50 ± 30.83	60.00 ± 48.99	3
FD together	0.95 ± 0.14	81.43 ± 15.71	69.50 ± 32.13	70.00 ± 45.83	2
all on-surface	0.95 ± 0.15	88.33 ± 14.06	89.00 ± 22.11	70.00 ± 45.83	2
all in-air	0.95 ± 0.16	**97.14 ± 5.71**	95.50 ± 9.07	100.00 ± 0.00	1
all together	0.95 ± 0.17	**97.14 ± 5.71**	95.50 ± 9.07	100.00 ± 0.00	1

MCC—Matthew's correlation coefficient; ACC—accuracy; SEN—sensitivity; SPE—specificity; feat.—number of features important for the trained model (i.e., feature importance of the feature > 0.0); The feature importances, as well as the exact names of these features, are summarized in the text.

With respect to the classification performance, the highest MCC achieved was 0.95 was for eight out of the total nine feature sets (with the exception being the feature set composed of conventional handwriting features computed for the on-surface movements). An interesting fact to note is that for all models based on conventional handwriting features, only a single feature was capable of providing the classification models with such a high discrimination power. In terms of the specific features important for the trained models, the following feature importances were returned by the models (feature importance quantifies the relative importance of the features in the ensemble of the trained XGBoost model [49]; therefore, the higher the value of the feature importance, the more important the feature for the prediction of the dependent variable): conventional on-surface (horizontal jerk (median) of repetitive loops), conventional in-air (horizontal velocity (median) of the sentence), conventional together (horizontal velocity (median) of the sentence), FD on-surface (jerk (max) $\alpha = 0.3$ of the letters *le*), FD in-air (vertical acceleration (mean) $\alpha = 0.6$ of the word *nepopadnout* (FI = 0.33), horizontal jerk (mean) $\alpha = 0.9$ of the word *nepopadnout* (FI = 0.33), horizontal jerk (mean) $\alpha = 0.2$ of the repetitive word *lektorka* (FI = 0.33)), FD together (jerk (max) $\alpha = 0.3$ of the letters *le* (on-surface; FI = 0.67), horizontal jerk (mean) $\alpha = 0.9$ of the word *nepopadnout* (in-air; FI = 0.33)), all on-surface (horizontal jerk (median) of repetitive loops (FI = 0.50), jerk (max) $\alpha = 0.3$ of the letters *le* (FI = 0.50)), all in-air (horizontal velocity (median) of the sentence), and all together (horizontal velocity (median) of the sentence (in-air)).

The results of multivariate regression analysis are summarized in Table 7. For this purpose, we used UPDRS V and PD duration as our target variables. As in the case of binary classification, we built and evaluated nine different regression models according to the same criteria. For each of the rating scales, the table shows the results achieved using the trained models and the associated feature importance values. All obtained results are discussed in the following section.

Table 7. Results of regression analysis for clinical data.

Feature Set	MAE	RMSE	EER [%]	Feat
		UPDRS V		
conventional on-surface	0.59 ± 0.29	0.71 ± 0.41	13.82 ± 6.71	1
conventional in-air	0.60 ± 0.30	0.72 ± 0.42	14.01 ± 6.98	1
conventional together	0.60 ± 0.31	0.73 ± 0.42	14.05 ± 6.90	1
FD on-surface	0.60 ± 0.32	0.65 ± 0.45	**12.51 ± 7.55**	1
FD in-air	0.60 ± 0.33	0.68 ± 0.43	13.49 ± 7.29	1
FD together	0.60 ± 0.34	0.66 ± 0.45	13.06 ± 7.55	2
all on-surface	0.60 ± 0.35	0.65 ± 0.45	**12.51 ± 7.55**	1
all in-air	0.60 ± 0.36	0.71 ± 0.43	13.72 ± 7.36	1
all together	0.60 ± 0.37	0.66 ± 0.45	13.06 ± 7.55	2
		PD duration		
conventional on-surface	4.29 ± 0.94	5.03 ± 1.09	24.52 ± 5.39	18
conventional in-air	4.91 ± 1.38	5.56 ± 1.50	28.03 ± 7.85	16
conventional together	4.14 ± 1.32	4.85 ± 1.52	**23.64 ± 7.55**	16
FD on-surface	4.45 ± 0.66	5.06 ± 0.85	25.40 ± 3.75	14
FD in-air	4.79 ± 0.73	5.48 ± 0.72	27.36 ± 4.20	19
FD together	4.55 ± 0.68	5.32 ± 0.78	26.00 ± 3.88	21
all on-surface	4.48 ± 0.86	5.12 ± 0.96	25.62 ± 4.92	16 (12 F, 4 C)
all in-air	4.95 ± 1.18	5.59 ± 1.17	28.30 ± 6.75	17 (13 F, 4 C)
all together	4.70 ± 1.10	5.45 ± 1.23	26.82 ± 6.30	17 (12 F, 6 C)

UPDRS V—Unified Parkinson's disease rating scale, part V: Modified Hoehn and Yahr staging score [40]; MAE—mean absolute error; RMSE—root mean squared error; EER—estimation error rate; F—FD-based features; C—conventional handwriting features; feat.—number of features important for the trained model (i.e., feature importance of the feature > 0.0); The feature importances, as well as the exact names of these features for models built to assess UPDRS V, are summarized in the text. In the case of PD duration, this data can be found in Table S1 provided in the Supplementary Material.

Considering EER as our performance evaluation metric, the following results are worth pointing out. In the case of UPDRS V, the lowest EER was achieved using a single FD-based feature—specifically, the standard deviation of vertical velocity ($\alpha = 0.1$) computed for the on-surface movements ($12.51 \pm 7.55\%$). The same feature was selected when both FD and conventional features were considered while building the model. In general, all models achieved an EER of around 12–13%. In comparison with the conventional features, the FD-based features performed better, with a difference of about 1%. In terms of the specific features important for the trained models, the following feature importances were returned by the models: conventional on-surface (vertical normalized jerk (mean) of the repetitive word *lektorka*), conventional in-air (vertical velocity (mean) of the sentence), conventional together (vertical velocity (mean) of the sentence), FD on-surface (vertical velocity (std) $\alpha = 0.1$ of the sentence), FD in-air (vertical velocity (median) $\alpha = 0.3$ of the sentence), FD together (vertical velocity (std) $\alpha = 0.1$ of the sentence (on-surface; FI $= 0.50$), vertical velocity (median) $\alpha = 0.3$ of the sentence (in-air; FI $= 0.50$)), all on-surface (vertical velocity (std) $\alpha = 0.1$ of the sentence), all in-air (vertical velocity (median) $\alpha = 0.3$ of the sentence), and all together (vertical velocity (std) $\alpha = 0.1$ of the sentence (on-surface; FI $= 0.50$), vertical velocity (median) $\alpha = 0.3$ of the sentence (in-air; FI $= 0.50$)). With respect to PD duration, the lowest EER was achieved using conventional handwriting features computed for both on-surface as well as in-air movements ($23.64 \pm 7.55\%$).

4. Discussion

To the best of our knowledge, except for our pilot work [20], there are no prior studies which integrate FD into a handwriting parameterization for quantitative PD dysgraphia analysis. Therefore, the results published in this paper are exploratory in nature.

In comparison with the conventional kinematic features, FD-based ones correlate more significantly with the clinical characteristics (UPDRS V and PD duration). We observed especially strong correlations for handwriting tasks based on the periodic repetition of specific movements (Archimedean spiral; repetitive letter *l*, syllable *le*, or word *les*). Although the levels of significance based on the conventional handwriting parameters are lower, similar handwriting tasks are involved in the most significant results. We hypothesize that this is due to their ability to highlight or better quantify the cardinal motor symptoms of PD. For example, the most significant relationship between handwriting performance and PD duration was identified in acceleration extracted from the Archimedean spiral. Rigidity combined with tremor and/or bradykinesia makes a PD patient's handwriting/drawing less fluent (increased changes in velocity and higher acceleration). This is highlighted in a task such as the spiral, where the proper coordination of the fingers, wrist, and arm is required. Generally, the observed problems with coordination are in line with the work of Dounskaia et al. [51] and Teulings et al. [52]. To better illustrate these manifestations, Figure 3 plots the velocity profiles of repetitive loops for a healthy control and a PD patient. As can be seen, the patient introduced more changes in velocity, and their drawing became much more non-fluent. To summarize these findings, FD features in combination with properly selected tasks provide a stronger relationship with the severity and progress of PD.

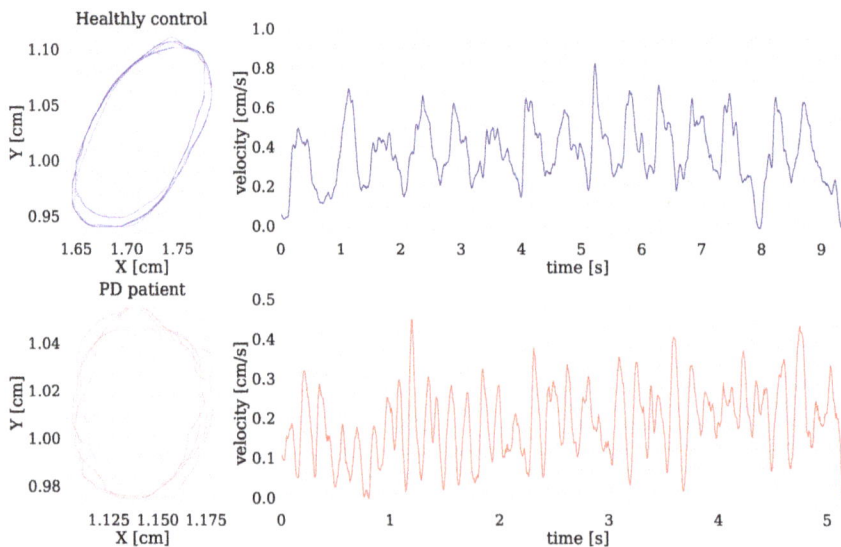

Figure 3. Handwriting samples of the repetitive loop task for HC and PD patients are on the left, and the resulting velocity profiles are on the right.

On the other hand, in terms of binary classification, the conventional parameters provided the best results. The classification performance is remarkable: ACC = 97.74%, SEN = 95.50%, and SPE = 100%. In fact, our results represent the highest classification accuracy that has ever been reported based on the PaHaW database (see Table 1). We hypothesize that the improvement was caused by the inclusion of the state-of-the-art XGBoost algorithm into our machine learning pipelines. As already mentioned, the result is based on one in-air feature: median horizontal velocity of a sentence. In comparison with the HC cohort, the PD patients exhibited much lower values of this measure, i.e., while writing the sentence, the PD patients were not able to perform horizontal transitions (movement between neighboring letters or words) as quickly as the HC could. This finding is in line with the work of Ma et al. [53], who observed that wrist extension stiffness in PD patients makes the handwriting in the horizontal direction more problematic. Therefore, scientists started to use the term *horizontal dysgraphia* [13]. Generally, vertical or horizontal dysgraphia may be considered a presymptomatic neurobehavioral biomarker of PD with possible significance in early PD diagnosis [13].

In [20], we proved that the FD features improved the accuracy of PD dysgraphia diagnosis in the Archimedean spiral drawing task by 10%. Contrary to our pilot results, in the frame of this work, these features did not lead to any improvements. After a deeper analysis, we found that this was caused by a combined task approach. Performance of the Archimedean spiral is a quasiparticle and continuous task with some repetitive patterns. It looks as though the FD features work especially well in these specific cases. Nevertheless, when combining these tasks with a complex handwriting task (such as a sentence), the measures quantifying in-air movement tend to be more discriminative (in our case, the median in-air horizontal velocity of a sentence). This brings us to the same conclusion that was given during the correlation analysis—the FD features advance the PD dysgraphia diagnosis only in some specific cases.

The best regression model, estimating the UPDRS V score with a 12.51% error, is based only on the standard deviation of on-surface vertical velocity ($\alpha = 0.1$) extracted from the sentence. This FD-based parameter was selected from the feature set combining all on-surface measures; therefore, we can confirm the positive influence of FC on the regression analysis performance. In fact, the FD features outperformed the conventional ones in all scenarios. To better understand this result, we plotted vertical velocity patterns of the sentence task for different orders of FD (see Figure 4). We can observe a big difference between $\alpha = 0.1$ and the rest of the orders, including the full derivative. This large distance means that we are working with completely new information that is far from that contained in the full derivative. Although it is difficult to clinically interpret this information, it is clear that FC opens new possibilities for monitoring PD severity.

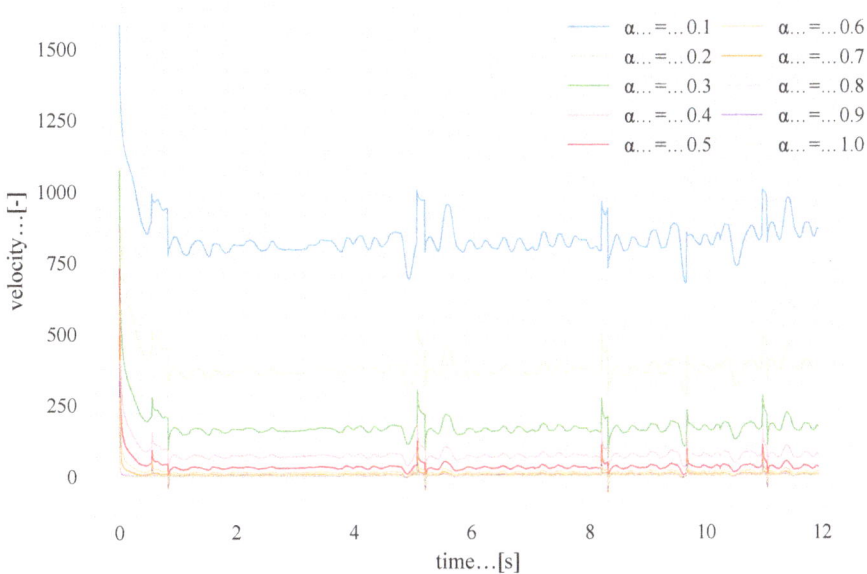

Figure 4. Vertical velocity patterns of the sentence task for different orders of fractional-order derivatives (FD).

Regarding the PD duration estimation results, the most successful model (EER = 23.46%) consists of 16 conventional on-surface/in-air features (all features' importance values can be found in Supplementary Table S1). The most frequent feature with the highest feature importance is the jerk extracted from several handwriting tasks. This probably means that as PD progresses, handwriting becomes more jerky and irregular. Vertical velocity is the second most frequent feature involved in the models, which is probably linked with micrographia. Generally, in the case of PD duration estimation, the FD-based features did not yield any improvement.

In conclusion, the FD-based features are better for modeling PD severity (in terms of UPDRS V score estimation), but they do not lead to an improvement in PD duration modeling. The progress of PD is nonlinear and very individual. This means that patients with the same PD duration can be in different stages of the disease. This fact supports our results: the estimation error of PD duration was generally much worse than the estimation error of the UPDRS V score. Since PD duration estimation is a difficult task with poor results, fine improvements based on FD parameters play no role.

5. Conclusions

This study deals with advanced approaches to PD dysgraphia diagnosis and monitoring based on FC integrated with online handwriting/drawing parameterization. To the best of our knowledge, it is the first work that performs a complex investigation into the possibilities for FC in online handwriting processing and proposes new advances in kinematic analyses based on FD. Although the conventional features provided better and very high classification accuracy, which is at the top of the state-of-the-art analyses based on the PaHaW database (ACC = 97.74%, SEN = 95.50%, and SPE = 100%), the newly designed parameters were proven to work better for specific tasks (continuous and/or repetitive, such as the Archimedean spiral) and for specific applications, i.e., PD severity estimation (EER = 12.51%). However, our results need to be confirmed by subsequent scientific research.

This study has several limitations and suggestions for further improvements. Since the dataset is small, to be able to generalize the results, bigger databases should be involved. On the other hand, it is common to have such small numbers of PD patients and HC samples in PD dysgraphia analysis, e.g., see our review in Table 1. Next, we considered only the kinematic measures. To better evaluate the discrimination power of the FD features and better evaluate their ability to estimate PD severity or progress, other feature types, such as temporal, spatial, and dynamic, should be included in future comparisons. Finally, the FD-based parameters could be further explored. For instance, we can consider other approximations (e.g., Caputo) or employ FC for other measures (e.g., entropies).

Supplementary Materials: The following are available online at http://www.mdpi.com/2076-3417/8/12/2566/s1, Table S1: Feature relevance from multivariate regression (modeling PD duration).

Author Contributions: Conceptualization, J.M. (Jan Mucha), J.M. (Jiri Mekyska) and M.F.-Z.; methodology, J.M. (Jan Mucha), M.F.-Z. and J.M. (Jiri Mekyska); software, J.M. (Jan Mucha) and J.M. (Jiri Mekyska); validation, J.M. (Jan Mucha); formal analysis, J.M. (Jan Mucha) and J.M. (Jiri Mekyska); investigation, J.M. (Jan Mucha), M.F.-Z., J.M. (Jiri Mekyska), I.R. and L.B.; resources, I.R. and L.B.; data curation, J.M. (Jan Mucha), J.M. (Jiri Mekyska), Z.G., V.Z., T.K., L.B. and I.R.; writing—original draft preparation, J.M. (Jan Mucha), J.M. (Jiri Mekyska), Z.G. and M.F.-Z.; visualization, J.M. (Jan Mucha) and Z.G.; supervision, J.M. (Jiri Mekyska), Z.S., I.R., M.F.-Z. and K.L.-d.-I.; project administration, J.M. (Jiri Mekyska), Z.S., I.R., M.F.-Z. and K.L.-d.-I.; funding acquisition, J.M. (Jiri Mekyska), Z.S., I.R., M.F.-Z. and K.L.-d.-I.

Funding: This project has received funding from the European Union's Horizon 2020 research and innovation programme under the Marie Skłodowska-Curie grant agreement No. 734718 (CoBeN). In addition, this work was supported by the grant of the Czech Science Foundation 18-16835S (Research of advanced developmental dysgraphia diagnosis and rating methods based on quantitative analysis of online handwriting and drawing) and the following projects: LO1401, FEDER and MEC, and TEC2016-77791-C4-2-R from the Ministry of Economic Affairs and Competitiveness of Spain. This article is based upon work from COST Action CA15225, a network supported by COST (European Cooperation in Science and Technology), and, for the research, infrastructure of the SIX Center was used.

Conflicts of Interest: The authors declare no conflict of interest.

Disclaimer: The article reflects only the authors' view, and the Research Executive Agency (REA) is not responsible for any use that may be made of the information it contains.

Abbreviations

The following abbreviations are used in this manuscript:

ACC	accuracy
ADA	AdaBoost
ANN	artificial neural network
ANOVA	analysis of variance
AUC	area under the ROC curve
CNN	convolutional neural network
EMD	empirical mode decomposition
EER	estimated error rate
FN	false negatives
FP	false positives
FC	fractional calculus
FD	fractional-order derivative
FI	feature importance
K-NN	K-nearest neighbors
LED	L-dopa equivalent daily dose
LDA	linear discriminant analysis
MCC	Matthew's correlation coefficient
max	maximum
MAE	mean absolute error
NB	naïve Bayes classifier
OPF	optimum path forest
PD	Parkinson's disease
RF	random forests
RMSE	root mean squared error
SEN	sensitivity
r_p	Pearson's correlation coefficient
r_s	Spearman's correlation coefficient
SPE	specificity
std	standard deviation
TN	true negatives
TP	true positives
SVM	support vector machine
UPDRS V	unified Parkinson's disease rating scale, part V: Modified Hoehn and Yahr staging score

References

1. Bellou, V.; Belbasis, L.; Tzoulaki, I.; Evangelou, E.; Ioannidis, J.P. Environmental risk factors and Parkinson's disease: An umbrella review of meta-analyses. *Parkinsonism Relat. Disord.* **2016**, *23*, 1–9. [CrossRef] [PubMed]
2. Berg, D.; Postuma, R.B.; Adler, C.H.; Bloem, B.R.; Chan, P.; Dubois, B.; Gasser, T.; Goetz, C.G.; Halliday, G.; Joseph, L.; et al. MDS research criteria for prodromal Parkinson's disease. *Mov. Disord.* **2015**, *30*, 1600–1611. [CrossRef] [PubMed]
3. Sapir, S.; Ramig, L.; Fox, C. Speech and swallowing disorders in Parkinson disease. *Curr. Opin. Otolaryngol. Head Neck Surg.* **2008**, *16*, 205–210. [CrossRef] [PubMed]
4. Hirsch, L.; Jette, N.; Frolkis, A.; Steeves, T.; Pringsheim, T. The incidence of Parkinson's disease: A systematic review and meta-analysis. *Neuroepidemiology* **2016**, *46*, 292–300. [CrossRef] [PubMed]
5. Hornykiewicz, O. Biochemical aspects of Parkinson's disease. *Neurology* **1998**, *51*, S2–S9. [CrossRef] [PubMed]
6. Elbaz, A.; Carcaillon, L.; Kab, S.; Moisan, F. Epidemiology of Parkinson's disease. *Rev. Neurol.* **2016**, *172*, 14–26, doi:10.1016/j.neurol.2015.09.012. [CrossRef] [PubMed]
7. Contreras-Vidal, J.L.; Poluha, P.; Teulings, H.L.; Stelmach, G.E. Neural dynamics of short and medium-term motor control effects of levodopa therapy in Parkinson's disease. *Artif. Intell. Med.* **1998**, *13*, 57–79. [CrossRef]

8. Drotar, P.; Mekyska, J.; Rektorova, I.; Masarova, L.; Smekal, Z.; Faundez-Zanuy, M. Evaluation of handwriting kinematics and pressure for differential diagnosis of Parkinson's disease. *Artif. Intell. Med.* **2016**, *67*, 39–46. [CrossRef]
9. Brabenec, L.; Mekyska, J.; Galaz, Z.; Rektorova, I. Speech disorders in Parkinson's disease: Early diagnostics and effects of medication and brain stimulation. *J. Neural Transm.* **2017**, *124*, 303–334. [CrossRef]
10. Mucha, J.; Galaz, Z.; Mekyska, J.; Kiska, T.; Zvoncak, V.; Smekal, Z.; Eliasova, I.; Mrackova, M.; Kostalova, M.; Rektorova, I.; et al. Identification of hypokinetic dysarthria using acoustic analysis of poem recitation. In Proceedings of the 2017 40th International Conference on Telecommunications and Signal Processing (TSP), Barcelona, Spain, 5–7 July 2017; pp. 739–742.
11. De Stefano, C.; Fontanella, F.; Impedovo, D.; Pirlo, G.; di Freca, A.S. Handwriting analysis to support neurodegenerative diseases diagnosis: A review. *Pattern Recognit. Lett.* **2018**, in press. [CrossRef]
12. Rosenblum, S.; Samuel, M.; Zlotnik, S.; Erikh, I.; Schlesinger, I. Handwriting as an objective tool for Parkinson's disease diagnosis. *J. Neurol.* **2013**, *260*, 2357–2361. [CrossRef]
13. Thomas, M.; Lenka, A.; Kumar Pal, P. Handwriting Analysis in Parkinson's Disease: Current Status and Future Directions. *Mov. Disord. Clin. Pract.* **2017**, *4*, 806–818. [CrossRef] [PubMed]
14. McLennan, J.; Nakano, K.; Tyler, H.; Schwab, R. Micrographia in Parkinson's Disease. *J. Neurol. Sci.* **1972**, *15*, 141–152. [CrossRef]
15. Letanneux, A.; Danna, J.; Velay, J.L.; Viallet, F.; Pinto, S. From micrographia to Parkinson's disease dysgraphia. *Mov. Disord.* **2014**, *29*, 1467–1475. [CrossRef] [PubMed]
16. Drotar, P.; Mekyska, J.; Smekal, Z.; Rektorova, I.; Masarova, L.; Faundez-Zanuy, M. Contribution of different handwriting modalities to differential diagnosis of Parkinson's Disease. In Proceedings of the 2015 IEEE International Symposium Medical Measurements and Applications (MeMeA), Turin, Italy, 7–9 May 2015; pp. 1–5.
17. Drotar, P.; Mekyska, J.; Rektorova, I.; Masarova, L.; Smekal, Z.; Zanuy, M.F. Decision support framework for Parkinson's disease based on novel handwriting markers. *IEEE Trans. Neural Syst. Rehabil. Eng.* **2015**, *23*, 508–516. [CrossRef] [PubMed]
18. Loconsole, C.; Trotta, G.F.; Brunetti, A.; Trotta, J.; Schiavone, A.; Tato, S.I.; Losavio, G.; Bevilacqua, V. Computer Vision and EMG-Based Handwriting Analysis for Classification in Parkinson's Disease. In *Intelligent Computing Theories and Application*; Springer International Publishing: Cham, Switzerland, 2017; pp. 493–503, doi:10.1007/978-3-319-63312-1_43.
19. Nackaerts, E.; Broeder, S.; Pereira, M.P.; Swinnen, S.P.; Vandenberghe, W.; Nieuwboer, A.; Heremans, E. Handwriting training in Parkinson's disease: A trade-off between size, speed and fluency. *PLoS ONE* **2017**, *12*, e0190223. [CrossRef] [PubMed]
20. Mucha, J.; Zvoncak, V.; Galaz, Z.; Faundez-Zanuy, M.; Mekyska, J.; Kiska, T.; Smekal, Z.; Brabenec, L.; Rektorova, I.; Lopez-de Ipina, K. Fractional Derivatives of Online Handwriting: A New Approach of Parkinsonic Dysgraphia Analysis. In Proceedings of the 2018 41st International Conference on Telecommunications and Signal Processing (TSP), Athens, Greece, 4–6 July 2018; pp. 214–217.
21. Kotsavasiloglou, C.; Kostikis, N.; Hristu-Varsakelis, D.; Arnaoutoglou, M. Machine learning-based classification of simple drawing movements in Parkinson's disease. *Biomed. Signal Process. Control* **2017**, *31*, 174–180. [CrossRef]
22. Taleb, C.; Khachab, M.; Mokbel, C.; Likforman-Sulem, L. Feature selection for an improved Parkinson's disease identification based on handwriting. In Proceedings of the 2017 1st International Workshop on Arabic Script Analysis and Recognition (ASAR), Nancy, France, 3–5 April 2017; pp. 52–56.
23. Moetesum, M.; Siddiqi, I.; Vincent, N.; Cloppet, F. Assessing visual attributes of handwriting for prediction of neurological disorders: A case study on Parkinson's disease. *Pattern Recognit. Lett.* **2018**. [CrossRef]
24. Heremans, E.; Nackaerts, E.; Vervoort, G.; Vercruysse, S.; Broeder, S.; Strouwen, C.; Swinnen, S.P.; Nieuwboer, A. Amplitude Manipulation Evokes Upper Limb Freezing during Handwriting in Patients with Parkinson's Disease with Freezing of Gait. *PLoS ONE* **2015**, *10*, e0142874. [CrossRef] [PubMed]
25. Pereira, C.R.; Pereira, D.R.; da Silva, F.A.; Hook, C.; Weber, S.A.; Pereira, L.A.; Papa, J.P. A step towards the automated diagnosis of parkinson's disease: Analyzing handwriting movements. In Proceedings of the 2015 IEEE 28th International Symposium on Computer-Based Medical Systems (CBMS), Sao Carlos, Brazil, 22–25 June 2015; pp. 171–176.

26. Heremans, E.; Nackaerts, E.; Broeder, S.; Vervoort, G.; Swinnen, S.P.; Nieuwboer, A. Handwriting Impairments in People With Parkinson's Disease and Freezing of Gait. *Neurorehabil. Neural Repair* **2016**, *30*, 911–919. [CrossRef]

27. Pereira, C.R.; Weber, S.A.; Hook, C.; Rosa, G.H.; Papa, J.P. Deep Learning-Aided Parkinson. In Proceedings of the 2016 29th SIBGRAPI Conference on Graphics, Patterns and Images (SIBGRAPI), Sao Paulo, Brazil, 4–7 October 2016; pp. 340–346.

28. Impedovo, D.; Pirlo, G.; Vessio, G. Dynamic Handwriting Analysis for Supporting Earlier Parkinson's Disease Diagnosis. *Information* **2018**, *9*, 247. [CrossRef]

29. Baleanu, D.; Jajarmi, A.; Bonyah, E.; Hajipour, M. New aspects of poor nutrition in the life cycle within the fractional calculus. *Adv. Differ. Equ.* **2018**, *2018*, 230. [CrossRef]

30. Khalili Golmankhaneh, A.; Baleanu, D. New Derivatives on the Fractal Subset of Real-Line. *Entropy* **2016**, *18*, 1. [CrossRef]

31. Podlubny, I. *Fractional Differential Equations an Introduction to Fractional Derivatives, Fractional Differential Equations, to Methods of Their Solution and Some of Their Applications*; Academic Press: San Diego, CA, USA, 1999.

32. Uchaikin, V.V. *Fractional Derivatives for Physicists and Engineers*; Springer: Berlin, Germany, 2013; Volume 2.

33. Samko, S.G.; Kilbas, A.A.; Marichev, O.I. *Fractional Integrals and Derivatives: Theory and Applications*; Gordon and Breach: Yverdon, Switzerland, 1993; p. 44.

34. Arshad, S.; Baleanu, D.; Bu, W.; Tang, Y. Effects of HIV infection on CD4+ T-cell population based on a fractional-order model. *Adv. Differ. Equ.* **2017**, *2017*, 92. [CrossRef]

35. Pinto, C.M.; Machado, J.T. Fractional model for malaria transmission under control strategies. *Comput. Math. Appl.* **2013**, *66*, 908–916. doi:10.1016/j.camwa.2012.11.017. [CrossRef]

36. Lopes, A.M.; Machado, J.T. Integer and fractional-order entropy analysis of earthquake data series. *Nonlinear Dyn.* **2016**, *84*, 79–90. [CrossRef]

37. Lopes, A.M.; Machado, J.T. Application of fractional techniques in the analysis of forest fires. *Int. J. Nonlinear Sci. Numer. Simul.* **2016**, *17*, 381–390. [CrossRef]

38. Scalas, E.; Gorenflo, R.; Mainardi, F. Fractional calculus and continuous-time finance. *Phys. A Stat. Mech. Its Appl.* **2000**, *284*, 376–384. [CrossRef]

39. Baleanu, D. *Fractional Calculus: Models and Numerical Methods*; World Scientific: Singapore, 2012; Volume 3.

40. Fahn, S.; Elton, R.L. UPDRS Development Committee (1987) Unified Parkinson's Disease Rating Scale. In *Recent Developments in Parkinson's Disease*; Macmillan: Florham Park, NJ, USA, 1987.

41. Lee, J.Y.; Kim, J.W.; Lee, W.Y.; Kim, J.M.; Ahn, T.B.; Kim, H.J.; Cho, J.; Jeon, B.S. Daily dose of dopaminergic medications in Parkinson's disease: Clinical correlates and a posteriori equation. *Neurol. Asia* **2010**, *15*, 137–143.

42. Sesa-Nogueras, E.; Faundez-Zanuy, M.; Mekyska, J. An information analysis of in-air and on-surface trajectories in online handwriting. *Cogn. Comput.* **2012**, *4*, 195–205. [CrossRef]

43. Alonso-Martinez, C.; Faundez-Zanuy, M.; Mekyska, J. A comparative study of in-air trajectories at short and long distances in online handwriting. *Cogn. Comput.* **2017**, *9*, 712–720. [CrossRef] [PubMed]

44. Drotar, P.; Mekyska, J.; Rektorova, I.; Masarova, L.; Smekal, Z.; Faundez-Zanuy, M. A new modality for quantitative evaluation of Parkinson's disease: In-air movement. In Proceedings of the 13th IEEE International Conference on BioInformatics and BioEngineering (BIBE 2013), Chania, Greece, 10–13 November 2013; pp. 1–4.

45. Mekyska, J.; Faundez-Zanuy, M.; Mzourek, Z.; Galaz, Z.; Smekal, Z.; Rosenblum, S. Identification and rating of developmental dysgraphia by handwriting analysis. *IEEE Trans. Hum.-Mach. Syst.* **2017**, *47*, 235–248. [CrossRef]

46. Tenreiro Machado, J. Fractional coins and fractional derivatives. *Abstr. Appl. Anal.* **2013**, *2013*, 205097. [CrossRef]

47. Kilbas, A.A.; Srivastava, H.M.; Trujillo, J.J. *Theory and Applications of Fractional Differential Equations*; Elsevier: Amsterdam, The Netherlands, 2006.

48. Scherer, R.; Kalla, S.L.; Tang, Y.; Huang, J. The Grünwald–Letnikov method for fractional differential equations. *Comput. Math. Appl.* **2011**, *62*, 902–917. [CrossRef]

49. Chen, T.; Guestrin, C. Xgboost: A scalable tree boosting system. In Proceedings of the 22nd ACM Sigkdd International Conference on Knowledge Discovery and Data Mining, San Francisco, CA, USA, 13–17 August 2016; pp. 785–794.

50. Matthews, B.W. Comparison of the predicted and observed secondary structure of T4 phage lysozyme. *Biochim. Biophys. Acta* **1975**, *405*, 442–451. [CrossRef]

51. Dounskaia, N.; Van Gemmert, A.W.; Leis, B.C.; Stelmach, G.E. Biased wrist and finger coordination in Parkinsonian patients during performance of graphical tasks. *Neuropsychologia* **2009**, *47*, 2504–2514. [CrossRef]

52. Teulings, H.L.; Contreras-Vidal, J.L.; Stelmach, G.E.; Adler, C.H. Parkinsonism Reduces Coordination of Fingers, Wrist, and Arm in Fine Motor Control. *Exp. Neurol.* **1997**, *146*, 159–170. [CrossRef]

53. Ma, H.I.; Hwang, W.J.; Chang, S.H.; Wang, T.Y. Progressive micrographia shown in horizontal, but not vertical, writing in Parkinson's disease. *Behav. Neurol.* **2013**, *27*, 169–174. [CrossRef]

applied
sciences

MDPI

Article

Changes in Phonation and Their Relations with Progress of Parkinson's Disease

Zoltan Galaz [1], Jiri Mekyska [1], Vojtech Zvoncak [1], Jan Mucha [1], Tomas Kiska [1], Zdenek Smekal [1], Ilona Eliasova [2,3], Martina Mrackova [2,3], Milena Kostalova [3,4], Irena Rektorova [2,3,*], Marcos Faundez-Zanuy [5], Jesus B. Alonso-Hernandez [6] and Pedro Gomez-Vilda [7]

[1]   Department of Telecommunications, Brno University of Technology, Technicka 10,
      616 00 Brno, Czech Republic; xgalaz00@stud.feec.vutbr.cz (Z.G.); mekyska@feec.vutbr.cz (J.M.);
      vojtech.zvoncak@gmail.com (V.Z.); xmucha05@stud.feec.vutbr.cz (J.M.); xkiska02@stud.feec.vutbr.cz (T.K.);
      smekal@feec.vutbr.cz (Z.S.)
[2]   First Department of Neurology, St. Anne's University Hospital, Pekarska 53, 656 91 Brno, Czech Republic;
      ilona.eliasova@fnusa.cz (I.E.); mrackova@fnusa.cz (M.M.)
[3]   Applied Neuroscience Research Group, Central European Institute of Technology, Masaryk University,
      Kamenice 5, 625 00 Brno, Czech Republic; Kostalova.Milena@fnbrno.cz
[4]   Department of Neurology, Faculty Hospital and Masaryk University, Jihlavska 20,
      639 00 Brno, Czech Republic
[5]   Escola Superior Politecnica, Tecnocampus, Avda. Ernest Lluch 32, 083 02 Mataro, Barcelona, Spain;
      faundez@tecnocampus.cat
[6]   Institute for Technological Development and Innovation in Communications (IDeTIC), University of Las
      Palmas de Gran Canaria, 35001 Las Palmas de Gran Canaria, Spain; jalonso@dsc.ulpgc.es
[7]   Neuromorphic Processing Laboratory (NeuVox Lab), Center for Biomedical Technology, Universidad
      Politecnica de Madrid, Campus de Montegancedo, s/n, Pozuelo de Alarcon, 28223 Madrid, Spain;
      pedro@fi.upm.es
[*]   Correspondence: irena.rektorova@fnusa.cz; Tel.: +420-543-182-639

Received: 17 October 2018; Accepted: 19 November 2018; Published: 22 November 2018

Abstract: Hypokinetic dysarthria, which is associated with Parkinson's disease (PD), affects several speech dimensions, including phonation. Although the scientific community has dealt with a quantitative analysis of phonation in PD patients, a complex research revealing probable relations between phonatory features and progress of PD is missing. Therefore, the aim of this study is to explore these relations and model them mathematically to be able to estimate progress of PD during a two-year follow-up. We enrolled 51 PD patients who were assessed by three commonly used clinical scales. In addition, we quantified eight possible phonatory disorders in five vowels. To identify the relationship between baseline phonatory features and changes in clinical scores, we performed a partial correlation analysis. Finally, we trained XGBoost models to predict the changes in clinical scores during a two-year follow-up. For two years, the patients' voices became more aperiodic with increased microperturbations of frequency and amplitude. Next, the XGBoost models were able to predict changes in clinical scores with an error in range 11–26%. Although we identified some significant correlations between changes in phonatory features and clinical scores, they are less interpretable. This study suggests that it is possible to predict the progress of PD based on the acoustic analysis of phonation. Moreover, it recommends utilizing the sustained vowel /i/ instead of /a/.

Keywords: phonation; acoustic analysis; follow-up study; hypokinetic dysarthria; Parkinson's disease

1. Introduction

Parkinson's disease (PD) is a frequent neurodegenerative disorder that is associated with a substantial reduction of dopaminergic neurons especially in substancia nigra pars compacta [1]. The primary motor symptoms of PD comprise tremor at rest, muscular rigidity, bradykinesia, and postural instability [1]. Patients with PD also develop a variety of non-motor symptoms [2] such as sleep disturbances, depression, cognitive impairment, etc. To diagnose, rate and monitor motor and non-motor symptoms of PD, various clinical rating scales such as Unified Parkinson's Disease Rating Scale (UPDRS) [3], Freezing Of Gait Questionnaire (FOG-Q) [4], or Addenbrooke's Cognitive Examination-Revised (ACE – R) [5] have been developed. Nevertheless, reliability of the assessment is often reduced by inter-rater variability [6].

Up to 90% [7] of patients with PD develop a multi-dimensional speech disorder named hypokinetic dysarthria (HD) [8], which is manifested in phonation, articulation, and prosody [9–11]. In the area of phonation, insufficient breath support, reduction in phonation time, increased acoustic noise, instability of articulatory organs, microperturbations of frequency/amplitude, and harsh breathy voice quality has been observed [9,12]. HD leads to serious complications in daily communication of patients with PD [13]. Generally, HD was found to be more severe in the advanced stages of PD [14].

As reported by the recent studies, acoustic analysis of HD can provide clinicians with non-invasive and reliable methodology of PD diagnosis, assessment and monitoring [9,15]. Moreover, this methodology has also been used to monitor the efficiency of PD treatment [10,16–18]. In the field of acoustic analysis of PD phonation, the authors mostly focused on the sustained vowel /a/ [9]. Conventional phonatory features such as jitter, shimmer, harmonic-to-noise ratio, degree of unvoiced segments, and formant-based parameters extracted from this vowel have been widely used to diagnose PD [12,19–23]. Although Hazan et al. [24] employed analysis of sustained phonation for diagnosis of PD even in its early stage, based on the recent review [9], most of the researchers find relevant applications of the phonatory analysis especially in moderate or severe stages of this disorder.

For example, the analysis of sustained phonation has been utilized during PD severity assessment. In 2010, Tsanas et al. [15] enrolled 42 PD patients and parameterized their sustained phonation of vowel /a/ by a set of conventional features that were consequently mapped to UPDRS, part III (motor examination) and the total score of this scale. Using classification and regression trees, they estimated the UPDRS III score with MAE (mean absolute error) equal to 5.95. The total UPDRS score was estimated with MAE = 7.52. A parametric version of this dataset has been made available for research purposes and other research teams further decreased the estimation error [25–27]. Another work that deals with the automatic clinical scores estimation was published by Mekyska et al. [21]. In this study, they acquired sustained phonation of vowels /a/, /e/, /i/, /o/, /u/ in 84 PD patients. Modeling conventional and advanced features by random forests provided the estimation of UPDRS III with MAE = 5.70. In addition, the authors estimated several other clinical scores such as UPDRS, part IV (complications of therapy) with MAE = 1.30 or Beck depression inventory (BDI) with MAE = 3.12.

Even though HD is one of the most problematic aspects of PD, the number of longitudinal studies investigating the evolution of HD in PD over time (based on the acoustic analysis) is very limited [28–31]. If we focus specifically on longitudinal monitoring of sustained phonation, then, in fact, we can identify only one study, which is published by Skodda et al. [31]. In this work, the authors repeatedly (with average time interval 32.50 months) acquired sustained vowel /a/ in 32 female and 48 male PD patients (age in session 1: 66.28 ± 8.11 years; PD duration in session 1: 6.10 ± 4.63 years; UPDRS III in session 1: 20.16 ± 10.96; UPDRS III in session 2: 19.58 ± 8.29). The voice was quantified by jitter, shimmer, noise-to-harmonic ratio, and mean fundamental frequency. Based on the paired *t*-test, the authors identified significant changes in shimmer and noise-to-harmonic ratio. In both cases, the values of these parameters increased. Another interesting finding is that, although some phonatory features significantly changed, UPDRS III was held widely stable over time. The authors provide two possible explanations: (1) voice impairment could be the result of an escalation of axial dysfunction too subtle to be mirrored by UPDRS III; (2) alterations of speech parameters could be completely independent of motor performance that may be based upon non-dopaminergic mechanisms. Inconsistencies in terms of the L-dopa effect on HD are further discussed in Brabenec et al. [9].

To sum it up, although the scientific community frequently addresses phonation in association with HD (especially when diagnosing or assessing PD), to the best of our knowledge, there is only one study that focuses on HD phonatory disorders from a longitudinal perspective. Moreover, the work deals with the analysis of phonation just partially, it considers only the sustained vowel /a/, and it does not explore a possibility of PD progress prediction based on a combination of acoustic analysis and machine learning. Therefore, in the frame of our two-year follow-up study, we are going much further with the following aims:

1. to identify phonatory acoustic features at baseline that are significantly correlated with changes in various clinical rating scales,
2. to investigate relationship between changes in the phonatory acoustic features and the clinical rating scales after the two-year follow-up,
3. to establish mathematical models that will estimate the change in clinical rating scales based on the change in acoustic measures,
4. to compare results based on five vowels: /a/, /e/, /i/, /o/, /u/.

The rest of this article is organized as follows: Section 2 describes a dataset of PD patients as well as methodology in terms of acoustic analysis, statistical analysis and machine learning. Results are reported in Section 3 and consequently discussed in Section 5. Finally, conclusions are given in Section 4.

2. Materials and Methods

2.1. Dataset

In this work, we enrolled 51 patients with idiopathic PD. All of them are Czech native speakers (17 females and 34 males; age: 65.47 ± 7.46 years; PD duration: 7.61 ± 4.01 years; mean LED (L-dopa equivalent daily dose) [32]: 1033.67 ± 567.96 mg/day) at the First Department of Neurology, St. Anne's University Hospital in Brno, Czech Republic. After two years, the patients were re-examined (age: 67.61 ± 7.38 years; PD duration: 9.57 ± 4.50 years; mean LED: 1115.11 ± 484.38 mg/day). All patients signed an informed consent form that has been approved (including the study) in 14 March 2016 by the Research Ethics Committee of Masaryk University (ref. no.: EKV-2016-004, project title: Effects of non-invasive brain stimulation on hypokinetic dysarthria, micrographia, and brain plasticity in patients with Parkinson's disease, investigator: Prof. MD. Irena Rektorova, PhD.).

None of the patients had a disease affecting the central nervous system other than PD. All patients were examined on their regular dopaminergic medication approximately 1 h after the L-dopa [32] dose. The following rating scales were used to evaluate the clinical symptoms of PD: UPDRS III and UPDRS IV [3], FOG-Q [4], REM sleep behavior disorder screening questionnaire (RBDSQ) [33], and ACE-R [5]. The full clinical characteristics of the dataset, i.e., mean $\pm$ sd values for the clinical rating scales in session 1, session 2, and session Δ (session 2 − session 1) can be seen in Table 1. Moreover, to identify statistically significant differences, the table reports p-values of the Wilcoxon signed-rank test between the data acquired in session 1 (baseline examination) and session 2 (two-year follow-up examination) too.

The clinical data from the Δ session were also used to generate descriptive visualizations (i.e., histograms, regression and residual plots) for the change in selected clinical rating scales, more specifically: LED, UPDRS III, UPDRS IV, FOG-Q, RBDSQ, ACE-R, see Figure 1. With this approach, it is possible to assess the improvement and/or decline in motor and non-motor deficits associated with PD in the horizon of two years as well as a relationship between the change in each of the scales relative to other scales in the selected set.

Table 1. Clinical characteristics of the patients.

Scale	Mean ± sd (s1)	Mean ± sd (s2)	Mean ± sd (Δ)	p (Wilcoxon)
LED	917.61 ± 544.78	1129.92 ± 477.50	212.31 ± −67.28	0.188
UPDRS III	22.49 ± 13.47	27.45 ± 12.68	4.96 ± −0.79	0.000
UPDRS IV	2.82 ± 2.58	3.44 ± 2.94	0.62 ± 0.36	0.632
FOG-Q	6.57 ± 5.40	8.33 ± 5.97	1.76 ± 0.57	0.000
RBDSQ	3.98 ± 3.25	3.78 ± 2.28	−0.2 ± −0.98	0.522
ACE-R	87.92 ± 7.62	85.89 ± 9.48	−2.03 ± 1.86	0.000

s1—first session; s2—second session; Δ—delta session (session 2 − session 1); p (Wilcoxon)—p-value for Wilcoxon signed-rank test (paired samples); LED—L-dopa equivalent daily dose (mg/day) [32]; UPDRS III—Unified Parkinson's Disease Rating Scale, part III: evaluation of motor function [3], UPDRS IV—Unified Parkinson's Disease Rating Scale, part IV: evaluation of complications of therapy [3]; FOG-Q—Freezing of gait questionnaire [4]; RBDSQ—The REM sleep behavior disorder screening questionnaire [33]; ACE-R—Addenbrooke's Cognitive Examination-Revised [5].

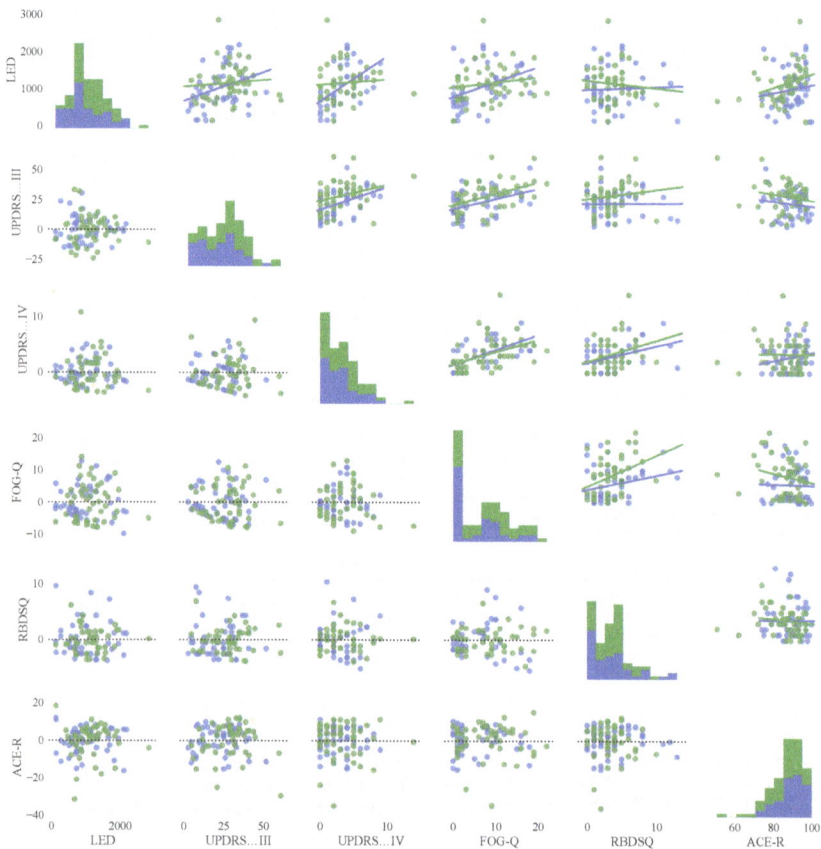

Figure 1. Descriptive statistical graphs of clinical characteristics of the PD patient dataset: on the main diagonal, histograms are visualized. Next, the upper triangular part of the graph-grid shows scatter plots with the fitted lines of linear regression models. Finally, the lower triangular part of the graph-grid is used to display residuals for the models shown in the upper grid. Color notation: the blue color represents data for session 1, and the green color represents data for session 2.

2.2. Vocal Tasks

To quantify the deterioration of phonation in patients with PD, we used a sustained phonation of vowels: /a/, /e/, /i/, /o/, /u/ as a basis for our experiments. The reason behind using all of the five vowels is to employ the analysis with the emphasis on quantifying all positions of a tongue during phonation. For more information, see the Hellwag (vowel) triangle [34]. In our view, using only a sustained phonation of the vowel /a/ is not fully justified as there is very little or no reason to assume that this particular position of the tongue can provide more information about phonatory disorders. In fact, as shown by previous studies, the analysis of other vowels is important for a more robust description of HD [19–21,23,24,35–37].

Sustained phonation of a vowel is a standard measure used to assess quality of phonation [9]. During this particular vocal task, a speaker is asked to sustain phonation of a vowel, attempting to maintain steady frequency and amplitude at a comfortable level [38]. The advantage of this task in comparison with other commonly used vocal tasks is its independence of articulatory and other linguistic confounds [38]. Moreover, it is also present in most of the databases and therefore the experiments proposed in our work are comparable with other commonly used databases [39,40].

The sustained phonation task used in this study is a part of a speech acquisition protocol derived from the standardized 3F Dysarthria Profile [41]. During the data acquisition, a large capsule cardioid microphone M-AUDIO Nova (Cumberland, RI, United States) mounted to a boom arm RODE PSA1 (Silverwater, Australia) and positioned at a distance of approximately 20 cm from the patient's mouth was used for the recording. Consequently, the signals were digitized by audio interface M-AUDIO Fast Track Pro (Cumberland, RI, United States) with the sampling frequency of 48 kHz (16-bit resolution) and checked by a trained acoustic engineer without having seen the patient's clinical data. Finally, the signals were parameterized using Praat [42] software as well as a set of MATLAB (MATLAB 9.4, MathWorks, Natick, MA, United States) parametrization functions [43] developed at the Brno University of Technology.

2.3. Acoustic Features

To describe a variety of phonatory disorders associated with HD, we quantified the following: (a) microperturbations in frequency of voice using period perturbation quotient (PPQ); (b) microperturbations in intensity of voice using amplitude perturbation quotient (APQ); (c) irregular pitch fluctuations using coefficient of variation of fundamental frequency (F0 (CV)); (d) irregular amplitude fluctuations using coefficient of variation of Teager–Kaiser operator (TKEO (CV)); (e) tremor of articulatory organs (such as jaw, tongue and lips), coefficient of variation of 1st formant (F1 (CV)), coefficient of variation of 2nd formant (F2 (CV)), coefficient of variation of 3rd formant (F3 (CV)); (f) increased acoustic noise using median of harmonic-to-noise ratio (HNR (Q2)), median of energy ratio (ER (Q2), energy ratio of bands 2000–4000 Hz and 70–900 Hz)), median of glottal-to-noise excitation ratio (GNE (Q2)), median of normalized noise energy (NNE (Q2)); (g) irregular acoustic noise fluctuations using standard deviation of harmonic-to-noise ratio (HNR (SD)), coefficient of variation of energy ratio (ER (CV)), standard deviation of glottal-to-noise excitation ratio (GNE (SD)), standard deviation of normalized noise energy (NNE (SD)); and (h) aperiodicity of voice using fraction of locally unvoiced frames (FLUF). All of these features are standard and clinically interpretable dysphonic measures and were selected based on a recommendation given in our recent review on acoustic analysis of voice/speech signals in patients suffering from HD [9]. For more information about the voice/speech parametrization, see [43].

2.4. Statistical Analysis

Before describing the analytical setup applied in this work, it is important to mention that the dataset did not contain any missing values, and therefore all data samples were used. Furthermore, even though we used six clinical rating scales when describing the dataset (see Section 2.1), only four of these scales were used for the analysis, specifically: UPDRS III, UPDRS IV, RBDSQ, and FOG-Q. The reason is that previous studies have already shown that non-motor manifestations of PD are not linked with the phonatory aspects of HD, but rather with the impairments of prosody and articulation [44] that are commonly being quantified using a sentence reading task, free speech (monologue), etc. Since this study is focused on the phonatory aspects of HD, clinical rating scales describing only motor symptoms of PD were used.

To reveal and assess the strength of a relationship between the computed acoustic features and patients' clinical data (UPDRS III, UPDRS IV, RBDSQ, and FOG-Q), Spearman's correlation coefficient was computed (the statistical assumptions for Spearman's correlation coefficient were satisfied as: (a) the acoustic features as well as the the clinical data are both variables that are measured on at least an ordinal scale, and (b) there is a monotonic relationship between the two variables). Since age, gender, and probably L-dopa, are manifested in a voice of PD patients [9], for the purpose of this work, we employed partial Spearman's correlation controlling for the effect of the following confounding factors (also known as covariates): patients' age, gender [29,45], and dopaminergic medication [32,46]. The significance level of correlation was set to 0.05. More specifically, two correlation scenarios were considered: (a) correlation between the acoustic features at the baseline and the change in values of the selected clinical rating scales, and (b) correlation between the change in the acoustic features and the change in the values of the selected clinical rating scales. With this approach, we aimed at identifying those acoustic features that are significantly correlated with the specific motor and non-motor symptoms assessed by the selected clinical rating scales in both scenarios.

Next, to evaluate the power of the acoustic features at the baseline to predict the change of the patients' clinical data in the horizon of two years, we used the acoustic features computed for the recordings acquired in session 1 (baseline examination) and built mathematical models predicting the change in the selected clinical rating scales (Δ). For this purpose, we employed Gradient Boosted Trees (more specifically, the famous XGBoost algorithm [47]) in a supervised learning setup: 10-fold cross-validation with 20 repetitions [48]. The XGBoost algorithm belongs to the state-of-the-art in machine learning, which is supported by the fact that it has been recently used to win competitions on Kaggle. It works well even on small datasets (where it outperforms deep learning approaches), it is robust to outliers and it is able to model complex interdependencies. For these reasons, it has been used by many researchers in various biomedical fields, e.g., [49–51], etc.

The performance of the models (precision of the predictions) was evaluated by MAE and estimation error rate (EER). These measures are defined as:

$$\text{MAE} = \frac{1}{n} \sum_{i=1}^{n} |y_i - \hat{y}_i|,$$

$$\text{EER} = \frac{1}{n \cdot r} \sum_{i=1}^{n} |y_i - \hat{y}_i| \cdot 100 \, [\%],$$

where y_i stands for the true label of i-th observation, $\hat{y}_i$ represents the predicted label of the i-th observation, n denotes the number of observations, and finally r stands for the range of values in the predicted clinical rating scale (not the range that can be theoretically reached, but the actual range of the values in the dataset). As can be seen, EER therefore describes a percentage of error predictions with respect to statistical properties of the dataset, which is particularly useful for easy interpretation of the results.

3. Results

The values of 16 acoustic features extracted from both sessions, as well as values of their differences (session 2 − session 1), are reported in Table 2. Based on the Wilcoxon signed-rank test, we can observe that none of the features extracted from vowel /a/ significantly changed after two years. Regarding vowel /e/, we identified significantly increased microperturbations in intensity of voice and also increased aperiodicity. The same significant changes were identified in vowel /i/ and /u/. In the case of vowel /u/, in addition, we monitored the increase of microperturbations in frequency of voice. The repeated acquisition of vowel /o/ was associated with increased aperiodicity and more dominant microperturbations in frequency of voice.

Table 2. Statistical description of acoustic features for all vocal tasks.

Feature	Mean ± sd (s1)	Mean ± sd (s2)	Mean ± sd (Δ)	p (Wilcoxon)
		vowel /a/		
PPQ	1.31 ± 1.28	1.89 ± 2.52	0.58 ± 1.24	0.069
APQ	10.69 ± 4.23	12.81 ± 6.66	2.12 ± 2.42	0.063
FLUF	3.87 ± 5.28	4.77 ± 5.61	0.90 ± 0.32	0.386
HNR (Q2)	13.35 ± 2.90	12.81 ± 3.73	−0.54 ± 0.83	0.390
HNR (SD)	4.17 ± 0.86	4.18 ± 0.89	0.02 ± 0.03	0.928
F1 (CV)	0.15 ± 0.06	0.18 ± 0.10	0.02 ± 0.04	0.180
F2 (CV)	0.22 ± 0.15	0.23 ± 0.15	0.00 ± 0.00	0.913
F3 (CV)	8.25 ± 40.44	4.50 ± 21.67	−3.75 ± −18.77	0.575
ER (Q2)	6.98 ± 33.97	7.40 ± 36.43	0.42 ± 2.46	0.954
ER (CV)	0.57 ± 0.26	0.57 ± 0.30	0.01 ± 0.05	0.880
F0 (CV)	0.30 ± 0.74	0.36 ± 1.15	0.06 ± 0.41	0.763
GNE (Q2)	−0.46 ± 1.85	−0.46 ± 1.85	0.00 ± −0.00	0.997
GNE (SD)	0.23 ± 0.74	0.30 ± 0.97	0.07 ± 0.23	0.679
TEO (CV)	−0.34 ± 1.71	−0.31 ± 1.75	0.03 ± 0.04	0.931
NNE (Q2)	−1.45 ± 7.41	−1.33 ± 7.08	0.13 ± −0.34	0.932
NNE (SD)	1.80 ± 0.76	1.79 ± 0.78	−0.01 ± 0.02	0.955
		vowel /e/		
PPQ	1.31 ± 1.17	1.80 ± 3.06	0.50 ± 1.89	0.269
APQ	11.21 ± 6.82	15.05 ± 9.95	3.84 ± 3.13	0.036
FLUF	2.62 ± 3.58	5.31 ± 6.02	2.69 ± 2.44	0.007
HNR (Q2)	14.36 ± 3.92	13.75 ± 4.40	−0.61 ± 0.48	0.394
HNR (SD)	4.19 ± 0.96	4.32 ± 1.18	0.13 ± 0.21	0.533
F1 (CV)	0.62 ± 0.23	0.56 ± 0.24	−0.05 ± 0.01	0.173
F2 (CV)	0.19 ± 0.07	0.19 ± 0.09	−0.00 ± 0.02	0.965
F3 (CV)	12.41 ± 86.28	7.06 ± 48.94	−5.36 ± −37.34	0.709
ER (Q2)	3.46 ± 23.30	3.39 ± 22.83	−0.07 ± −0.47	0.989
ER (CV)	0.59 ± 0.28	0.61 ± 0.37	0.02 ± 0.10	0.702
F0 (CV)	0.42 ± 1.47	0.19 ± 0.50	−0.24 ± −0.97	0.293
GNE (Q2)	−0.36 ± 1.77	−0.17 ± 1.18	0.19 ± −0.59	0.536
GNE (SD)	0.08 ± 0.36	0.06 ± 0.30	−0.02 ± −0.07	0.726
TEO (CV)	−0.07 ± 0.50	−0.19 ± 1.31	−0.11 ± 0.80	0.571
NNE (Q2)	−0.19 ± 1.35	−0.21 ± 1.44	−0.01 ± 0.10	0.961
NNE (SD)	1.67 ± 0.56	1.83 ± 0.61	0.17 ± 0.05	0.055
		vowel /i/		
PPQ	1.26 ± 1.68	1.92 ± 3.08	0.66 ± 1.40	0.196
APQ	10.49 ± 5.31	14.83 ± 9.65	4.35 ± 4.34	0.005
FLUF	2.13 ± 3.73	4.60 ± 6.89	2.46 ± 3.15	0.013
HNR (Q2)	17.16 ± 3.38	16.17 ± 5.32	−0.98 ± 1.95	0.212
HNR (SD)	4.51 ± 1.12	4.53 ± 1.40	0.02 ± 0.28	0.921
F1 (CV)	0.67 ± 0.48	0.58 ± 0.40	−0.09 ± −0.08	0.170
F2 (CV)	0.14 ± 0.07	0.16 ± 0.08	0.02 ± 0.00	0.088
F3 (CV)	0.09 ± 0.21	38.22 ± 207.82	38.14 ± 207.61	0.205
ER (Q2)	0.09 ± 0.08	8.35 ± 33.15	8.27 ± 33.07	0.087
ER (CV)	0.61 ± 0.37	0.56 ± 0.44	−0.05 ± 0.07	0.490
F0 (CV)	0.36 ± 1.39	0.35 ± 1.17	−0.01 ± −0.22	0.961
GNE (Q2)	−0.30 ± 1.49	−0.30 ± 1.48	0.01 ± −0.01	0.980
GNE (SD)	0.07 ± 0.32	0.12 ± 0.40	0.06 ± 0.08	0.449
TEO (CV)	0.78 ± 0.55	0.36 ± 2.47	−0.42 ± 2.47	0.238
NNE (Q2)	−0.01 ± 0.18	−0.20 ± 11.53	−0.19 ± 1.15	0.228
NNE (SD)	1.44 ± 0.55	1.40 ± 0.70	−0.05 ± 0.15	0.727

Table 2. *Cont.*

Feature	Mean ± sd (s1)	Mean ± sd (s2)	Mean ± sd (Δ)	*p* (Wilcoxon)
		vowel /o/		
PPQ	1.14 ± 1.06	1.68 ± 2.06	0.54 ± 0.99	0.047
APQ	11.09 ± 4.41	14.16 ± 9.94	3.06 ± 5.53	0.051
FLUF	2.64 ± 4.32	5.63 ± 7.24	2.99 ± 2.91	0.008
HNR (Q2)	15.49 ± 3.30	15.28 ± 4.79	−0.22 ± 1.49	0.773
HNR (SD)	4.93 ± 1.14	4.57 ± 1.36	−0.36 ± 0.21	0.149
F1 (CV)	0.24 ± 0.19	0.31 ± 0.22	0.06 ± 0.03	0.091
F2 (CV)	0.14 ± 0.10	0.14 ± 0.10	0.00 ± 0.01	0.867
F3 (CV)	8.75 ± 35.36	23.86 ± 96.34	15.11 ± 60.97	0.317
ER (Q2)	8.17 ± 32.52	8.07 ± 32.42	−0.10 ± −0.10	0.989
ER (CV)	0.64 ± 0.39	0.61 ± 0.36	−0.03 ± −0.02	0.705
F0 (CV)	0.40 ± 0.91	0.63 ± 1.81	0.24 ± 0.89	0.434
GNE (Q2)	−0.91 ± 2.72	−1.25 ± 3.11	−0.35 ± 0.39	0.582
GNE (SD)	0.30 ± 0.75	0.39 ± 0.90	0.09 ± 0.15	0.580
TEO (CV)	−0.74 ± 3.27	−0.28 ± 1.30	0.46 ± −1.97	0.376
NNE (Q2)	−0.19 ± 0.79	−0.11 ± 0.47	0.09 ± −0.31	0.530
NNE (SD)	1.62 ± 0.83	1.58 ± 0.91	−0.04 ± 0.08	0.843
		vowel /u/		
PPQ	1.35 ± 1.17	2.60 ± 3.07	1.26 ± 1.90	0.009
APQ	12.66 ± 5.60	17.03 ± 9.50	4.37 ± 3.91	0.007
FLUF	2.77 ± 5.13	8.52 ± 9.83	5.75 ± 4.69	0.001
HNR (Q2)	15.28 ± 4.23	14.32 ± 5.22	−0.96 ± 0.99	0.270
HNR (SD)	5.40 ± 1.56	5.08 ± 1.46	−0.32 ± −0.10	0.252
F1 (CV)	0.69 ± 0.44	0.71 ± 0.34	0.03 ± −0.10	0.667
F2 (CV)	0.17 ± 0.09	0.18 ± 0.09	0.01 ± −0.00	0.468
F3 (CV)	10.28 ± 52.21	15.04 ± 104.80	4.76 ± 52.58	0.779
ER (Q2)	8.27 ± 33.82	3.63 ± 25.29	−4.64 ± −8.53	0.453
ER (CV)	0.67 ± 0.39	0.74 ± 0.43	0.06 ± 0.04	0.431
F0 (CV)	0.23 ± 0.67	0.18 ± 0.10	−0.05 ± −0.57	0.586
GNE (Q2)	−0.19 ± 1.31	0.00 ± 0.00	0.19 ± −1.31	0.323
GNE (SD)	0.09 ± 0.40	0.10 ± 0.70	0.01 ± 0.30	0.966
TEO (CV)	−0.51 ± 2.69	−0.07 ± 0.51	0.43 ± −2.18	0.275
NNE (Q2)	−1.27 ± 7.09	−0.31 ± 2.190	0.96 ± −4.91	0.373
NNE (SD)	1.57 ± 0.73	1.68 ± 0.49	0.11 ± −0.24	0.379

The results of Spearman's partial correlation between the baseline acoustic features (session 1) and change in clinical data (Δ) can be seen in Table 3. None of the features significantly correlated with UPDRS, part III. On the other hand, in the case of part IV, we can observe negative correlation with aperiodicity (FLUF, vowels /e/, /i/, /o/, /u/), i.e., low aperiodicity at the baseline resulted in increased complications with therapy. Similarly, we identified negative correlation with tremor of jaw (F2 (CV), vowel /a/), but positive correlation with the tremor of lips (F3 (CV), vowel /o/). Another positive correlations were observed with median of energy ratio (vowels /o/, /u/), irregular pitch fluctuations (F0 (CV), vowel /a/), and variability of voice quality (GNE (SD), vowel /a/). Change in UPDRS IV negatively correlated with irregular amplitude fluctuations (TEO (CV), vowel /u/), acoustic noise (NNE (Q2), vowel /u/) and its variation (NNE (SD), vowel /a/). Results linked with the acoustic noise quantified by the median GNE are not consistent.

RBDSQ significantly and positively correlated with microperturbations in frequency of voice (PPQ, vowel /u/) and microperturbations of its intensity (APQ, vowel /a/), i.e., increased microperturbations in frequency/amplitude at the baseline resulted in deterioration of sleep. In addition, RBDSQ negatively correlated with the variation of voice quality (HNR (SD), vowel /o/).

Regarding gait difficulties, as assessed by FOG-Q, we can observe two positive correlations with tremor of jaw (F1 (CV), vowel /i/) and irregular pitch fluctuations (F0 (CV), vowel /a/). The total score of this questionnaire negatively correlates with variation of acoustic noise (NNE (SD), vowel /o/).

Table 3. Spearman's correlation coefficients between baseline acoustic features and Δ of clinical data.

Feature	/a/	/e/	/i/	/o/	/u/	/a/	/e/	/i/	/o/	/u/
	UPDRS III					UPDRS IV				
PPQ	−0.07	−0.08	0.26	−0.09	−0.10	−0.11	−0.26	−0.08	−0.17	−0.08
APQ	−0.06	−0.05	0.17	−0.10	−0.09	0.08	0.12	−0.01	0.06	−0.09
FLUF	0.10	0.08	0.11	0.10	0.16	−0.23	−0.47 **	−0.33 *	−0.34 *	−0.32 *
HNR (Q2)	0.16	0.07	−0.05	0.17	0.09	−0.02	−0.08	0.07	0.11	0.16
HNR (SD)	0.10	0.11	−0.04	−0.05	−0.15	0.07	−0.02	0.20	−0.22	−0.09
F1 (CV)	0.04	0.10	−0.17	−0.02	0.19	−0.27	−0.07	0.08	−0.32 *	0.08
F2 (CV)	−0.15	0.20	−0.10	−0.11	0.11	−0.37 *	−0.11	−0.12	−0.23	0.04
F3 (CV)	−0.24	0.25	−0.23	−0.17	0.11	0.22	0.28	0.10	0.32 *	0.10
ER (Q2)	−0.25	0.25	−0.14	−0.17	0.17	0.16	0.28	0.12	0.34 *	0.32 *
ER (CV)	0.12	−0.09	0.22	−0.03	0.11	−0.17	−0.14	−0.22	−0.18	−0.06
F0 (CV)	0.28	−0.12	0.04	−0.08	0.17	0.33 *	−0.04	−0.24	−0.16	−0.21
GNE (Q2)	−0.14	0.11	−0.15	0.06	−0.17	−0.35 *	0.25	0.15	0.27	0.33 *
GNE (SD)	−0.16	−0.08	−0.12	−0.22	0.24	0.30 *	−0.01	−0.04	0.22	0.11
TEO (CV)	0.28	−0.25	0.11	0.15	−0.25	−0.23	−0.28	−0.28	−0.23	−0.30 *
NNE (Q2)	0.28	−0.25	−0.09	0.14	−0.23	−0.15	−0.28	−0.22	−0.26	−0.31 *
NNE (SD)	0.09	−0.25	0.12	0.11	−0.26	−0.43 **	−0.17	−0.28	−0.12	−0.12
	RBDSQ					FOG-Q				
PPQ	0.15	0.11	0.21	−0.13	0.33 *	−0.20	−0.09	0.15	−0.09	0.08
APQ	0.29 *	0.27	0.21	0.27	0.20	−0.18	−0.14	−0.16	−0.16	−0.12
FLUF	−0.06	0.06	−0.22	−0.17	−0.24	0.20	0.15	0.13	0.17	0.23
HNR (Q2)	−0.20	−0.17	−0.19	−0.27	−0.20	0.09	0.03	−0.03	0.09	−0.05
HNR (SD)	−0.17	−0.16	0.21	−0.36 *	−0.27	0.07	0.24	0.21	0.18	0.12
F1 (CV)	0.08	−0.02	0.24	0.09	0.11	0.16	−0.22	0.33 *	−0.20	−0.20
F2 (CV)	−0.08	−0.09	−0.26	−0.17	−0.27	0.18	−0.09	−0.03	−0.13	0.26
F3 (CV)	−0.10	−0.04	−0.11	0.07	−0.08	0.03	0.22	0.14	0.08	−0.11
ER (Q2)	−0.19	−0.21	−0.26	−0.12	−0.16	−0.13	0.04	0.25	0.10	−0.05
ER (CV)	0.08	0.28	0.20	−0.28	−0.11	0.04	−0.13	−0.11	0.10	0.06
F0 (CV)	0.21	0.13	0.19	0.25	0.10	0.37 *	0.20	−0.03	−0.10	0.19
GNE (Q2)	−0.16	−0.17	−0.28	−0.18	−0.16	−0.29	−0.25	0.05	−0.19	−0.19
GNE (SD)	−0.05	−0.09	−0.25	0.09	−0.20	0.08	0.14	−0.05	0.12	0.15
TEO (CV)	0.05	0.04	0.27	−0.17	0.15	−0.05	−0.22	−0.07	−0.25	−0.01
NNE (Q2)	0.24	0.28	0.18	0.20	0.21	−0.04	0.21	0.17	0.16	0.20
NNE (SD)	−0.27	−0.05	−0.19	−0.16	0.11	−0.23	−0.12	0.16	−0.22 **	0.05

*—*p*-value of Spearman's correlation coefficient <0.05; **—*p*-value of Spearman's correlation coefficient <0.01.

The results of Spearman's partial correlation between the change of baseline acoustic features (Δ) and the change in clinical data (Δ) can be seen in Table 4. Regarding the change of UPDRS III, it negatively correlated with the change of microperturbations in frequency of voice (PPQ, vowel /i/), aperiodicity (FLUF, vowels /e/, /o/), tremor of tongue (F1 (CV), vowels /a/, /u/), tremor of jaw (F2 (CV), vowel /e/), irregular pitch fluctuations (F0 (CV), vowels /a/, /u/), and variation of acoustic noise (NNE (SD), vowel /i/). Significant positive correlations were identified with the change of lips tremor (F3 (CV), vowel /a/), acoustic noise (ER (Q2), vowel /a/), and variation of voice quality (GNE (SD), vowel /e/).

In the case of UPDRS IV, we identified seven significant positive correlations with the change of microperturbations in frequency of voice (PPQ, vowel /e/), tremor of jaw (F2 (CV), vowel /a/), irregular amplitude fluctuations (TEO (CV), vowels /a/, /u/), and acoustic noise (NNE (Q2), vowels /o/, /u/). The change in UPDRS IV significantly negatively correlated with the change of acoustic noise (ER (Q2), vowel /u/), and its variation (ER (CV), vowel /e/).

Changes in RBDSQ significantly negatively correlated with the change of microperturbations in frequency of voice (PPQ, vowel /u/), microperturbations of its intensity (APQ, vowels /e/, /i/, /u/), tremor of lips (F3 (CV), vowel /o/), acoustic noise (NNE (Q2), vowel /e/), and its variation (ER (CV), vowel /e/). Positive correlations were identified with the change in voice quality (HNR (Q2), all vowels) and its variability (HNR (SD), vowels /e/, /o/, /u/). The similar results can be observed when assessing the quality by GNE (vowel /e/).

Finally, in terms of changes in FOG-Q, we identified significant negative correlations with the change in aperiodicity (FLUF, vowels /a/, /e/, /o/, /u/), tremor of jaw (F1 (CV), vowel /i/), tremor of tongue (F2 (CV), vowel /u/), and variation of acoustic noise (ER (CV), vowels /e/, /i/). One significant positive correlation can be observed with the change in acoustic noise variation (NNE (SD), vowel /o/). The results based on irregular amplitude fluctuations (TEO (CV)) are not consistent.

Table 4. Spearman's correlation coefficients between Δ of acoustic features and Δ of clinical data.

Feature	/a/	/e/	/i/	/o/	/u/	/a/	/e/	/i/	/o/	/u/
			UPDRS III					UPDRS IV		
PPQ	−0.12	−0.20	−0.31 *	0.10	−0.13	0.09	0.40 **	0.25	0.21	0.10
APQ	−0.17	−0.14	−0.26	−0.06	−0.15	0.06	0.08	0.10	0.06	0.03
FLUF	−0.27	−0.30 *	−0.15	−0.40 **	−0.25	0.17	0.21	0.28	0.04	0.21
HNR (Q2)	−0.04	0.08	0.22	−0.07	0.09	−0.02	−0.03	−0.14	−0.14	−0.17
HNR (SD)	0.12	−0.18	−0.25	−0.05	0.19	−0.10	0.08	−0.21	−0.08	0.13
F1 (CV)	−0.35 *	−0.26	−0.04	−0.28	−0.38 **	0.05	−0.04	−0.07	−0.02	−0.11
F2 (CV)	0.13	−0.34 *	−0.22	−0.07	−0.23	0.39 **	0.07	−0.07	0.28	0.21
F3 (CV)	0.29 *	−0.19	0.22	0.24	−0.09	−0.16	−0.12	−0.21	−0.21	−0.10
ER (Q2)	0.31 *	−0.15	0.04	0.26	−0.13	−0.11	−0.07	−0.16	−0.27	−0.29 *
ER (CV)	−0.23	−0.05	−0.16	−0.17	0.05	−0.08	−0.33 *	0.12	−0.04	0.14
F0 (CV)	−0.32 *	0.18	0.21	−0.19	−0.30 *	−0.15	0.12	0.15	−0.13	0.15
GNE (Q2)	0.27	−0.22	−0.22	0.20	0.17	0.17	−0.18	−0.15	0.11	−0.27
GNE (SD)	0.16	0.32 *	0.23	0.14	−0.25	−0.18	0.10	−0.07	−0.25	−0.05
TEO (CV)	−0.23	−0.18	−0.11	−0.13	0.23	0.29 *	−0.13	0.28	0.28	0.30 *
NNE (Q2)	−0.25	0.14	−0.12	−0.19	0.20	0.10	0.16	0.24	0.30 *	0.31 *
NNE (SD)	−0.12	0.05	−0.42 **	−0.06	0.11	0.29 *	0.15	0.16	0.24	0.16
			RBDSQ					FOG-Q		
PPQ	−0.23	−0.19	−0.16	−0.17	−0.37 *	0.15	−0.17	−0.18	0.12	−0.10
APQ	−0.28	−0.37 *	−0.38 **	−0.29	−0.32 *	0.08	0.08	0.06	0.14	0.09
FLUF	0.05	−0.06	0.06	0.10	−0.18	−0.41 **	−0.29 *	−0.10	−0.35 *	−0.40 **
HNR (Q2)	0.29 *	0.35 *	0.31 *	0.36 *	0.41 **	0.15	0.11	0.12	0.02	0.04
HNR (SD)	0.23	0.36 *	0.10	0.40 **	0.30 *	−0.06	0.07	−0.18	−0.20	0.01
F1 (CV)	−0.23	0.04	−0.12	0.06	−0.16	−0.29	−0.04	−0.42 **	0.12	0.13
F2 (CV)	−0.06	0.08	0.16	0.17	0.06	−0.27	−0.08	−0.20	−0.09	−0.37 *
F3 (CV)	0.07	0.15	−0.25	−0.40 **	−0.10	−0.07	0.18	0.09	−0.08	0.20
ER (Q2)	0.17	0.23	−0.27	−0.27	0.09	0.12	0.26	−0.06	−0.21	0.17
ER (CV)	−0.10	−0.46 **	0.06	0.28	−0.11	−0.00	−0.30 *	−0.34 *	0.08	−0.19
F0 (CV)	−0.17	−0.14	−0.17	−0.12	−0.11	−0.23	−0.17	0.20	−0.05	−0.23
GNE (Q2)	0.13	0.30 *	0.25	0.13	0.15	0.20	0.17	−0.21	0.07	0.27
GNE (SD)	0.04	0.40 **	−0.23	−0.23	0.25	−0.07	0.14	0.24	−0.03	0.14
TEO (CV)	−0.06	−0.25	0.24	0.25	−0.18	0.06	−0.37 *	−0.17	0.34 *	−0.20
NNE (Q2)	−0.15	−0.42 **	0.24	0.27	−0.23	−0.20	−0.28	−0.05	0.25	−0.13
NNE (SD)	0.25	0.02	0.20	0.24	−0.15	0.18	0.10	−0.25	0.06 **	−0.15

*—*p*-value of Spearman's correlation coefficient <0.05; **—*p*-value of Spearman's correlation coefficient <0.01.

The results of the clinical scales' estimation are reported in Table 5. Using the acoustic analysis of sustained phonation of the baseline vowel /e/ in combination with mathematically modeling based on the XGBoost algorithm, we estimated the change in UPDRS III score with 25.7% error (MAE = 7.3, range(UPDRS III Δ) = 29). The change in UPDRS IV was estimated with the lowest error equal to 11.3% (MAE = 1.7, range(UPDRS IV Δ) = 15) when employing acoustic analysis of the baseline vowel /o/. The change in RBDSQ was estimated with 16.3% error (MAE = 2.0, range(RBDSQ Δ) = 13) based on phonatory analysis of vowel /i/. Finally, the lowest error of FOG-Q change estimation is 13.2% (MAE = 2.8, range(FOG-Q Δ) = 22). In this case, the acoustic analysis of vowel /u/ outperformed the other ones.

Table 5. Results of the clinical scales' estimation.

VT	MAE	EER [%]	MAE	EER [%]	MAE	EER [%]	MAE	EER [%]
	UPDRS III		UPDRS IV		RBDSQ		FOG-Q	
/a/	8.2 ± 2.6	29.1 ± 9.2	1.9 ± 0.6	12.9 ± 4.1	2.2 ± 1.0	17.6 ± 8.1	3.1 ± 0.8	14.7 ± 3.8
/e/	7.3 ± 2.0	25.7 ± 7.0	1.8 ± 0.7	12.2 ± 4.8	2.0 ± 0.8	16.4 ± 6.9	3.4 ± 1.0	16.1 ± 5.0
/i/	7.4 ± 2.7	26.3 ± 9.4	1.9 ± 0.8	12.9 ± 5.5	2.0 ± 0.7	16.3 ± 6.3	2.9 ± 0.7	13.6 ± 3.6
/o/	7.9 ± 2.1	28.2 ± 7.7	1.7 ± 0.7	11.3 ± 4.8	2.1 ± 0.8	16.8 ± 6.3	3.3 ± 0.5	15.4 ± 2.6
/u/	7.7 ± 2.5	27.2 ± 8.8	2.0 ± 0.8	13.8 ± 5.5	2.1 ± 0.9	17.3 ± 7.2	2.8 ± 0.9	13.2 ± 4.5

VT—vocal task; MAE—mean absolute error; EER—estimation error rate.

Due to inter-rater variability as well as intra-rater variability [52–54], consistent scoring of PD using the commonly used clinical rating scales is not an easy task. Automatic scoring, i.e., the estimation of the values of the clinical rating scales must be viewed as a tool that can provide clinicians with an additional, unbiased, and objective information that can help them with their decision-making, not as a tool that will substitute the work of clinicians. With this in mind, the predictions made by the trained XGBoost models can be considered rather reasonable as the error of 10–20% is comparable with a deviation caused by inter/intra-rater variability. Moreover, each clinical rating scale is different. On one hand, there are complex scales such as UPDRS III describing various motor aspects of PD, and, on the other hand, there are scales specifically focusing on a subset of its symptoms, e.g., FOG-Q (gait difficulties), RBDSQ (sleep disorders), etc. This information must be taken into account when evaluating the prediction errors because, the more complex the scale is, the more difficult it becomes to predict its values. This can be seen in our results as well. The most complex of the scales was predicted with the largest prediction error.

Feature importances of the SGBoost models are visualized in Figure 2. The figure shows the feature importances for all of the trained models. Feature importances quantify a relative importance of the features in the ensemble of the trained XGBoost model [47]. Therefore, the higher the value of the feature importance, the more important the feature is for the prediction of the dependent variable. With this in mind, the rationale behind this visualization is to show which features are important, and how strong that importance is, for the trained models in direction of predicting the change in the particular clinical rating scales in the horizon of two years given the acoustic features at the baseline.

Based on these graphs, we can conclude that the estimation of UPDRS III change requires a complex parametrization because, in all scenarios, at least 13 acoustic features were employed. In this case, especially median NNE was not frequently used. Although the models estimate the change of UPDRS IV with the lowest error, they usually use just a few phonatory parameters. In fact, in the case of vowel /o/, we observed 11.3% estimation error based on the following three phonatory features: GNE (Q2), ER (Q2), and FLUF. Generally, these features quantify quality of voicing. The best estimation of the RBDSQ change is based on eight phonatory parameters extracted from vowel /i/. The most important features quantify tremor of jaw (F1 (CV)), aperiodicity (FLUF), and microperturbations in intensity (APQ). Finally, based on the feature importances, we can observe that the most important role in FOG-Q change estimation was played by formant frequencies quantifying tremor of the articulatory organs.

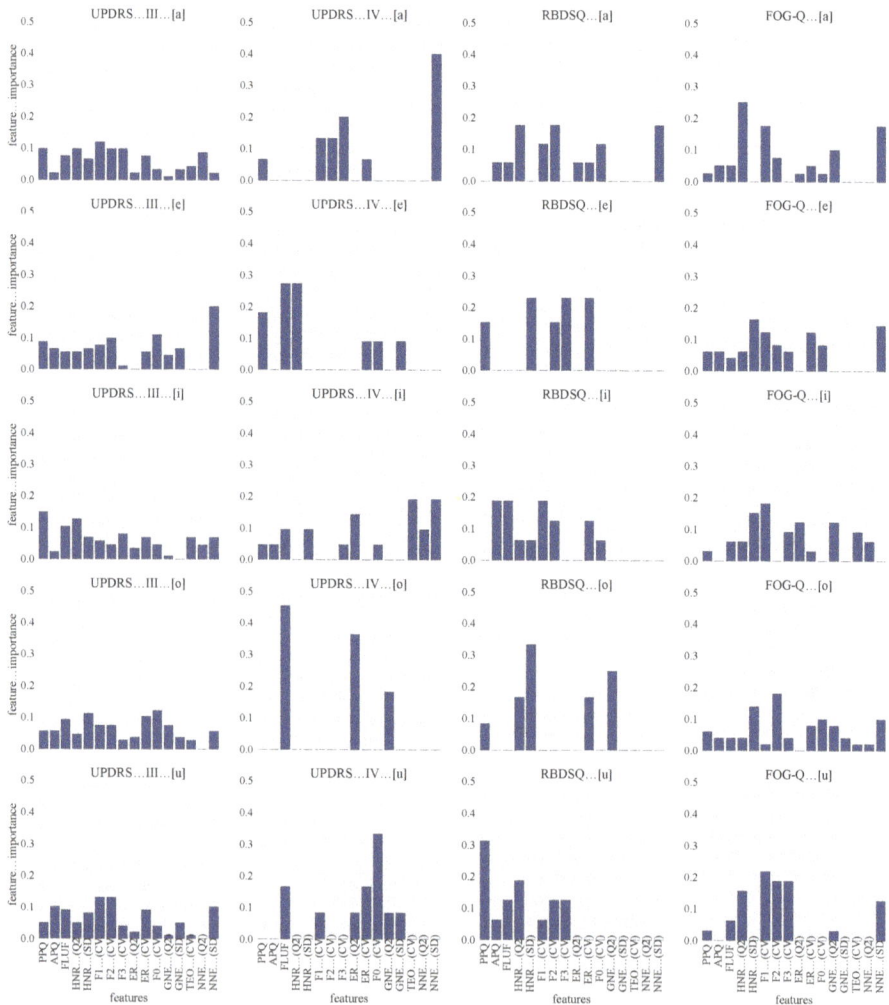

Figure 2. Feature importance graphs for trained XGBoost models. Each column shows graphs for the models trained to estimate the change (Δ) in values of a particular clinical rating scale (UPDRS III, UPDRS IV, RBDSQ, FOG-G). Each row shows graphs for the models trained using the features extracted from the recordings of a particular vowel phonation (/a/, /e/, /i/, /o/, /u/). The scale of the graphs is unified so that it is easier to compare the values among the models.

4. Discussion

Although the only existing longitudinal study [31] is different in the interval between sessions (32.5 vs. 24.0 months), we are going to compare our findings with the results reported by these authors. In contrary to Skodda et al., who observed significant change in shimmer of the sustained phonation of vowel /a/, we have not identified any significant differences in this vowel. Nevertheless, we identified significant changes in the same feature extracted from vowels /e/, /i/, /u/. In addition, we monitored some significant changes in jitter and FLUF. Based on these results, we can conclude that, for two years, patients' voices became more aperiodic with increased microperturbations of frequency and amplitude.

None of the acoustic features at baseline significantly correlated with a change in UPDRS III, which supports the results of the clinical scales' estimation where the lowest estimation error was above 25%. However, we identified some significant correlations between changes of phonatory features and the clinical scale. Surprisingly, except tremor of lips (F3 (CV)), acoustic noise (ER (Q2)), and variation of voice quality (GNE (SD)), worsening in UPDRS III (motor performance) was associated with improvement in phonatory characteristics. This could be explained by the fact that HD belongs to axial symptoms [9,31] that do not play significant part in UPDRS III. In other words, although several significant correlations were identified, we hypothesize that some underlying pathophysiological mechanism are involved and a direct interpretation is not possible.

Regarding the change in complications of therapy (as assessed by UPDRS IV), although the most significant correlations were observed with baseline features extracted from the vowel /a/, the lowest estimation error (11%) was based on vowel /o/. In this case, low aperiodicity, but increased lips tremor and increased acoustic noise at baseline, was associated with increased complications in the follow-up examination.

Only three significant correlations are reported between baseline acoustic parameters (quantifying microperturbations of frequency/amplitude and variation in voice quality) and change in RBDSQ. Although we have not identified any significant correlations based on vowel /i/, the XGBoost algorithm reached the lowest error (16%) including features calculated from this vowel. This result could originate from the ability of XGBoost to model complex interdependencies that are not evident at first sight [47]. Regarding the partial correlations between changes in RBDSQ and phonatory features, we can conclude that mainly changes in voice aperiodicity and voice quality are linked with changes in sleep disorders.

HD and freezing of gait (FOG) are both axial symptoms of PD [55]. In our recent work, we have found out that these symptoms share some pathophysiological mechanism [56]. More specifically, we proved that FOG is mainly linked with improper articulation, disturbed speech rate and with intelligibility. We did not identify any significant relations between FOG and phonatory features. On the other hand, we analyzed only the sustained vowel /a/ and partial correlations were calculated only with some baseline FOG-Q sub-scores. The current study provides deeper and more complex results in terms of FOG and phonatory features relations. The first correlation analysis (baseline features vs. ΔFOG-Q) identified just a few significant correlations. However, based on mainly formant frequencies extracted from vowel /u/, the XGBoost model estimated the change in FOG-Q with 13% error. Generally, the significant impact of formants in this specific task is in line with our previous study [56]. The second correlation analysis (Δ of the baseline features vs. ΔFOG-Q) revealed some relations between changes in FOG and changes in aperiodicity, tremor of jaw/tongue, and acoustic noise.

Although most of the studies dealing with the acoustic analysis of phonation in PD patients focus on sustained vowel /a/, it is not sufficiently explained why this corner vowel is more important than the other two, i.e., /i/ or /u/. Looking at the Hellwag (vowel) triangle [34], we can see that, during phonation of vowel /a/, the tongue is in its lowest position from a vertical point of view, and in its central position from a horizontal one. In other words, a speaker does not have to make an effort to keep the tongue in a limit position (the tongue is almost relaxed). Therefore, some phonatory disorders could not be accented. This limitation is not present in vowels /i/ or /o/, where the speaker has to exert a force in both directions. On the other hand, the lowest limit position of jaw is reached during the phonation of vowel /a/. In summary, although some research teams employed a more complex set of vowels in their experiments [19–21,23,24,35–37], the vowel /a/ is still the most frequently used one. However, this choice should be supported by a complex, robust, and multilingual study (theoretically, the effect of culture and language plays no role here, but this should be proven as well). Based on these assumptions, we have decided to explore significance of all five Czech vowels. In addition, the results suggest that the progress of PD is reflected in each vowel differently. Moreover, each vowel differently correlates with changes in scores of clinical scales. Finally, in our case, the best prediction of the change in the clinical rating scales under the focus have never been based on phonatory parameters of the vowel /a/. If we have to choose one optimal candidate for considered clinical scores changes prediction (see Table 5), it would be the corner vowel /i/, where the tongue is in limit position in both directions.

In our previous works, we proved that HD shares some pathophysiological mechanisms with other motor/non-motor features of PD. For instance, based on a combination of acoustic analysis and machine learning approaches, it is possible to predict cognitive deficits or gait disorders [44,56]. Although in the frame of this research we explored only the field of phonation, our results confirm the ability of acoustic HD analysis to predict the progress of PD. These findings and conclusions could have practical applications in eHealth, mHealth and generally Health 4.0 systems that could be used to remotely monitor and assess motor/non-motor deficits in PD patients.

5. Conclusions

This study deals with a quantitative analysis of changes in sustained phonation that has been acquired twice (with a two-year interval) in 51 PD patients. These changes are linked with progress of PD as assessed by three commonly used clinical scales. Finally, it explores a possibility of PD progress prediction based on a combination of acoustic analysis and machine learning modeling.

Based on the reported results, we conclude that, for two years, patients' voices became more aperiodic with increased microperturbations of frequency and amplitude. Although we did not identify many significant correlations between baseline values of phonatory features and changes in clinical scores, the XGBoost algorithm was able to predict these changes with errors ranging from 11% (in the case of UPDRS IV) to 26% (in the case of UPDRS III). These results accent the impact of acoustic HD analysis in Health 4.0 systems. Next, we identified significant correlations between changes in phonatory features and changes in clinical scores; however, probably due to some underlying pathophysiological mechanisms and complex interdependencies, these relations are less interpretable. Finally, our results suggest that the researchers should consider acoustic analysis of corner vowel /i/ instead of the corner vowel /a/.

Admittedly, the main limitation of this study is the small size of patient cohort. On the other hand, longitudinal studies of PD patients are very time-consuming (the patients are usually examined by several experts such as neurologists, clinical psychologists, and clinical speech therapists), physically demanding (PD is a movement disorder, therefore it requires patients to make significant effort to get into a hospital), and it is difficult to assess a large number of patients due to a low prevalence which is estimated to 1.5% for people aged over 65 years [57]. In fact, as far as we know, this is the first complex study analyzing changes in phonation and their relations with progress of PD based on such a big dataset. Moreover, it is the first study employing acoustic analysis of phonation in combination with machine learning modeling in order to predict the progress of PD. Nevertheless, our findings should be confirmed by further scientific research that will include bigger cohorts.

Author Contributions: Conceptualization, Z.G., J.M. and I.R.; Methodology, Z.G. and J.M.; Software, Z.G. and J.M.; Validation, Z.G.; Formal Analysis, Z.G.; Investigation, Z.G., J.M., i.e., M.M., M.K. and I.R.; Resources, Z.S., i.e., M.M., M.K. and I.R.; Data Curation, J.M., T.K., i.e., M.M., M.K., I.R.; Writing—Original Draft Preparation, Z.G., J.M., V.Z., J.M., T.K.; Visualization, Z.G.; Supervision, J.M., Z.S., I.R., M.F.-Z., JB.A.-H. and P.G.-V.; Project Administration, J.M. and I.R.; Funding Acquisition, J.M., I.R., M.F.-Z., JB.A.-H. and P.G.-V.

Funding: This research was funded by the grant of the Czech Ministry of Health 16-30805A (Effects of non-invasive brain stimulation on hypokinetic dysarthria, micrographia, and brain plasticity in patients with Parkinson's disease) and the following projects: LO1401, FEDER and MEC, TEC2016-77791-C4-2-R, TEC2016-77791-C4-1-R, CENIE_TECA-PARK_55_02 INTERREG V-A Spain—Portugal (POCTEP), and TEC2016-77791-C4-4-R from the Ministry of Economic Affairs and Competitiveness of Spain. For the research, infrastructure of the SIX Center was used.

Conflicts of Interest: The authors declare no conflict of interest.

Abbreviations

The following abbreviations are used in this manuscript:

ACE-R	Addenbrooke's cognitive examination-revised
APQ	amplitude perturbation quotient
CV	coefficient of variation
EER	estimation error rate
ER	energy ratio
F0	fundamental frequency
F*i*	*i*th formant
FLUF	fraction of locally unvoiced frames
FOG	freezing of gait
FOG-Q	freezing of gait questionnaire
GNE	glottal-to-noise excitation ratio
HD	hypokinetic dysarthria
HNR	harmonic-to-noise ratio
LED	L-dopa equivalent daily dose
MAE	mean absolute error
NNE	normalized noise energy
PD	Parkinson's disease
PPQ	period perturbation quotient
Q2	second quartile (median)
RBDSQ	REM sleep behavior disorder screening questionnaire
SD	standard deviation
TKEO	Teager–Kaiser energy operator
UPDRS III	Unified Parkinson's disease rating scale, part III: evaluation of motor functions
UPDRS IV	Unified Parkinson's disease rating scale, part IV: evaluation of complications of therapy

References

1. Hornykiewicz, O. Biochemical aspects of Parkinson's disease. *Neurology* **1998**, *51*, S2–S9. [CrossRef] [PubMed]
2. Hoehn, M.M.; Yahr, M.D. Parkinsonism: Onset, Progression, and Mortality. *Neurology* **1967**, *17*, 427–442. [CrossRef] [PubMed]
3. Fahn, S.; Elton, R.L. *UPDRS Development Committee (1987) Unified Parkinson's Disease Rating Scale*; Recent Developments in Parkinson's Disease; Macmillan: Florham Park, NJ, USA, 1987.
4. Giladi, N.; Shabtai, H.; Simon, E.S.; Biran, S.; Tal, J.; Korczyn, S.D. Construction of freezing of gait questionnaire for patients with Parkinsonism. *Parkinsonism Relat. Disord.* **2000**, *6*, 165–170. [CrossRef]
5. Larner, A.J. Addenbrooke's cognitive examination-revised (ACE-R) in day-to-day clinical practice. *Age Ageing* **2007**, *36*, 685–686. [CrossRef] [PubMed]
6. Ramaker, C.; Marinus, J.; Stiggelbout, A.M.; Van Hilten, B.J. Systematic evaluation of rating scales for impairment and disability in Parkinson's disease. *Mov. Disord.* **2002**, *17*, 867–876, doi:10.1002/mds.10248. [CrossRef] [PubMed]
7. Ramig, L.O.; Fox, C.; Sapir, S. Speech treatment for Parkinson's disease. *Expert Rev. Neurother.* **2008**, *8*, 297–309. [CrossRef] [PubMed]
8. Darley, F.L.; Aronson, A.E.; Brown, J.R. Differential Diagnostic Patterns of Dysarthria. *J. Speech Lang. Hear. Res.* **1969**, *12*, 246–269. [CrossRef]
9. Brabenec, L.; Mekyska, J.; Galaz, Z.; Rektorova, I. Speech disorders in Parkinson's disease: Early diagnostics and effects of medication and brain stimulation. *J. Neural Transm.* **2017**, *124*, 303–334. [CrossRef] [PubMed]
10. Eliasova, I.; Mekyska, J.; Kostalova, M.; Marecek, R.; Smekal, Z.; Rektorova, I. Acoustic evaluation of short-term effects of repetitive transcranial magnetic stimulation on motor aspects of speech in Parkinson's disease. *J. Neural Transm.* **2013**, *120*, 597–605. [CrossRef] [PubMed]

11. Elfmarkova, N.; Gajdos, M.; Mrackova, M.; Mekyska, J.; Mikl, M.; Rektorova, I. Impact of Parkinson's disease and levodopa on resting state functional connectivity related to speech prosody control. *Parkinsonism Relat. Disord.* **2016**, *22* (Suppl. 1), S52–S55. [CrossRef] [PubMed]
12. Gómez-Vilda, P.; Mekyska, J.; Ferrández, J.M.; Palacios-Alonso, D.; Gómez-Rodellar, A.; Rodellar-Biarge, V.; Galaz, Z.; Smekal, D.; Rektorova, I.; Eliasova, I.; et al. Parkinson Disease Detection from Speech Articulation Neuromechanics. *Front. Neuroinform.* **2017**, *11*, 56. [CrossRef] [PubMed]
13. Lirani-Silva, C.; Mourão, L.F.; Gobbi, L.T.B. Dysarthria and Quality of Life in neurologically healthy elderly and patients with Parkinson's disease. *CoDAS* **2015**, *27*, 248–254. [CrossRef] [PubMed]
14. Ho, A.K.; Iansek, R.; Marigliani, C.; Bradshaw, J.L.; Gates, S. Speech Impairment in a Large Sample of Patients with Parkinson's disease. *J. Behav. Neurol.* **1999**, *11*, 131–137. [CrossRef]
15. Tsanas, A.; Little, M.; McSharry, P.; Ramig, L. Accurate telemonitoring of Parkinson's Disease progression by noninvasive speech tests. *IEEE Trans. Bio-Med. Eng.* **2010**, *57*, 884–893. [CrossRef] [PubMed]
16. Harel, B.T.; Cannizzaro, M.S.; Cohen, H.; Reilly, N.; Snyder, P.J. Acoustic characteristics of Parkinsonian speech: A potential biomarker of early disease progression and treatment. *J. Neurolinguist.* **2004**, *17*, 439–453. [CrossRef]
17. Rusz, J.; Cmejla, R.; Ruzickova, H.; Klempir, J.; Majerova, V.; Picmausova, J.; Roth, J.; Ruzicka, E. Evaluation of speech impairment in early stages of Parkinson's disease: A prospective study with the role of pharmacotherapy. *J. Neural Transm.* **2013**, *120*, 319–329. [CrossRef] [PubMed]
18. Skodda, S.; Grönheit, W.; Schlegel, U.; Südmeyer, M.; Schnitzler, A.; Wojtecki, L. Effect of subthalamic stimulation on voice and speech in Parkinson's disease: For the better or worse? *Front. Neurol.* **2014**, *4*, 218. [CrossRef] [PubMed]
19. Orozco-Arroyave, J.R.; Hönig, F.; Arias-Londoño, J.D.; Vargas-Bonilla, J.F.; Daqrouq, K.; Skodda, S.; Rusz, J.; Nöth, E. Automatic detection of Parkinson's disease in running speech spoken in three different languages. *J. Acoust. Soc. Am.* **2016**, *139*, 481–500. [CrossRef] [PubMed]
20. Mekyska, J.; Smekal, Z.; Galaz, Z.; Mzourek, Z.; Rektorova, I.; Faundez-Zanuy, M.; López-de Ipiña, K. Perceptual features as markers of Parkinson's Disease: the issue of clinical interpretability. In *Recent Advances in Nonlinear Speech Processing*; Springer: New York, NY, USA, 2016; pp. 83–91.
21. Mekyska, J.; Galaz, Z.; Mzourek, Z.; Smekal, Z.; Rektorova, I.; Eliasova, I.; Kostalova, M.; Mrackova, M.; Berankova, D.; Faundez-Zanuy, M.; et al. Assessing progress of Parkinson's disease using acoustic analysis of phonation. In Proceedings of the 2015 4th International Work Conference on Bioinspired Intelligence (IWOBI), San Sebastian, Spain, 10–12 June 2015; pp. 111–118.
22. Arora, S.; Venkataraman, V.; Zhan, A.; Donohue, S.; Biglan, K.; Dorsey, E.; Little, M. Detecting and monitoring the symptoms of Parkinson's disease using smartphones: A pilot study. *Parkinsonism Relat. Disord.* **2015**, *21*, 650–653. [CrossRef] [PubMed]
23. Villa-Cañas, T.; Orozco-Arroyave, J.; Vargas-Bonilla, J.; Arias-Londoño, J. Modulation spectra for automatic detection of Parkinson's disease. In Proceedings of the 2014 XIX Symposium on Image, Signal Processing and Artificial Vision (STSIVA), Armenia, Colombia, 17–19 September 2014; pp. 1–5.
24. Hazan, H.; Hilu, D.; Manevitz, L.; Ramig, L.O.; Sapir, S. Early diagnosis of Parkinson's disease via machine learning on speech data. In Proceedings of the 2012 IEEE 27th Convention of Electrical & Electronics Engineers in Israel (IEEEI), Eilat, Israel, 14–17 November 2012; pp. 1–4.
25. Eskidere, Ö.; Ertaş, F.; Hanilçi, C. A comparison of regression methods for remote tracking of Parkinson's disease progression. *Expert Syst. Appl.* **2012**, *39*, 5523–5528. [CrossRef]
26. Castelli, M.; Vanneschi, L.; Silva, S. Prediction of the Unified Parkinson's Disease Rating Scale assessment using a genetic programming system with geometric semantic genetic operators. *Expert Syst. Appl.* **2014**, *41*, 4608–4616. [CrossRef]
27. Naranjo, L.; Pérez, C.J.; Martín, J. Addressing voice recording replications for tracking Parkinson's disease progression. *Med. Biol. Eng. Comput.* **2017**, *55*, 365–373. [CrossRef] [PubMed]
28. Skodda, S.; Rinsche, H.; Schlegel, U. Progression of dysprosody in Parkinson's disease over time—A longitudinal study. *Mov. Disord.* **2009**, *24*, 716–722. [CrossRef] [PubMed]
29. Skodda, S.; Flasskamp, A.; Schlegel, U. Instability of syllable repetition as a marker of disease progression in Parkinson's disease: A longitudinal study. *Mov. Disord.* **2011**, *26*, 59–64. [CrossRef] [PubMed]
30. Skodda, S.; Grönheit, W.; Schlegel, U. Impairment of vowel articulation as a possible marker of disease progression in Parkinson's disease. *PLoS ONE* **2012**, *7*, e32132. [CrossRef] [PubMed]

31. Skodda, S.; Gronheit, W.; Mancinelli, N.; Schlegel, U. Progression of Voice and Speech Impairment in the Course of Parkinson's Disease: A Longitudinal Study. *Parkinson's Dis.* **2013**, *2013*, 389195. [CrossRef] [PubMed]
32. Lee, J.Y.; Kim, J.W.; Lee, W.Y.; Kim, J.M.; Ahn, T.B.; Kim, H.J.; Cho, J.; Jeon, B.S. Daily dose of dopaminergic medications in Parkinson's disease: clinical correlates and a posteriori equation. *Neurol. Asia* **2010**, *15*, 137–143.
33. Stiasny-Kolster, K.; Mayer, G.; Schafer, S.; Muller, J.C.; Heinzel-Gutenbrunner, M.; Oertel, W.H. The REM sleep behavior disorder screening questionnaire—A new diagnostic instrument. *Mov. Disord.* **2007**, *22*, 2386–2393. [CrossRef] [PubMed]
34. Mol, H. Lossfree Twin-Tube Resonator and the Vowel Triangle of Hellwag. *J. Acoust. Soc. Am.* **1965**, *37*, 1186. [CrossRef]
35. Orozco-Arroyave, J.R.; Hönig, F.; Arias-Londoño, J.D.; Vargas-Bonilla, J.; Skodda, S.; Rusz, J.; Nöth, E. Automatic detection of Parkinson's disease from words uttered in three different languages. In Proceedings of the Fifteenth Annual Conference of the International Speech Communication Association, Singapore, 14–18 September 2014; pp. 1573–1577.
36. Rusz, J.; Cmejla, R.; Tykalova, T.; Ruzickova, H.; Klempir, J.; Majerova, V.; Picmausova, J.; Roth, J.; Ruzicka, E. Imprecise vowel articulation as a potential early marker of Parkinson's disease: Effect of speaking task. *J. Acoust. Soc. Am.* **2013**, *134*, 2171–2181. [CrossRef] [PubMed]
37. Rusz, J.; Cmejla, R.; Ruzickova, H.; Ruzicka, E. Quantitative acoustic measurements for characterization of speech and voice disorders in early untreated Parkinson's disease. *J. Acoust. Soc. Am.* **2011**, *129*, 350–367. [CrossRef] [PubMed]
38. Titze, I.R. *Principles of Voice Production*; Prentice Hall: Englewood Cliffs, NJ, USA, 1994.
39. Harar, P.; Alonso-Hernandezy, J.B.; Mekyska, J.; Galaz, Z.; Burget, R.; Smekal, Z. Voice Pathology Detection Using Deep Learning: A Preliminary Study. In Proceedings of the 2017 International Conference and Workshop on Bioinspired Intelligence (IWOBI), Funchal, Portugal, 10–12 July 2017; pp. 45–48.
40. Harar, P.; Galaz, Z.; Alonso-Hernandez, J.B.; Mekyska, J.; Burget, R.; Smekal, Z. Towards robust voice pathology detection. *Neural Comput. Appl.* **2018**. [CrossRef]
41. Kostalova, M.; Mrackova, M.; Marecek, R.; Berankova, D.; Eliasova, I.; Janousova, E.; Roubickova, J.; Bednarik, J.; Rektorova, I. The 3F Test Dysarthric Profile—Normative Speech Values in Czech. *Ceska Slovenska Neurologie Neurochirurgie* **2013**, *76*, 614–618.
42. Boersma, P.; Weenink, D. Praat, a system for doing phonetics by computer. *Glot Int.* **2002**, *5*, 341–345.
43. Mekyska, J.; Janousova, E.; Gomez-Vilda, P.; Smekal, Z.; Rektorova, I.; Eliasova, I.; Kostalova, M.; Mrackova, M.; Alonso-Hernandez, J.B.; Faundez-Zanuy, M.; et al. Robust and complex approach of pathological speech signal analysis. *Neurocomputing* **2015**, *167*, 94–111. [CrossRef]
44. Rektorova, I.; Mekyska, J.; Janousova, E.; Kostalova, M.; Eliasova, I.; Mrackova, M.; Berankova, D.; Necasova, T.; Smekal, Z.; Marecek, R. Speech prosody impairment predicts cognitive decline in Parkinson's disease. *Parkinsonism Relat. Disord.* **2016**, *29*, 90–95. [CrossRef] [PubMed]
45. Arias-Vergara, T.; Vásquez-Correa, J.C.; Orozco-Arroyave, J.R. Parkinson's Disease and Aging: Analysis of Their Effect in Phonation and Articulation of Speech. *Cogn. Comput.* **2017**, *9*, 731–748. [CrossRef]
46. Rusz, J.; Tykalova, T.; Klempir, J.; Cmejla, R.; Ruzicka, E. Effects of dopaminergic replacement therapy on motor speech disorders in Parkinson's disease: Longitudinal follow-up study on previously untreated patients. *J. Neural Transm.* **2016**, *123*, 379–387. [CrossRef] [PubMed]
47. Chen, T.; Guestrin, C. Xgboost: A scalable tree boosting system. In Proceedings of the 22nd ACM SIGKDD International Conference on Knowledge Discovery and Data Mining, San Francisco, CA, USA, 13–17 August 2016; pp. 785–794.
48. Breiman, L.; Friedman, J.H.; Olshen, R.A.; Stone, C.J. *Classification and Regression Trees*; Wadsworth and Brooks: Monterey, CA, USA, 1984.
49. Torlay, L.; Perrone-Bertolotti, M.; Thomas, E.; Baciu, M. Machine learning—XGBoost analysis of language networks to classify patients with epilepsy. *Brain Inform.* **2017**, *4*, 159. [CrossRef] [PubMed]
50. Chen, Y.; Wang, X.; Jung, Y.; Abedi, V.; Zand, R.; Bikak, M.; Adibuzzaman, M. Classification of short single lead electrocardiograms (ECGs) for atrial fibrillation detection using piecewise linear spline and XGBoost. *Physiol. Meas.* **2018**, *39*, 104006. [CrossRef] [PubMed]

51. Zhong, J.; Sun, Y.; Peng, W.; Xie, M.; Yang, J.; Tang, X. XGBFEMF: An XGBoost-based Framework for Essential Protein Prediction. *IEEE Trans. NanoBiosci.* **2018**, *17*, 243–250. [CrossRef] [PubMed]

52. Palmer, J.L.; Coats, M.A.; Roe, C.M.; Hanko, S.M.; Xiong, C.; Morris, J.C. Unified Parkinson's Disease Rating Scale-Motor Exam: Inter-rater reliability of advanced practice nurse and neurologist assessments. *J. Adv. Nurs.* **2010**, *66*, 1382–1387. [CrossRef] [PubMed]

53. Baggio, J.A.O.; Curtarelli, M.B.; Rodrigues, G.R.; Tumas, V. Validity of the Brazilian version of the freezing of gait questionnaire. *Arquivos de Neuro-Psiquiatria* **2012**, *70*, 599–603. [CrossRef] [PubMed]

54. Santos, D.G.; Macías, M.A. Inter-rater variability in motor function assessment in Parkinson's disease between experts in movement disorders and nurses specialising in PD management. *Neurologia* **2017**. [CrossRef]

55. Jankovic, J. Parkinson's disease: Clinical features and diagnosis. *J. Neurol. Neurosurg. Psychiatry* **2008**, *79*, 368–376. [CrossRef] [PubMed]

56. Mekyska, J.; Galaz, Z.; Kiska, T.; Zvoncak, V.; Mucha, J.; Smekal, Z.; Eliasova, I.; Kostalova, M.; Mrackova, M.; Fiedorova, D.; et al. Quantitative Analysis of Relationship Between Hypokinetic Dysarthria and the Freezing of Gait in Parkinson's Disease. *Cogn. Comput.* **2018**. [CrossRef]

57. Berg, D.; Postuma, R.B.; Adler, C.H.; Bloem, B.R.; Chan, P.; Dubois, B.; Gasser, T.; Goetz, C.G.; Halliday, G.; Joseph, L.; et al. MDS research criteria for prodromal Parkinson's disease. *Mov. Dis.* **2015**, *30*, 1600–1611. [CrossRef] [PubMed]

applied
sciences

MDPI

Article

An Efficient Method to Learn Overcomplete Multi-Scale Dictionaries of ECG Signals

David Luengo [1,*], David Meltzer [2] and Tom Trigano [3]

[1] Department of Signal Theory and Communications, Escuela Técnica Superior de Ingeniería y Sistemas de Telecomunicación, Universidad Politécnica de Madrid, C/Nikola Tesla s/n, 28031 Madrid, Spain

[2] Department of Telematic and Electronic Engineering, Escuela Técnica Superior de Ingeniería y Sistemas de Telecomunicación, Universidad Politécnica de Madrid, C/Nikola Tesla s/n, 28031 Madrid, Spain; david.meltzer@upm.es

[3] Department of Electrical and Electronics Engineering, Shamoon College of Engineering, Ashdod 77245, Israel; tom.trigano@gmail.com or thomast@sce.ac.il

* Correspondence: david.luengo@upm.es; Tel.: +34-9106-73380

Received: 7 November 2018; Accepted: 22 November 2018; Published: date

Abstract: The electrocardiogram (ECG) was the first biomedical signal for which digital signal processing techniques were extensively applied. By its own nature, the ECG is typically a sparse signal, composed of regular activations (QRS complexes and other waveforms, such as the P and T waves) and periods of inactivity (corresponding to isoelectric intervals, such as the PQ or ST segments), plus noise and interferences. In this work, we describe an efficient method to construct an overcomplete and multi-scale dictionary for sparse ECG representation using waveforms recorded from real-world patients. Unlike most existing methods (which require multiple alternative iterations of the dictionary learning and sparse representation stages), the proposed approach learns the dictionary first, and then applies a fast sparse inference algorithm to model the signal using the constructed dictionary. As a result, our method is much more efficient from a computational point of view than other existing algorithms, thus becoming amenable to dealing with long recordings from multiple patients. Regarding the dictionary construction, we located first all the QRS complexes in the training database, then we computed a single average waveform per patient, and finally we selected the most representative waveforms (using a correlation-based approach) as the basic atoms that were resampled to construct the multi-scale dictionary. Simulations on real-world records from Physionet's PTB database show the good performance of the proposed approach.

Keywords: electrocardiogram (ECG); Least Absolute Shrinkage and Selection Operator (LASSO); overcomplete multi-scale dictionary construction; signal representation; sparse inference

1. Introduction

Since the development of the first practical apparatus for the recoding of the electrocardiogram (ECG) by Willem Einthoven in 1903, ECGs have been widely used by physicians to diagnose and monitor many cardiac disorders. Indeed, the use of the ECG has become so widespread that it is nowadays routinely used in both clinical and ambulatory settings to obtain a series of indicators related to the health status of patients [1]. This ubiquitous presence of ECGs in the medical field has been greatly enabled by digital signal processing (DSP) techniques: ECGs were the first biomedical signals where DSP algorithms were extensively applied to remove noise and interferences, detect and characterize the different waveforms contained in the ECG, extract the signals of interest (e.g., the fetal ECG) from the composite ECG, etc. [1,2].

By its own nature, the ECG is typically a sparse signal, composed of regular activations (the QRS complexes and other waveforms, such as the P and T waves) and periods of inactivity (corresponding to isoelectric intervals, such as the PQ or ST segments), as well as noise and interferences (baseline wander, powerline interference, electromyographic noise, motion artifacts, etc.) [1]. Since the introduction of the Least Absolute Shrinkage and Selection Operator (LASSO) regularizer by Tibshirani in 1996 [3], many sparse inference and representation techniques have been developed and successfully applied for all kinds of signals [4]: images, sound/audio recordings, biomedical waveforms, etc. However, to obtain a good sparse model for a given signal, it is essential to have an adequate dictionary composed of atoms that properly represent the significant waveforms contained in the observed signals. This has led to the development of many families of dictionaries (based on wavelets and wavelet packets, curvelets, contourlets, etc.) for different applications, as well as several on-line dictionary learning algorithms (e.g., see [5–7] for reviews of different dictionary learning methods for sparse inference) that typically require multiple alternative iterations of the dictionary learning and sparse representation stages.

In electrocardiographic signal processing, many approaches have been proposed for the sparse representation of single-channel and multi-channel ECGs using different types of simple analytical waveforms: Gaussians [8–10], generalized Gaussians and Gabor dictionaries [11], several families of wavelets (e.g., the Mexican hat or the coiflet4) [12,13], etc. Although these approaches can lead to good practical results, the resulting models usually contain many spurious activations that must be removed to obtain physiologically interpretable signals, for instance by means of a post-processing stage [9,13] or through the minimization of a complex non-convex cost function [14]. Conversely, a customized dictionary, built from real-world signals, will provide a better performance in terms of the reconstruction error obtained for a given level of sparsity. Consequently, several *on-line* dictionary learning approaches have also been applied, both in the context of sparse inference and compressed sensing (CS), to ECG signals: the K-SVD algorithm in [15], the shift-invariant K-SVD in [16], and the method of optimal directions in [17]. Unfortunately, all these methods have a high computational cost (due to their need to iterate between the dictionary learning and sparse approximation stages) and lead to dictionaries whose atoms do not correspond to real-world signals (thus reducing the interpretability of the sparse model, as well as the ability to easily locate the relevant waveforms). Alternatively, an *off-line* dictionary construction methodology (where a dictionary with real-world waveforms is initially built and then directly used for CS and sparse modeling without any further modification) was recently proposed by Fira et al. [12,18–20]. However, the atoms of the dictionary are either selected randomly from segments of the signal or taken directly from the first half of the ECG without any attempt to determine the most relevant waveforms.

In this work, we describe an efficient method to construct an overcomplete and multi-scale dictionary for sparse ECG representation using waveforms recorded from real-world patients. Unlike on-line dictionary methods (which require multiple alternative iterations of the dictionary learning and sparse representation stages), the proposed approach learns the dictionary first, and then applies a fast sparse inference algorithm, Convolutional Sparse Approximation (CoSA) [21], to model the signal using the constructed dictionary. As a result, our method is much more efficient from a computational point of view than other existing algorithms, thus becoming amenable to deal with long recordings from multiple patients. Regarding the dictionary construction, we locate all the QRS complexes in the training database first, then we compute a single average waveform per patient, and finally we select the most representative waveforms (using a correlation-based approach) as the basic atoms that will be resampled to construct the multi-scale dictionary. With respect to the approach of Fira et al., our method selects the optimal atoms to construct the dictionary, thus resulting in a much more compact solution. Numerical simulations demonstrate that the proposed approach is able to obtain a very sparse representation without missing any QRS complex or introducing spurious activations. Note that a preliminary version of this paper, where a single waveform was used to construct the overcomplete and multi-scale dictionary, was published in [22]. From a theoretical point of view, the main extension with respect to [22] is the proposal of a precise and novel procedure to incorporate multiple waveforms in the construction of the dictionary. Additional improvements, such as the simple and effective approach to remove the edge effects that appear after each resampling stage, have also been introduced in the pre-processing stages. Finally, a much more detailed literature review has been performed and many more numerical simulations, including both patients and channels (leads) not used to derive the dictionary, have been performed to characterize the behavior of the novel scheme.

The rest of the paper is organized as follows. Section 2 formulates the sparse representation problem of ECGs, emphasizing the importance of an appropriate dictionary. Then, in Section 3, we describe in detail the procedure followed to derive a multi-scale dictionary from real-world signals: the database used, the pre-processing steps, and the actual dictionary construction. Finally, Section 4 validates the proposed approach (focusing on the capability of the derived dictionary to model different ECG leads from multiple patients), and the paper is closed by the conclusions and future lines in Section 5. Throughout the paper, we concentrate on the description of the proposed method without focusing on any particular application. However, note that the constructed dictionary can be useful in many practical applications: lossy compression of ECG signals for their storage and transmission [17,20], denoising of ECGs contaminated by different types of interferences using sparse inference techniques [23], compressed sampling and sparse inference for heart rate variability analysis [24], sparse coding for atrial fibrillation (AF) classification [25], etc.

2. Problem Formulation

Let us assume that we have a single discrete-time ECG, $x[n]$, that has been obtained from a properly filtered and amplified continuous-time ECG, $x_c(t)$, through uniform sampling with a sampling period $T_s = 1/f_s$, i.e., $x[n] = x_c(nT_s)$. An *external ECG* captures the electrical activity occurring within the heart that triggers the mechanical cycle (systole and diastole) of the heart. Consequently, it is composed of a set of waveforms that reflect the different stages of the electrical cycle of the heart [1]: atrial depolarization (P waveform), ventricular depolarization (QRS complex) and ventricular repolarization (T waveform). Note that atrial repolarization cannot be observed in external ECGs, since it is masked by the ventricular depolarization, which happens simultaneously and produces a much stronger signal. All of these waveforms repeat themselves regularly during the heart's electrical cycle (thus leading to the well-known P-QRS-T cycle), although important changes in morphology, as well as fluctuations in amplitude, duration and interarrival times can be observed both for intra-patient and inter-patient recordings. Figure 1 shows an example of a single cycle from a clean synthetic ECG generated using the ECGSYN waveform generator [26] downloaded from Physionet [27], where all the relevant P-QRS-T waveforms, as well as the QRS onset and offset (also known as μ and j points [28]) can be clearly identified.

On top of the relevant electrical activity, the ECG also contains several types of noise and interference signals [1]: additive white Gaussian noise introduced by the electronic equipment used to acquire the ECGs (sensors, amplifiers and filters), baseline wander caused by the patient's respiration, powerline interference arising from the electrical network, electromyographic noise, motion artifacts, electrode contact noise, etc. Mathematically, this situation can be modeled as the superposition of the waveforms of interest (QRS complexes, P and T waveforms) as well as all the noise and interferences:

$$x[n] = \sum_{k=-\infty}^{\infty} E_k \Phi_k(t_n - T_k) + \epsilon[n], \quad n = 0, \ldots, N-1, \tag{1}$$

where T_k denotes the arrival time of the kth electrical pulse; E_k its amplitude; Φ_k is the associated, unknown pulse shape corresponding to QRS complexes, P and T waveforms; and $\epsilon[n]$ the noise and interference term. Note that, in real-world applications, Φ_k, T_k, and E_k are not precisely known. However, for the ECG, the typical shapes and durations of the Φ_k are known for all of the relevant waveforms. Therefore, they can be approximated by a time-shifted, multi-scale dictionary of known waveforms with finite support $M \ll N$ that can then be used to infer the E_k and T_k.

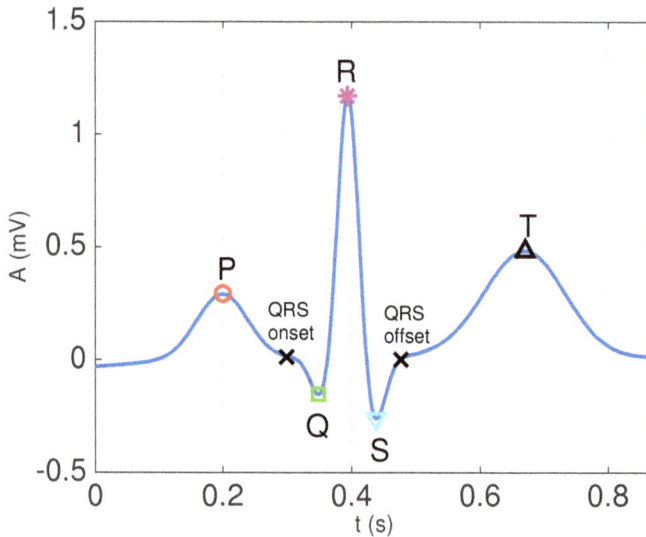

Figure 1. Example of a clean P-QRS-T cycle, with the peaks of all the relevant waveforms marked. The signal has been generated in Matlab using the ECGSYN waveform generator [26] with the following command: [s, ipeaks] = ecgsyn(1000,60,0,60,1,0.5,1000);.

More precisely, let us define a set of P candidate waveforms, Γ_p for $p = 1,\dots,P$, with a finite support of M_p samples such that $M_1 < M_2 < \cdots < M_P$ and $M = \max_{p=1,\dots,P} M_p = M_P$. If properly chosen, these waveforms can provide a good approximation of the local behavior of the signal around each sampling point, thus allowing us to approximate Equation (1) through the following model:

$$x[n] = \sum_{k=0}^{N-M-1} \sum_{p=1}^{P} \beta_{k,p} \Gamma_p[n-k] + \varepsilon[n], \quad n = 0,\dots,N-1, \tag{2}$$

where the $\beta_{k,p}$ are coefficients that indicate the amplitude of the pth waveform shifted to the kth time instant, $t_k = kT_s$; and $\varepsilon[n]$ includes also the additional approximation error associated to using Equation (2) instead of Equation (1), as well as all the noise and interferences already contained in $\epsilon[n]$. Let us now group all the candidate waveforms into a single matrix $\mathbf{A} = [\mathbf{A}_0 \ \mathbf{A}_1 \ \cdots \ \mathbf{A}_{N-M-1}]$, where the $N \times P$ matrices $\mathbf{A}_k$ (for $k = 0,\dots,N-M-1$) have column entries equal to $\Gamma_p[m-k]$ for $m = k,\dots,k+M-1$ and 0 otherwise. Then, the model of Equation (2) can be expressed more compactly in matrix form as follows:

$$\mathbf{x} = \mathbf{A}\boldsymbol{\beta} + \boldsymbol{\varepsilon}, \tag{3}$$

where $\mathbf{x} = [x[0],\dots,x[N-1]]^\top$ is an $N \times 1$ vector with all the ECG samples, $\boldsymbol{\beta} = [\beta_{0,1},\dots,\beta_{0,P},\beta_{1,1},\dots,\beta_{1,P},\dots,\beta_{N-M-1,1},\dots,\beta_{N-M-1,P}]^\top$ is an $(N-M)P \times 1$ coefficients vector, and $\boldsymbol{\varepsilon} = [\varepsilon[0],\dots,\varepsilon[N-1]]^\top$ is the $N \times 1$ noise vector.

Note that the matrix $\mathbf{A}$ can be considered as a global dictionary composed of $N - M$ sub-dictionaries, $\mathbf{A}_k$ for $k = 0,1,\dots,N-M-1$, that contain replicas of the candidate waveforms time shifted to $t_0 = 0$, $t_1 = T_s$, $\dots$, $t_{N-M-1} = (N-M-1)T_s$. In practice, the usual approach is to either use a single or several different waveforms with different time scales (*atoms*) to cope with the uncertainty about the shape and duration of the pulses that can be found in $x[n]$. Hence, as a result, we obtain an *overcomplete dictionary* (as the number of columns is larger than the number of rows, i.e., $(N-M)P > N$) composed of time-shifted, multi-scale waveforms which resemble the relevant electrical impulses that can be observed in the recorded ECGs. Now, two key questions arise:

1. Which is the optimal dictionary to model external ECG signals and how can we construct it?
2. Given an overcomplete dictionary, how can we obtain the optimal set of coefficients that represent only the relevant signal components?

Regarding the second question, let us remark that the only unknown term in Equation (3) is β when the dictionary is fixed. A classical solution to obtain this set of coefficients β is then minimizing the L_2 norm of the error between the model and the observed signal, thus obtaining the least squares (LS) solution:

$$\hat{\beta}_{LS} = \arg\min_{\beta} \|x - A\beta\|_2^2, \tag{4}$$

with $\|\cdot\|_2$ denoting the L_2 norm of a vector. The solution of Equation (4) is not unique, as it requires solving an overdetermined system of linear equations, but the standard solution (i.e., the solution that minimizes the L_2 norm of the obtained coefficients) is $\hat{\beta}_{LS} = A^{\sharp}x$, where $A^{\sharp} = (A^{\top}A)^{-1}A^{\top}$ denotes the pseudoinverse of A. (Note that, even though $\hat{\beta}_{LS}$ can be computed analytically from a theoretical point of view, it requires inverting an $(N-M)P \times (N-M)P$ matrix. Hence, we can easily encounter computational or numerical problems when $(N-M)P$ is large and/or A is ill-conditioned.) However, the LS approach leads to a solution where all the coefficients in $\hat{\beta}_{LS}$ are likely to be non-zero. This solution does not take into account the sparse nature of the relevant waveforms in $x[n]$ and results in overfitting, as part of the noise and interference terms are also implicitly modeled by the first term in Equation (3). Hence, a better alternative in this case is explicitly enforcing sparsity in β by applying the so-called LASSO approximation [3], which minimizes a cost function composed of the L_2 norm of the reconstruction error and the L_1 norm of the coefficient vector:

$$\hat{\beta} = \arg\min_{\beta} \|x - A\beta\|_2^2 + \lambda\|\beta\|_1, \tag{5}$$

where $\|\cdot\|_1$ denotes the L_1 norm, and λ is a parameter defining the trade-off between the sparsity of β and the precision of the estimation: the larger is the value of λ, the sparser is the solution obtained, but also the larger is the mean squared error.

Regarding the first question, it is obvious that a good dictionary, tailored to the shapes of the relevant waveforms in the ECG, will lead to a sparser representation and thus a better temporal localization of those waveforms. In the particular case of ECG modeling, many families of waveforms have been proposed within the related fields of sparse inference and compressed sensing, as discussed in the Introduction. However, there is increasing evidence that the best dictionaries are those constructed using atoms directly extracted from the signals to be modeled [15,17,20]. In the following, we describe a novel approach to construct a single overcomplete and multi-scale dictionary by learning the most representative waveforms from multiple patients. The goal of this paper is then investigating whether the resulting dictionary is able to model the outputs from multiple patients with a single set of representative waveforms.

3. Multi-Scale Dictionary Derivation

In this section, we describe the novel approach for off-line construction of a single overcomplete and multi-scale dictionary using QRS complexes extracted from multiple ECGs recorded from healthy patients. The database used to construct the dictionary is described first in Section 3.1, and the method is described next: the pre-processing stage in Section 3.2 and the dictionary creation stage in Section 3.3. Finally, the obtained dictionary was stored and applied to attain a sparse reconstruction of the desired ECGs (which may be in the database or not) using the LASSO approach, as described in Section 4.

3.1. Database

To construct the dictionary, we used the Physikalisch-Technische Bundesanstalt (PTB) database, compiled by the National Metrology Institute of Germany for research, algorithmic benchmarking and teaching purposes [29]. The ECGs were collected from healthy volunteers and patients with different heart diseases by Prof. Michael Oeff, at the Dep. of Cardiology of Univ. Clinic Benjamin Franklin in Berlin (Germany), and can be freely downloaded from Physionet [27]. (https://www.physionet.org/physiobank/database/ptbdb/). The database contains 549 records from 290 subjects (aged 17–87 years) composed of 15 simultaneously measured signals: the 12 standard leads plus the 3 Frank lead ECGs [1,2]. Each signal lasts approximately 2 min and is digitized using a sampling frequency $f_s = 1000$ Hz with a 16 bit resolution. Out of the 268 subjects for which the clinical summary is available, we selected channel 10 (lead V4) of the first recording of the $Q = 51$ healthy patients available in order to build the dictionary.

3.2. Pre-Processing

The block diagram of the pre-processing stage is shown in Figure 2. (Let us remark that we focus here on the QRS complexes because they are the most relevant waveforms that can be found in the ECGs. However, the proposed approach can also be applied to construct dictionaries of typical P and T waveforms.) Firstly, all the QRS complexes were extracted separately from each of the Q available ECGs, and those patients for which a significant number of QRS complexes cannot be reliably obtained wree removed from subsequent stages. After resampling to the maximum length of all the QRS complexes found for each of the remaining $Q' \leq Q$ patients, an individual average QRS complex was obtained per patient. Then, a second resampling stage was applied to the average QRS complexes of all the Q' valid patients to ensure that they have the same length, followed by a windowing stage to obtain initial and final samples equal to zero, and a normalization to remove the mean and enforce unit energy on all the signals. Finally, these Q' waveforms were stored in a QRS complexes database and used to construct the desired overcomplete dictionary. In the sequel, we provide a detailed description of each of the blocks in Figure 2.

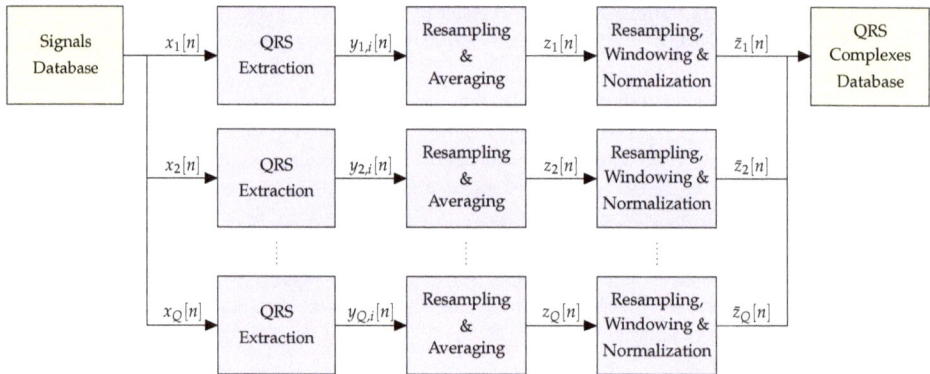

Figure 2. Block diagram of the pre-processing stage applied before the creation of the dictionary.

3.2.1. QRS Extraction

The first pre-processing step consists in extracting all the QRS complexes from each of the Q ECGs, $x_q[n]$ for $q = 1,\ldots,Q$. To attain this goal, we followed the approach described in [30]:

1. Apply a 4th order Butterworth bandpass filter with cut-off frequencies $f_{c1} = 1$ Hz and $f_{c2} = 40$ Hz to remove noise and interferences. Forward–backward filtering, with an appropriate choice of the initial state to remove transients [31], is used to avoid phase distortion.
2. Locate the positions of the R waveforms using the Pan–Tompkins QRS detector [32].
3. Determine the fiducial points that mark the beginning and end of the QRS complexes by tracking backwards and forward from the R peaks, estimating the QRS onset and offset points using the minimum *radius of curvature* technique, as described in Section 4.2 of [30].

This approach resulted in $Q' = 44$ valid subjects (i.e., $\approx 84.6\%$ out of the $Q = 51$ available individuals), for which a significant and variable number of QRS complexes ($y_{q,i}[n]$ for $1 \leq q \leq Q'$, $1 \leq i \leq P_q$ and $0 \leq n \leq L_{q,i} - 1$), with a variable length of samples for each of them, were extracted for the database used. A total of 6266 QRS complexes were extracted, implying an average of 142.4 QRS complexes per patient (with a maximum of 194 QRS complexes obtained from a single subject) with lengths from 90 to 124 samples (i.e., from 90 to 124 ms).

3.2.2. Resampling and Averaging

To compute an average QRS complex for each individual, we need to work with QRS complexes that have a fixed length. (Note that extracting a single waveform per patient can be a limitation. However, since the recordings used in this work correspond to healthy patients and are rather short (less than 2 min), a single waveform is often enough to represent the average QRS complex for each patient. Developing an efficient method to extract multiple waveforms from each patient is a challenging issue that will be considered in future works.) The easiest solution to achieve this goal is resampling the extracted QRS complexes to the maximum length for each patient, $L_{\max}^q = \max_{1 \leq i \leq P_q} L_{q,i} \leq L_{\max} = \max_{1 \leq q \leq Q'} L_{\max}^q = 124$ samples. A change in the sampling rate of discrete time signals can be accomplished by means of interpolation and decimation [33]. If the ith QRS complex ($i = 1, \ldots, P_q$) has a length $L_{q,i} \leq L_{\max}^q$ samples and $M_{q,i} = \text{GCD}(L_{q,i}, L_{\max}^q)$, with the Greatest Common Divisor (GCD) being the largest positive integer that divides each of the two integer numbers, then we need first to interpolate by a factor $L_{\max}^q / M_{q,i}$ and then to decimate by a factor $L_{q,i} / M_{q,i}$ (i.e., the fractional resampling rate is $L_{\max}^q / L_{q,i} \geq 1$).

The aforementioned approach includes a digital lowpass antialiasing filter between the interpolator and the decimator, with a cut-off frequency $w_c^{q,i} = \pi M_{q,i} / L_{\max}^q$ rad, which assumes that the sequence to process starts and ends with sequences of zeroes. Although the recorded ECGs have been initially bandpass filtered to remove baseline wander and other artifacts, the QRS complexes cannot be assumed to start and end with the required zero samples. In fact, the set of available QRS complexes (from the QRS onset to the QRS offset) always contain negative starting and ending values: from -0.0219 to -0.2349 mV for the initial sample, and from -0.0318 to -0.1796 mV for the final sample. As the starting and ending samples of all the QRS complexes are not equal to zero, resampling produces an undesired effect (*edge effect*), which consists on deviations from the expected values in the starting and ending values of the resampled signals. To remove the edge effect, we propose the following simple and effective approach:

1. From the ith QRS complex signal, $y_{q,i}[n]$ for $n = 0, 1, \ldots, L_{q,i} - 1$, we first constructed the following two sequences:

$$y_\ell^{(q,i)}[n] = y_{q,i}[n] - y_{q,i}[0],$$
$$y_r^{(q,i)}[n] = y_{q,i}[n] - y_{q,i}[L_{q,i} - 1],$$

   which are not likely to be affected by the edge effect on their leftmost and rightmost samples, respectively.

2. We performed the resampling by the factor $L_{q,i} / L_{\max}^i$ separately on $y_\ell^{(q,i)}[n]$ and $y_r^{(q,i)}[n]$, obtaining two resampled sequences $\tilde{y}_\ell^{(q,i)}[n]$ and $\tilde{y}_r^{(q,i)}[n]$.

3. The desired resampled sequence is finally given by

$$\tilde{y}_{q,i}[n] = \begin{cases} \tilde{y}_\ell^{(q,i)}[n], & 0 \leq n \leq \left\lfloor \frac{N}{2} \frac{L_{\max}^q}{L_{q,i}} \right\rfloor - 1; \\ \tilde{y}_r^{(q,i)}[n], & \left\lfloor \frac{N}{2} \frac{L_{\max}^q}{L_{q,i}} \right\rfloor \leq n \leq \left\lfloor N \frac{L_{\max}^q}{L_{q,i}} \right\rfloor - 1. \end{cases} \tag{6}$$

Figure 3 shows the leftmost and rightmost samples corresponding to one of the resampled QRS complexes (from patient 214 in the PTB database) when resampling is performed directly on $y_{q,i}[n]$. The edge effect on the left and right parts of the signals (i.e., the deviation of the red line with respect to the desired values indicated by the black dots) is evident in this case. On the other hand, when the proposed approach was applied, the resampled sequence ($\tilde{y}_{q,i}[n]$) is not affected by the edge effect on either its left or its right side, as also seen in Figure 3.

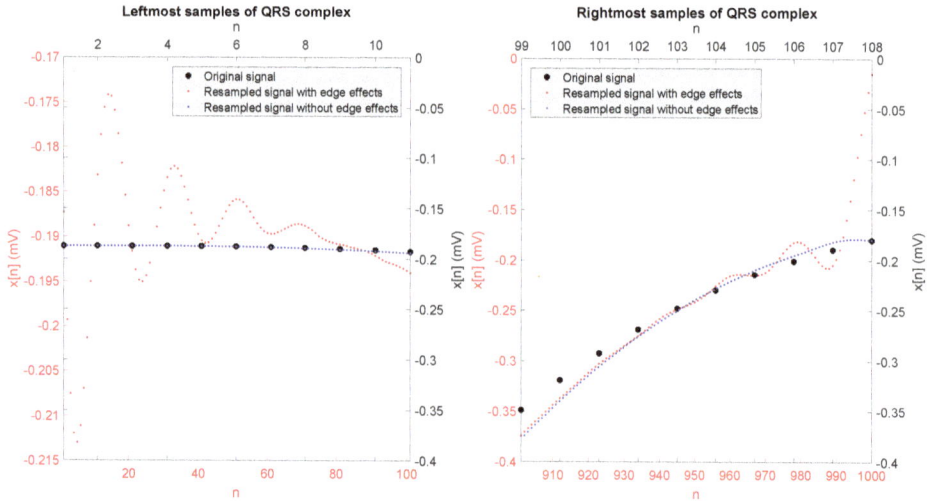

Figure 3. Leftmost and rightmost samples of an original QRS complex (from patient 214 in the PTB database) and its resampled version with and without edge effects.

Finally, the averaged QRS complex for each patient was obtained simply by computing the sample mean for each time instant:

$$z_q[n] = \frac{1}{P_q} \sum_{i=1}^{P_q} \tilde{y}_{q,i}[n], \tag{7}$$

with P_q denoting the number of QRS complexes found in the qth ECG.

3.2.3. Windowing and Normalization

After averaging, all the averaged QRS complexes were resampled again to obtain Q' signals with the same number of samples, $L_{max} = \max_{1 \leq q \leq Q'} L^q_{max} = 124$, using a resampling factor L_{max}/L^q_{max}. Then, the resulting signals were windowed to ensure a smooth decay of the QRS complexes towards zero, and normalized by removing their mean and dividing by their standard deviation. The signals that were finally stored in the QRS complex database are

$$\tilde{z}_q[n] = \frac{\check{z}_q[n]w[n] - \mu_q}{\sigma_q}, \tag{8}$$

where $\check{z}_q[n]$ are the averaged QRS complexes after the second resampling stage (i.e., after ensuring that their sample length is equal to L_{max}), μ_q and σ_q are their sample mean and standard deviations, respectively,

$$\mu_q = \frac{1}{L_{\max}} \sum_{n=0}^{L_{\max}-1} \tilde{z}_q[n]w[n], \tag{9}$$

$$\sigma_q = \sqrt{\frac{1}{L_{\max}-1} \sum_{n=0}^{L_{\max}-1} (\tilde{z}_q[n]w[n] - \mu_q)^2}, \tag{10}$$

and $w[n]$ is the window used in this case, a window that follows the spectral shape of the raised cosine filter, widely used in digital communications [34], in the time domain: $w[n] = w_c(nT_s)$ for $n = -(L_{\max}-1)/2, \dots, -1, 0, 1, \dots, (L_{\max}-1)/2$, with

$$w_c(t) = \begin{cases} 1, & |t| \le (1-\alpha)T_0; \\ \frac{1}{2}\left[1 + \cos\left(\frac{\pi}{2\alpha T_0}|t - (1-\alpha)T_0|\right)\right], & (1-\alpha)T_0 < |t| < (1+\alpha)T_0, \end{cases} \tag{11}$$

$T_0 = \frac{L_{\max}-1}{1+\alpha}\frac{T_s}{2}$, and α denoting the roll-off factor that controls the decay of $w_c(t)$ towards zero: for $\alpha = 0$, we have a rectangular window that abruptly goes to zero at $\pm T_0$, whereas for $\alpha = 1$ the window is bell-shaped and starts decaying smoothly towards zero immediately after $|t| > 0$. This window, whose time-domain shape is shown in Figure 4 for several values of α, ensures that the central samples of the QRS complexes remain undistorted, while their amplitudes quickly decay towards zero at the borders. Throughout the paper, we have always used $\alpha = 0.25$.

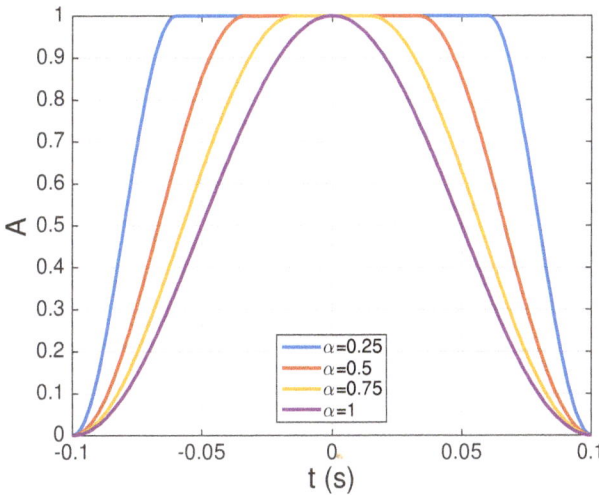

Figure 4. Several examples of the raised cosine window of Equation (11) for $T_s = 1$ ms, $L_{\max} = 201$, and different values of the parameter α.

3.3. Dictionary Construction

After the pre-processing described in the previous section, we have $Q' \le Q$ waveforms from different patients stored in the QRS complexes database. These waveforms ($\tilde{z}_q[n]$ for $1 \le q \le Q'$ and $0 \le n \le L_{\max} - 1$) could be directly used to build the sub-dictionaries. However, they are highly correlated and thus the resulting dictionary would provide a poor performance and lead to a high computation time. Therefore, to obtain a reduced dictionary composed of distinct shapes, we performed the procedure described in the following sections.

3.3.1. Selection of the First Atom

For the first atom of the dictionary, we sought the most representative waveform in the QRS complexes database. Indeed, if we intend to construct a sparse model for all the ECGs of all the patients using a single basis signal, then we should choose a waveform which resembles as closely as possible the set of QRS complexes (the most relevant part of the ECGs) in the different patients. To achieve this goal, using the average signals stored in the QRS complexes database, we followed two steps:

1. Compute a $Q' \times Q'$ correlation matrix $\mathbf{C}$, whose $(Q')^2$ elements correspond to Pearson's correlation coefficient among each pair of waveforms in the QRS complexes database (in practice, only $Q'(Q'-1)/2$ coefficients have to be computed, since $\rho_{ii} = 1$ and $\rho_{ij} = \rho_{ji}\ \forall i,j \in \{1,\dots,Q'\}$).

$$\rho_{ij} = \frac{C_{ij}}{\sqrt{C_{ii}C_{jj}}} = C_{ij}, \tag{12}$$

where C_{ij} denotes the cross-covariance between the ith and jth waveforms at lag 0 (i.e., without any time shift), and the last expression ($\rho_{ij} = C_{ij}$) is due to the energy normalization described in Section 3.2.3, which implies that $C_{ii} = 0\ \forall i$.

2. Select the waveform with the highest average correlation (in absolute value) with respect to all the other candidate waveforms, i.e.,

$$\ell_0 = \arg\max_{i=1,\dots,Q'} \sum_{j=1}^{Q'} |\rho_{ij}|, \tag{13}$$

which corresponds to the most representative waveform of all the candidate waveforms.

3.3.2. Selection of Additional Atoms

Additional atoms can be incorporated to the dictionary in order to increase its flexibility in representing different ECGs. These atoms should be constructed using highly correlated waveforms with respect to the remaining candidates (to obtain representative dictionary atoms), as well as with low absolute correlation with respect to already selected waveforms (to avoid similar atoms). Here, we propose to use the following procedure to select a total of K representative waveforms:

1. Set the number of accepted atoms equal to one ($k = 1$), the pool of candidate waveforms as $C = \{1,\dots,\ell_0 - 1, \ell_0 + 1,\dots,Q'\}$ (i.e., all the waveforms except for the one selected for the first atom), the pool of accepted waveforms as $\mathcal{A} = \{\ell_0\}$, and construct a reduced correlation matrix by removing the row and column corresponding to the first atom selected from the global correlation matrix $\mathbf{C}$:

$$\mathbf{C}_r = \begin{bmatrix} \mathbf{C}_{1:\ell_0-1,1:\ell_0-1} & \mathbf{C}_{1:\ell_0-1,\ell_0+1:Q'} \\ \mathbf{C}_{\ell_0+1:Q',1:\ell_0-1} & \mathbf{C}_{\ell_0+1:Q',\ell_0+1:Q'} \end{bmatrix}. \tag{14}$$

2. WHILE $k < K$:

    (a) Select, from the remaining waveforms in the pool of candidates, the one with the highest average correlation (in absolute value) with respect to all the other candidate waveforms in the pool, i.e.,

$$\tilde{\ell}_k = \arg\max_{i=1,\dots,Q'-k} \sum_{j=1}^{Q'-k} |\mathbf{C}_r(i,j)|, \tag{15}$$

and obtain the associated index, ℓ_k, in the original set of candidate waveforms.

(b)   Compute the maximum correlation (in absolute value) between the selected candidate and all the already accepted atoms,

$$\rho_{max} = \max_{i=0,\dots,k-1} |\mathbf{C}(\ell_k, \ell_i)|. \tag{16}$$

(c)   Remove the ℓ_kth waveform from the pool of candidates (i.e., set $\mathcal{C} = \mathcal{C} \setminus \{\ell_k\}$), and construct a new reduced matrix $(\mathbf{C}_r)$ by removing the $\tilde{\ell}_k$th row and column from the current $\mathbf{C}_r$.

(d)   IF $\rho_{max} < \gamma$ (with $0 \le \gamma \le 1$ denoting a pre-defined maximum correlation threshold), THEN add the selected waveform to the pool of accepted atoms (i.e., $\mathcal{A} = \mathcal{A} \cup \{\ell_k\}$), and set $k = k + 1$.

END

Note that the value of γ sets an upper bound on the number of waveforms that can be accepted for a given set of candidate waveforms. Therefore, K is the maximum number of waveforms that can be accepted in the previous algorithm, but the number of waveforms selected can actually be smaller than K. See Section 4.1 for a detailed description of the values of γ and K used in this work.

3.3.3. Construction of the Multi-Scale Dictionary

Finally, the K selected waveforms were resampled to obtain an overcomplete and multi-scale sub-dictionary composed of R different time scales. Note that the total number of atoms is thus $P = KR$, corresponding to K different waveforms with R distinct time scales for each one. In this case, we used $R = 11$ different time scales spanning a time frame slightly wider than the typical durations of QRS complexes: $60, 70, \dots, 160$ ms. Note also that the global dictionary is simply obtained by performing $N - M$ different time shifts of the resulting sub-dictionary [13].

4. Numerical Results

In this section, we first detail the construction of the dictionary and then present the application of the resulting dictionary to perform the sparse reconstruction of the signals from different patients.

4.1. Dictionary Construction

As mentioned earlier, to construct the dictionary, we used channel 10 (lead V4) from the first register of the $Q = 51$ healthy patients in the PTB database: Patients 104, 105, 116, 117, 121, 122, 131, 150, 155, 156, 166, 169, 170, 172–174, 180, 182, 184, 185, 198, 214, 229, 233–248, 251, 252, 255, 260, 263, 264, 266, 267, 276, 277, 279, and 284. From this whole set of patients, we were able to obtain reliable average QRS complexes for $Q' = 44$ Patients: 104, 105, 117, 121, 122, 131, 150, 155, 156, 169, 170, 174, 180, 182, 184, 185, 198, 214, 229, 234–248, 251, 252, 260, 263, 264, 267, 276, 277, 279, and 284. The remaining $Q - Q' = 7$ patients (116, 166, 172, 173, 233, 255, and 266), where the extraction of the QRS complexes fails, were used as the test set. The average waveforms for the $Q' = 44$ patients, after resampling to $L = 124$ samples (i.e., 124 ms), windowing and normalization, can be seen in Figure 5. Note that, as expected, there is a large degree of similarity among all the waveforms, since they all correspond to regular heartbeats from healthy patients. This similarity is illustrated also by the color plot of the absolute value of the correlation coefficient in Figure 6. Note the large correlation (corresponding to dark red points) among most waveforms. For this reason, in [22], we decided to extract a single waveform in order to construct the multi-scale and overcomplete dictionary. However, in Figure 6, we also notice that some waveforms exhibit low correlation values (as shown by blue points). This motivates us to explore here the performance as more than one waveform is extracted from the pool of average QRS complexes shown in Figure 5 in order to construct the dictionary.

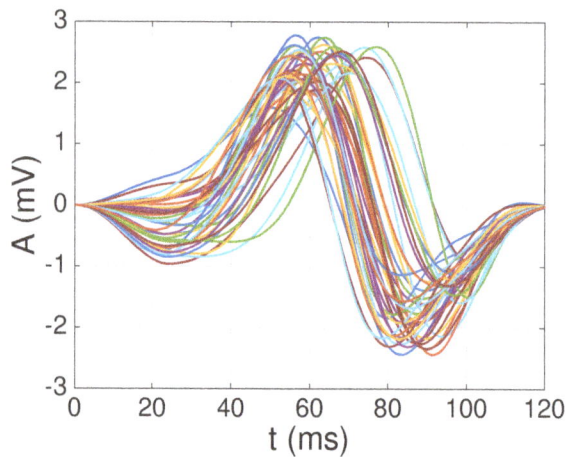

Figure 5. $Q' = 44$ reliable average QRS complexes extracted from the $Q = 51$ healthy patients in the PTB database [29] after resampling, windowing and normalization.

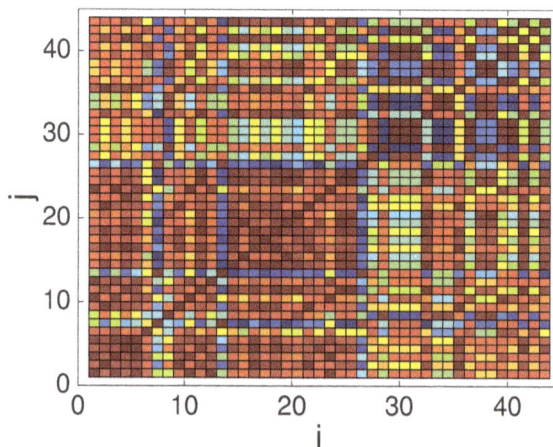

Figure 6. Color map showing the absolute value of the correlation coefficient, $|\rho_{ij}|$, for the $Q' = 44$ average QRS waveforms. Dark red colors indicate values close to 1, whereas dark blue colors indicate values closer to 0.

To build dictionaries composed of different waveforms, we tested several threshold levels: $\gamma = 0.1, 0.2, \ldots, 0.9$. Note that the larger is the value of γ the less restrictive is the condition to incorporate a new atom to the dictionary, and thus the larger is the number of final atoms (K) used: for $\gamma = 0$, we would always obtain $K = 1$ atoms (since no new atoms can be incorporated after the first one), whereas, for $\gamma = 1$, we would obtain $K = Q'$ (as all waveforms would be considered valid). Following this approach, we obtained $K = 1$ atoms for $\gamma \leq 0.2$, $K = 2$ atoms for $0.3 \leq \gamma \leq 0.6$, $K = 3$ atoms when $\gamma = 0.7$, $K = 4$ atoms when $\gamma = 0.8$, and $K = 6$ atoms when $\gamma = 0.9$. Figure 7 shows the six atoms selected for $\gamma = 0.9$. Note that, since we followed a deterministic procedure, the first atom was the one obtained for $K = 1$ (i.e., when $\gamma \leq 0.2$), the first and second ones were those obtained for $K = 2$ (i.e., for $0.3 \leq \gamma \leq 0.6$), and so on.

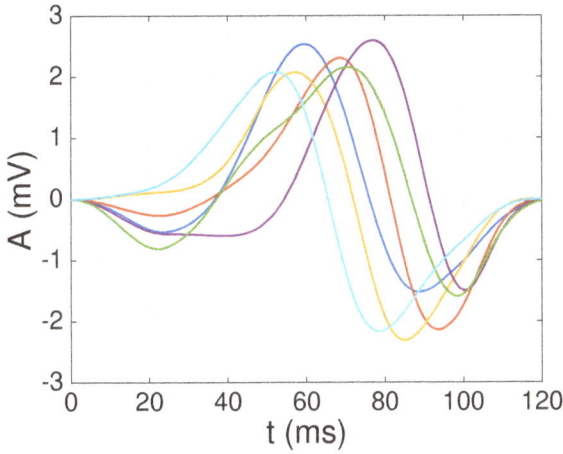

Figure 7. $K = 6$ atoms selected from the $Q' = 44$ reliable average QRS complexes obtained from the $Q = 51$ healthy patients in the PTB database, using a threshold $\gamma = 0.9$.

Finally, the selected waveforms for $K = 1, \ldots, 6$ were resampled in such a way that their duration ranged from 60 ms to 160 ms (with a time step of 10 ms). The resulting base dictionary, composed of $P = 11 \times K = 11, 22, \ldots, 66$ atoms, is shown in Figure 8 for $K = 6$. The final multi-scale and overcomplete dictionary consists of these P waveforms time shifted to the $N - M$ locations of the sampled ECGs, implying that its size is $N \times P(N - M)$. For instance, for $N = 115,200$ samples (a typical signal size in the PTB database) and $M = 160$ samples (the maximum support of the selected waveforms), the size of the matrix dictionary ranges from $N \times 11(N - M) = 115,200 \times 1,265,440$ for $K = 1$ up to $N \times 66(N - M) = 115,200 \times 7,592,640$ for $K = 6$.

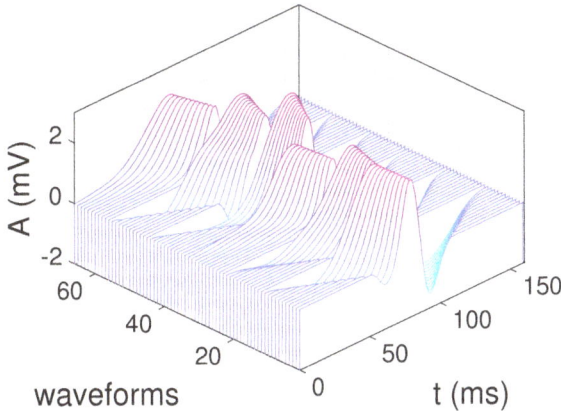

Figure 8. Multi-scale dictionary constructed using the $K = 6$ most representative waveforms shown in Figure 7.

4.2. Sparse ECG Representation

We next tested the constructed dictionary on 21 recordings corresponding to 17 healthy patients from the PTB database. Our main goal was determining whether the constructed dictionary (which is not patient-specific) can be effectively used to perform a sparse representation of ECG signals from multiple patients. To solve Equation (5), we used the CoSA algorithm recently proposed in [21], which allows us to process the $N \approx 115{,}200$ samples (almost 2 min of recorded time) of the whole signal at once (i.e., without having to partition it into several segments that have to be processed separately) in a reasonable amount of time. Since several signals showed a significant degree of baseline wander, before applying CoSA all signals were filtered using a third-order high-pass IIR (infinite impulse response) Butterworth filter designed using Matlab's `filterDesigner` tool: stop-band frequency $f_{stop} = 0.1$ Hz; pass-band frequency $f_{pass} = 1$ Hz; minimum stop-band attenuation $A_{stop} = 40$ dB; and maximun pass-band attenuation $A_{pass} = 1$ dB. Forward–backward filtering was applied again to avoid phase distortion. An example of the reconstructed signal using $K = 1$ and $\lambda = 20$ is shown in Figure 9. Note that all the QRS complexes (our main goal here) are properly represented by the sparse model. By increasing the number of signals (K), and especially by decreasing the sparsity factor (λ), a better model that also includes the P and T waveforms can be obtained. However, let us remark that this is not our main goal. Indeed, a better option to model the P and T waveforms would be constructing specific dictionaries of P and T waveforms: either using synthetic waveforms (e.g., Gaussians) or applying the proposed approach to construct real-world dictionaries of P and T waveforms. We intend to explore this issue in future works, constructing mega-dictionaries of P-QRS-T waveforms which are able to model all the relevant activations in the ECGs.

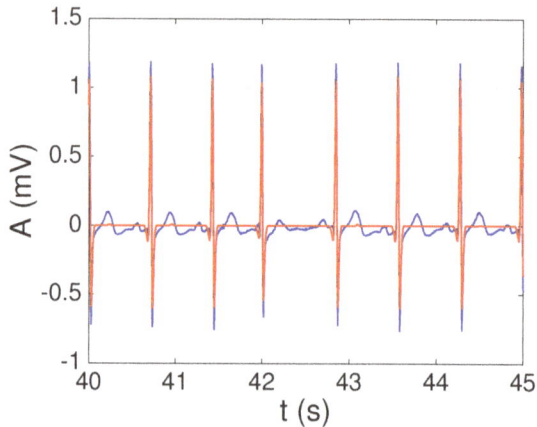

Figure 9. Example of sparse ECG representation with the derived dictionary for a segment of signal 121 from the PTB database. Real signal in blue; sparse representation (QRS complexes) in red.

To measure the effectiveness of the proposed approach, we used several performance metrics that measure both the model's sparsity and its accuracy in representing the original signal. On the one hand, the sparsity was gauged by the coefficient sparsity (C-Sp),

$$\text{C-Sp (\%)} = \frac{\|\boldsymbol{\beta}\|_0}{(N - M)P} \times 100 \tag{17}$$

which measures how many non-null coefficients (out of the total ones) are required to represent the signal, and the signal sparsity (S-Sp),

$$\text{S-Sp (\%)} = \frac{\|\mathbf{x}\|_0}{N} \times 100 \tag{18}$$

which measures how many samples of the signal's approximation (out of the total number of samples) are not equal to zero. Note that $\|\cdot\|_0$ indicates the L_0 "norm" of a vector, i.e., its number of non-null elements. On the other hand, the reconstruction error was measured by the normalized mean squared error (NMSE) and its logarithmic counterpart, the reconstruction signal to noise ratio (R-SNR):

$$\text{NMSE (\%)} = \frac{\|\mathbf{x} - \mathbf{A}\boldsymbol{\beta}\|_2^2}{\|\mathbf{x}\|_2^2} \times 100 \tag{19}$$

$$\text{R-SNR (dB)} = 10 \cdot \log_{10}(\text{NMSE}), \tag{20}$$

where $\| \cdot \|_2^2$ denotes the squared L_2 norm of a vector.

The results, for the four performance measured previously described, are displayed in Figure 10. On the left hand side, we show the results (mean value and standard deviation) for the first 10 patients in the training set: Patients 104, 105, 117, 121, 122, 131, 150, 155, 156, and 169. On the right hand side, we show the results (mean value and standard deviation again) for the 11 signals in the test set: Patients 116, 166, 172, 173, 233 (5 recordings), 255, and 266. Note the large sparsity attained in all cases (especially as λ increases), and the good reconstruction error for small/moderate values of λ (i.e., $\lambda \leq 5$). Note also the improvement in performance when incorporating additional waveforms to the dictionary (e.g., for $\lambda = 2$ there is a 3.32 dB improvement in mean R-SNR for the patients in the test set when using $K = 6$ instead of $K = 1$), although this comes at the expense of a reduction in the signal sparsity (e.g., for the same case, the mean S-Sp goes down by 4.2% when using $K = 6$ instead of $K = 1$). Finally, let us remark that there are no substantial differences between the performance on signals from the test set (i.e., signals used to build the dictionary) and signals from the training set (i.e., signals not used to build the dictionary), as evidenced by the similarity of curves on the left hand side and right hand side of Figure 10.

As a last performance check, we applied the Pan–Tompkins algorithm to the reconstructed signal in order to test whether the sparse model introduced some distortion in the location of the QRS complexes. As a result, we found that all the QRS complexes were always properly detected and located within two samples (i.e., ± 2 ms) of the QRS complexes found in the original signal, even in the sparsest case (i.e., with $K = 1$ and $\lambda = 20$). Therefore, if we are only interested in the QRS complexes or some analysis derived from them (e.g., heart rate variability studies), the proposed model is a very good option to construct a sparse model that keeps all the relevant information.

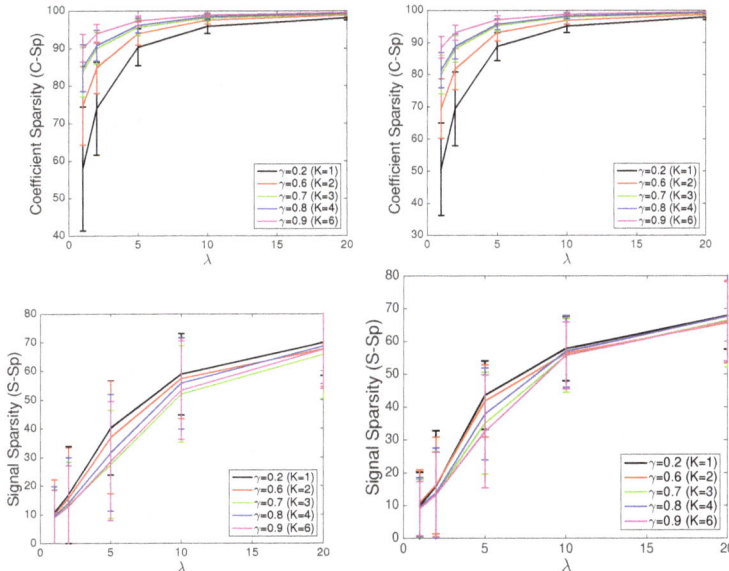

Figure 10. *Cont.*

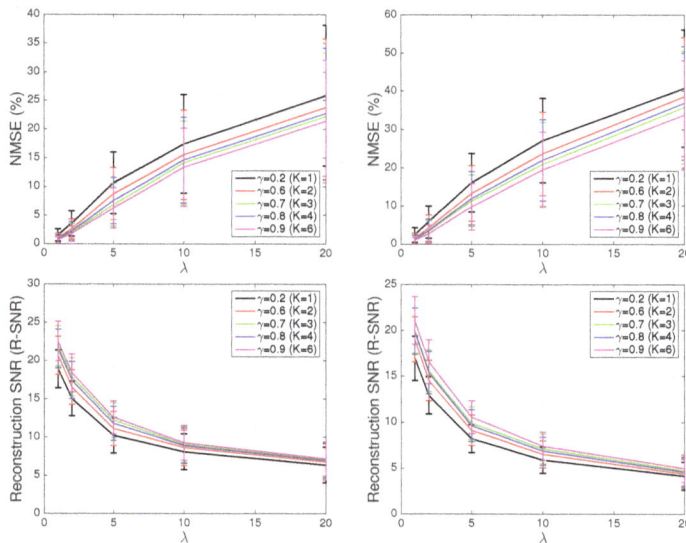

Figure 10. Different performance metrics: (**Left**) results on signals from training set; and (**Right**) results on signals from test set (i.e., signals not used to build the dictionary).

4.3. Sparse ECG Representation of Other Channels

In this section, we investigate the feasibility of using the dictionary learnt on lead V4 to represent ECGs recorded in other leads. To do so, we used the constructed dictionary to model all 15 channels (leads) available for Patient 104. Let us emphasize that no signals at all from any of the other 14 leads available were used to derive the dictionary, since all samples used from all the patients during the dictionary construction stage correspond to lead V4. Table 1 displays the results for $\lambda = 1$ and $K = 2$, showing that good results are obtained in general for most of the leads. Indeed, the performance for several leads (II, aVR, V5, V6 and Vx) in terms R-SNR is better than for lead v4, which was used to construct the dictionary, and poor R-SNR results were only attained for leads aVL and Vy. Similar conclusions were obtained when considering other values of λ and K. Overall, this shows the feasibility of constructing a single multi-scale and overcomplete dictionary (possibly using QRS complexes extracted from several leads) for multiple channels.

Table 1. Performance of the constructed dictionary (using waveforms extracted from lead V4) on other leads from Patient 104 not used to construct the dictionary.

Channel	Lead	C-Sp (%)	S-Sp (%)	NMSE (%)	R-SNR (dB)
1	I	86.5245	11.0095	4.1476	13.8220
2	II	83.6901	5.1302	1.8310	17.3732
3	III	92.0191	38.1580	5.1706	12.8646
4	aVR	85.4093	6.0408	2.5404	15.9510
5	aVL	92.9162	61.9314	10.2057	9.9116
6	aVF	88.1383	12.0894	3.0323	15.1823
7	V1	87.4629	5.5182	3.1652	14.9960
8	V2	90.7240	51.0356	3.0689	15.1302
9	V3	86.5196	40.1319	2.3935	16.2097
10	V4	83.1699	26.8689	2.9814	15.2557
11	V5	80.9004	5.1181	2.1296	16.7170
12	V6	81.0859	3.4931	1.6270	17.8862
13	Vx	80.1663	4.3333	1.7768	17.5037
14	Vy	93.0671	62.7804	11.4420	9.4150
15	Vz	94.0583	69.9991	5.8457	12.3316

5. Conclusions

In this paper, we have described a novel mechanism to derive a realistic, multi-scale and overcomplete dictionary from recorded real-world ECG signals. The dictionary was constructed offline, thus avoiding the computational burden of on-line approaches and ensuring the scalability of the proposed methodology for large datasets with many individuals and/or sample sizes. The obtained dictionary was been used to perform an accurate sparse representation of several ECGs recorded from healthy patients, showing that it can properly capture all the QRS complexes without introducing false alarms. Potential future lines include testing the proposed approach on a larger number of patients (especially including subjects with cardiac pathologies), the construction of dictionaries composed of multiple waveforms (e.g., P and T waveforms), and the combination of waveforms extracted from real patients with synthetic waveforms.

Author Contributions: Conceptualization, D.L.; methodology, D.L. and D.M.; software, D.L., D.M. and T.T.; validation, D.L.; data curation, D.M.; original draft preparation, D.L.; and review and editing, D.M. and T.T.

Funding: This research was funded by Ministerio de Economía y Competitividad (Spain) through the MIMOD-PLC project (grant number TEC2015-64835-C3-3-R).

Conflicts of Interest: The authors declare no conflict of interest. The funders had no role in the design of the study; in the collection, analyses, or interpretation of data; in the writing of the manuscript, or in the decision to publish the results.

References

1. Sörnmo, L.; Laguna, P. *Bioelectrical Signal Processing in Cardiac and Neurological Applications*; Academic Press: Cambridge, MA, USA, 2005.
2. Clifford, G.; Azuaje, F.; McSharry, P. (Eds.) *Advanced Methods and Tools for ECG Data Analysis*; Artech House: Norwood, MA, USA, 2009.
3. Tibshirani, R. Regression shrinkage and selection via the Lasso. *J. R. Stat. Soc. Ser. B (Methodol.)* **1996**, *58*, 267–288. [CrossRef]
4. Elad, M. *Sparse and Redundant Representations: From Theory to Applications in Signal and Image Processing*; Springer: Berlin, Germany, 2010.
5. Kreutz-Delgado, K.; Murray, J.F.; Rao, B.D.; Engan, K.; Lee, T.W.; Sejnowski, T.J. Dictionary learning algorithms for sparse representation. *Neural Comput.* **2003**, *15*, 349–396. [CrossRef] [PubMed]
6. Rubinstein, R.; Bruckstein, A.M.; Elad, M. Dictionaries for sparse representation modeling. *Proc. IEEE* **2010**, *98*, 1045–1057. [CrossRef]
7. Tosic, I.; Frossard, P. Dictionary learning. *IEEE Signal Process. Mag.* **2011**, *28*, 27–38. [CrossRef]
8. Billah, M.; Mahmud, T.; Snigdha, F.; Arafat, M. A novel method to model ECG beats using Gaussian functions. In Proceedings of the 2011 4th International Conference on Biomedical Engineering and Informatics (BMEI), Shanghai, China, 15–17 October 2011; Volume 2, pp. 612–616.
9. Monzón, S.; Trigano, T.; Luengo, D.; Artés-Rodríguez, A. Sparse spectral analysis of atrial fibrillation electrograms. In Proceedings of the 2012 IEEE International Workshop on Machine Learning for Signal Processing, Santander, Spain, 23–26 September 2012; pp. 1–6.
10. Trigano, T.; Kolesnikov, V.; Luengo, D.; Artés-Rodríguez, A. Grouped sparsity algorithm for multichannel intracardiac ECG synchronization. In Proceedings of the 2014 22nd European Signal Processing Conference (EUSIPCO), Lisbon, Portugal, 1–5 September 2014; pp. 1537–1541.
11. Divorra-Escoda, O.; Granai, L.; Lemay, M.; Hernandez, J.M.; Vandergheynst, P.; Vesin, J.M. Ventricular and atrial activity estimation through sparse ECG signal decompositions. In Proceedings of the 2006 IEEE International Conference on Acoustics Speech and Signal Processing Proceedings, Toulouse, France, 14–19 May 2006; Volume II, pp. 1060–1063.
12. Fira, M.; Goras, L.; Barabasa, C.; Cleju, N. On ECG compressed sensing using specific overcomplete dictionaries. *Adv. Electr. Comput. Eng.* **2010**, *10*, 23–28. [CrossRef]
13. Luengo, D.; Monzón, S.; Trigano, T.; Vía, J.; Artés-Rodríguez, A. Blind analysis of atrial fibrillation electrograms: A sparsity-aware formulation. *Integr. Comput.-Aided Eng.* **2015**, *22*, 71–85. [CrossRef]
14. Luengo, D.; Vía, J.; Monzón, S.; Trigano, T.; Artés-Rodríguez, A. Cross-products LASSO. In Proceedings of the 2013 IEEE International Conference on Acoustics, Speech and Signal Processing, Vancouver, BC, Canada, 26–31 May 2013; pp. 6118–6122.

15. Wang, C.; Liu, J.; Sun, J. Compression algorithm for electrocardiograms based on sparse decomposition. *Front. Electr. Electron. Eng. China* **2009**, *4*, 10–14. [CrossRef]
16. Mailhé, B.; Gribonval, R.; Bimbot, F.; Lemay, M.; Vandergheynst, P.; Vesin, J.M. Dictionary learning for the sparse modelling of atrial fibrillation in ECG signals. In Proceedings of the 2009 IEEE International Conference on Acoustics, Speech and Signal Processing, Taipei, Taiwan, 19–24 April 2009; pp. 465–468.
17. Polania, L.F.; Barner, K.E. Multi-scale dictionary learning for compressive sensing ECG. In Proceedings of the IEEE Digital Signal Processing and Signal Processing Education Meeting (DSP/SPE), Napa, CA, USA, 11–14 August 2013; pp. 36–41.
18. Fira, M.; Goras, L.; Barabasa, C.; Cleju, N. ECG compressed sensing based on classification in compressed space and specified dictionaries. In Proceedings of the 19th European Signal Processing Conference (EUSIPCO), Barcelona, Spain, 29 August–2 September 2011; pp. 1573–1577.
19. Fira, M.; Goras, L.; Barabasa, C. Reconstruction of compressed sensed ECG signals using patient specific dictionaries. In Proceedings of the International Symposium on Signals, Circuits and Systems ISSCS2013, Iasi, Romania, 11–12 July 2013; pp. 1–4.
20. Fira, M.; Goras, L. On projection matrices and dictionaries in ECG compressive sensing— A comparative study. In Proceedings of the 12th Symposium on Neural Network Applications in Electrical Engineering (NEUREL), Belgrade, Serbia, 25–27 November 2014; pp. 3–8.
21. Trigano, T.; Shevtsov, I.; Luengo, D. CoSA: An accelerated ISTA algorithm for dictionaries based on translated waveforms. *Signal Process.* **2017**, *139*, 131–135. [CrossRef]
22. Luengo, D.; Meltzer, D.; Trigano, T. Sparse ECG Representation with a Multi-Scale Dictionary Derived from Real-World Signals. In Proceedings of the 41st International Conference on Telecommunications and Signal Processing (TSP), Athens, Greece, 4–6 July 2018; pp. 1–5.
23. Satija, U.; Ramkumar, B.; Manikandan, M.S. Noise-aware dictionary-learning-based sparse representation framework for detection and removal of single and combined noises from ECG signal. *Healthc. Technol. Lett.* **2017**, *4*, 2–12. [CrossRef] [PubMed]
24. Faust, O.; Acharya, U.R.; Ma, J.; Min, L.C.; Tamura, T. Compressed sampling for heart rate monitoring. *Comput. Methods Prog. Biomed.* **2012**, *108*, 1191–1198. [CrossRef] [PubMed]
25. Whitaker, B.M.; Rizwan, M.; Aydemir, V.B.; Rehg, J.M.; Anderson, D.V. AF classification from ECG recording using feature ensemble and sparse coding. *Computing* **2017**, *44*, 1.
26. McSharry, P.E.; Clifford, G.D.; Tarassenko, L.; Smith, L.A. A dynamical model for generating synthetic electrocardiogram signals. *IEEE Trans. Biomed. Eng.* **2003**, *50*, 289–294. [CrossRef] [PubMed]
27. Goldberger, A.; Amaral, L.A.; Glass, L.; Hausdorff, J.M.; Ivanov, P.C.; Mark, R.G.; Mietus, J.E.; Moody, G.B.; Peng, C.K.; Stanley, H.E. Physiobank, physiotoolkit, and physionet. *Circulation* **2000**, *101*, e215–e220. [CrossRef] [PubMed]
28. Hu, X.; Liu, J.; Wang, J.; Xiao, Z. Detection of onset and offset of QRS complex based a modified triangle morphology. In *Frontier and Future Development of Information Technology in Medicine and Education*; Springer: Berlin, Germany, 2014; pp. 2893–2901.
29. Bousseljot, R.; Kreiseler, D.; Schnabel, A. Nutzung der EKG-Signaldatenbank CARDIODAT der PTB über das Internet. *Biomed. Tech./Biomed. Eng.* **1995**, *40*, 317–318. [CrossRef]
30. Israel, S.; Irvine, J.M.; Cheng, A.; Wiederhold, M.D.; Wiederhold, B.K. ECG to identify individuals. *Pattern Recognit.* **2005**, *38*, 133–142. [CrossRef]
31. Gustafsson, F. Determining the initial states in forward–backward filtering. *IEEE Trans. Signal Process.* **1996**, *44*, 988–992. [CrossRef]
32. Pan, J.; Tompkins, W.J. A real-time QRS detection algorithm. *IEEE Trans. Biomed. Eng.* **1985**, *32*, 230–236. [CrossRef] [PubMed]
33. Oppenheim, A.V.; Schafer, R.W. *Discrete-Time Signal Processing*; Pearson Education: London, UK, 2014.
34. Proakis, J.G. *Digital Communications*; McGraw-Hill: New York, NY, USA, 1995.

applied sciences

MDPI

Article

Optimized High Resolution 3D Dense-U-Net Network for Brain and Spine Segmentation [†]

Martin Kolařík [1,*], Radim Burget [1], Václav Uher [1], Kamil Říha [1] and Malay Kishore Dutta [2]

[1] Department of Telecommunications, Brno University of Technology, Brno 616 00, Czech Republic; burgetrm@feec.vutbr.cz (R.B.); xuherv00@stud.feec.vutbr.cz (V.U.); rihak@feec.vutbr.cz (K.Ř.)

[2] Centre for Advanced Studies, Dr. A.P.J. Abdul Kalam Technical University, Lucknow 226031, India; malaykishoredutta@gmail.com

[*] Correspondence: martin.kolarik@vutbr.cz or xkolar54@stud.feec.vutbr.cz

[†] This paper is an extended version of our paper published in 3D Dense-U-Net for MRI Brain Tissue Segmentation Published in 2018 41st International Conference on Telecommunications and Signal Processing (TSP).

Received: 7 November 2018; Accepted: 17 January 2019; Published: 25 January 2019

Abstract: The 3D image segmentation is the process of partitioning a digital 3D volumes into multiple segments. This paper presents a fully automatic method for high resolution 3D volumetric segmentation of medical image data using modern supervised deep learning approach. We introduce 3D Dense-U-Net neural network architecture implementing densely connected layers. It has been optimized for graphic process unit accelerated high resolution image processing on currently available hardware (Nvidia GTX 1080ti). The method has been evaluated on MRI brain 3D volumetric dataset and CT thoracic scan dataset for spine segmentation. In contrast with many previous methods, our approach is capable of precise segmentation of the input image data in the original resolution, without any pre-processing of the input image. It can process image data in 3D and has achieved accuracy of 99.72% on MRI brain dataset, which outperformed results achieved by human expert. On lumbar and thoracic vertebrae CT dataset it has achieved the accuracy of 99.80%. The architecture proposed in this paper can also be easily applied to any task already using U-Net network as a segmentation algorithm to enhance its results. Complete source code was released online under open-source license.

Keywords: 3D segmentation; brain; deep learning; neural network; open-source; semantic segmentation; spine; u-net

1. Introduction

Image segmentation is an important process of automated image processing based on the principle of partitioning the input image into areas sharing common features and therefore extract the information that the input image contains. The segmentation itself can be described as a method for labelling each pixel, or in the case of 3D data each voxel, with a corresponding class and is used nowadays as one of the basic image processing method for understanding the content of the input image in many areas of the computer vision.

Together with a rising availability of modern medical image scanning systems such as the magnetic resonance imaging (MRI) or computed tomography (CT) comes the need for automated processing of the scanned data. Evaluation of the results gathered by these scanning methods is usually done by hand by doctors and can be a repetitive and time–consuming task even for an experienced radiologist. The automation of this process is therefore very valuable and can help doctors to determine the correct diagnose faster when they are presented with precisely segmented scanned data within few seconds.

Problem with automatic segmentation of any tissue in medicine is that the method must be reliable and at least as precise as when doctor would have done the same task by hand.

The majority of work in medical image data segmentation focuses on segmenting abnormal tissue regions to determine correct diagnose and the progress of cancer tumours. The problem we are addressing with this paper is the semantic segmentation and subsequent modelling of 3D volumetric segmentations of brain and spine of the patient from MRI and CT scans. Example of MRI and CT scan slices can be seen in Figure 1. Current technology in additive manufacturing and virtual reality brings the doctors new possibilities in examining the patient before operation in high level of detail. To help automate the process of creating accurate 3D models of different parts of human body we propose an optimized neural network architecture evaluated on both MRI and CT images of soft and bone tissue capable of processing data in its original resolution and accelerated on graphical process unit (GPU) for faster parallel computation.

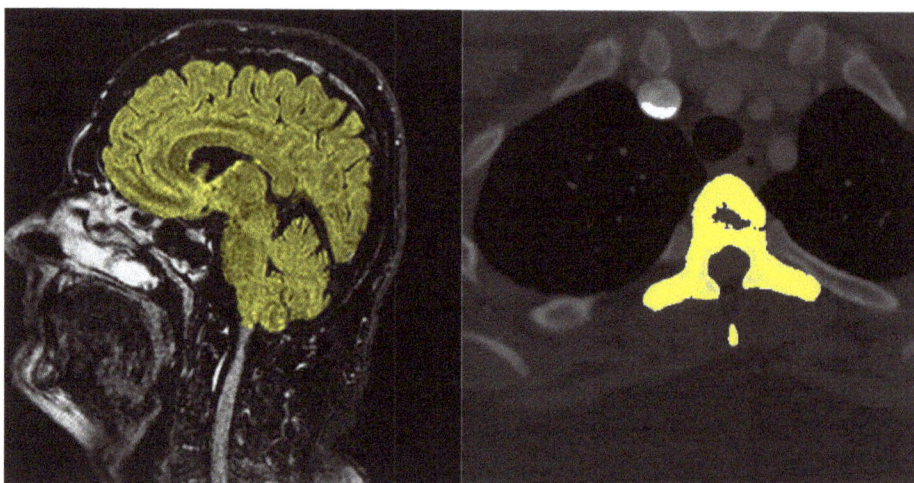

Figure 1. Example of MRI sagittal brain scan slice (**left**) and CT transversal thoracic scan slice (**right**)—tissue segmented with our system is highlighted in yellow.

The Paper is structured into four main sections—Introduction, Materials and Methods, Results and Conclusions. We also added Appendix A. in which we describe how to implement our published source code to segment image data other than the datasets used in this paper. Section 1 covers Introduction and Related works in the field of segmentation in general and also in focus with segmenting medical image data. We also describe specifics of segmenting the brain and spine image data. In the Section 2—Materials and Methods, we describe used datasets, neural network architectures and training process. This section provides all the information needed in order to replicate our experiment. In the Section 3, Results, we publish the results of our experiment measured in five different metrics and also provide visualisation of segmented 3D models of brain and spine created by our algorithm. In the Section 4, Conclusions, we summarize the results of this paper and provide future ways where we want to aim our work in medical image segmentation using deep densely conected neural networks.

1.1. Background and Related Work

1.1.1. Deep Learning Segmentation Methods

Current state of the art methods for segmenting image data are deep learning based. A comparison of most currently used network architectures is described in [1]. Shelhamer et al. [2] proposed the first fully convolutional architecture for segmentation. This work laid foundations for most currently used segmentation neural network architectures, because using only convolutional layers made a significant rise in accuracy. However, this architecture needs a large dataset to be fully trained and therefore it is not usually suitable for medical image segmentation. Using only convolutional layers combined with skip connections became one of the most popular architectures the autoencoder type of network called U-Net [3]. It consists of downsampling, bridge and upsampling blocks and is widely used for segmenting medical data thanks to its ability to be properly trained using little data comparing to its competitors such as [2]. Development in the area of classification networks using data interconnections from coarse to fine layers of the network as in residual networks [4] and followed by densely connected networks [5] led to application of this principle also in segmentation networks. Popular segmentation architecture using these principles is Tiramisu network [6]. All cited architectures are designed for 2D data. Reason for 2D data processing is that these networks were primarily published to be used on general image data and also because the training is implemented for GPU. GPU is used for parallel data processing and can achieve 20 or more times faster computation, but with limited memory use. With the latest advancements in GPU technology we are able to train larger architectures due to increased GPU RAM memory and utilize information in the third dimension for higher accuracy, because most medical data that are used in clinical practice consist of 3D image volumes.

Implementation of U-Net network for 3D data processing was done in [7] and achieved higher accuracy on testing data than original U-Net. Another design of 3D segmentation U-Net type network was done in [8]. Combining 2D and 3D data processing in a hybrid densely connected neural network architecture was done in [9] These methods however are designed to process medical images in lower resolution and therefore are not suitable for processing high resolution thoracic and other upper body medical images. The implementation and evaluation of residual and dense interconnections in the 3D U-Net segmentation model is the main proposal of this paper. We tested both Residual-U-Net and Dense-U-Net architectures and optimized them for medical image processing in 3D.

In contrast with previously mentioned architectures, our goal was to design U-Net architecture with densely connected layers for 3D data processing optimized for processing data in high resolution and compare its accuracy with the results of the original U-Net implementation and U-Net with added residual connections.

1.1.2. Brain Segmentation

The problem of automatic brain tissue segmentation has been very well explored before deep learning algorithms became a standard for semantic segmentation. Despotović et al. [10] covers an overview of older methods such as thresholding, clustering or some form of simpler machine learning algorithms. These methods also rely heavily on image preprocessing as in [11] therefore they are not automatic. One of quite frequently used preprocessing method is some form of skull-stripping algorithm, which removes the bone tissue of skull from the input picture before processing as in [12]. One of the most notable works in the brain segmentation field is the FMRIB's Automated Segmentation Tool (FSL) [13]. This method is based on a hidden Markov random field model and an associated Expectation-Maximization algorithm. We used this tool to evaluate our brain segmentation results to a method that does not use deep neural networks. Since deep learning methods achieve higher accuracy even without any preprocessing of input image, most of the ongoing research including this paper is now using some form of deep neural network for brain segmentation. An overview of the current state-of-the-art in the field of brain segmentation using deep learning can be found in [14]. This is an extended version of our paper 3D Dense-U-Net for MRI Brain Tissue Segmentation which was focused only on brain segmentation [15].

1.1.3. Spine Segmentation

Segmentation and volumetric 3D modelling of individual vertebrae or the complete human spine is an important task for surgeon pre-operation preparation. Combined with 3D printing or virtual reality systems for spine model examination, the surgeon gets much deeper understanding of the patient's problem. As the input images are usually thoracic scans, simple methods like thresholding are unable to distinguish between all bones in the thoracic region and spine segmentation is a difficult task. Spine segmentation is a challenging problem also because the input images contain less distinctive features compared to brain segmentation. Also when scanning younger children with not fully ossified bones, the contrast is very low and the spine tissue is not easily distinguishable from surrounding tissues.

Recent work using deep learning U-Net architecture [16] achieved highly precise results and by combining more neural network architectures in a chain even a higher precision can be achieved [17].

2. Data and Segmentation Methods

This section covers overview and implementation details of used segmentation architectures. Subsection dataset contains all the information about data we used for this paper and their preparation process.

2.1. Dataset

Modern convolutional deep learning architectures require large and properly annotated datasets divided into three parts—training, validation and testing data. Training is the part of the dataset that is used for learning the feature representation of the input data. Validation data are used during training process to control the progress of training accuracy. Testing data are used for validating accuracy of the algorithm after the training process. They are not used during the training process and should represent the production data which the algorithm will be used for.

2.1.1. Medical Image Data Formats

Processing 3D medical image data formats includes converting the provided input data from labelled medical data formats into simple stack of images and can be a confusing task for someone who does not know any details of different medical formats. Medical image represents an internal structure of an anatomic region in the form of an array of elements called pixels in 2D or voxels in 3D. It is a result of a sampling/reconstruction process mapping numerical values to positions in the space. Medical image data formats include common information such as pixel depth, metadata, pixel data and photometric interpretation. These formats also store more than one 3D data scans and therefore can be used for 4D image processing. All data used in this paper consist of only one 3D scan per patient [18].

Common medical data formats include:

- NRRD—Nearly raw raster data, general medical image data format, data suffixes differ from versions with attached header (.nrrd) and detached header (.raw/.mhd) where metadata are stored separately from the image data.
- Nifti—Neuroimaging Informatics Technology Initiative, this format is usually used for brain imaging data and uses suffix .nii.
- Dicom—Digital Imaging and Communications in Medicine, general image data format and most commonly used for different medical image data, uses suffix .dcm.
- Analyze—Analyze 7.5 format which uses suffixes .img/.hdr, also a detached header format.

An important part of processing medical images is their representation in the 3D vector space. This can be quite challenging and during this process loss of part of the image information can occur. During our data preparation process we converted NRRD 3D scans into image slices and lost the information about image spacing in the third axis. This information had to be recovered and added in order to generate 3D models of the segmented tissue.

2.1.2. Brain Dataset

Dataset consisted of 22 MRI T1 weighted brain scans from different patients and each scan contained 257 sagittal slices of human brain. The pictures had resolution 400×400 pixels and were provided in PNG format. MRI data were provided by the Department of Radiology from The University Hospital Brno and all original slices were labelled by two independent human experts resulting into two sets of ground truth masks suitable for semantic segmentation labeling all pixels either as brain tissue (white) or non-brain tissue (black). Patients involved were in the age of 35–55 years old both male and female and were in good physical condition. Data have been anonymized and approved for research and scientific purposes.

The first expert labelled the data very precisely and accurately and these labels are used as a reference data for training and evaluation. The second expert labelled the data as it is done regularly on everyday basis in medical praxis. These data are used for comparing the algorithm accuracy to accuracy of a human expert segmentation. An example of reference ground truth mask and mask labelled by our system can be seen in Figure 2.

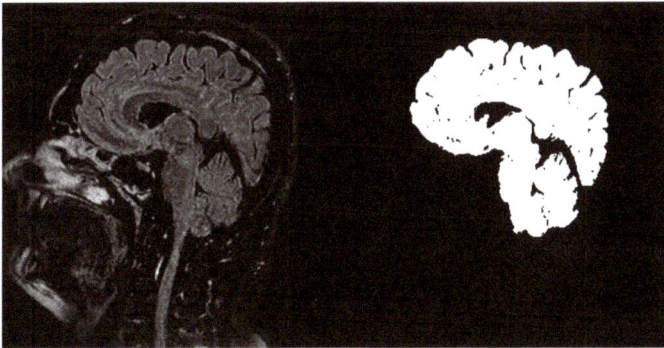

Figure 2. Example of brain dataset sagittal image slice (**left**) and according ground truth mask (**right**).

2.1.3. Spine Dataset

Spine dataset consists of 10 CT scans of different patients in the age 16–35 years old. The pictures had resolution 512×512 pixels and were provided in NRRD format. Number of slices in each scan was in range from 520 to 600 slices in the third dimension. Scans cover lumbar and thoracic spine region and were acquired without intravenous contrast. Slice thickness is 1 mm per slice and the in-plane resolution is between 0.31 and 0.45 mm. The data have been acquired at the Department of Radiological Sciences, University of California, Irvine, School of Medicine and scanners used include Philips or Siemens multidetector. Data were published as a part of 2014 CSI workshop challenge of the web http://spineweb.digitalimaginggroup.ca. Dataset can be used for development, training and evaluation of spine segmentation algorithms. Image data are provided in NRRD format. An example image slice and according ground truth mask can be seen in Figure 3.

Ground truth segmentation masks have been semi-automatically segmented and verified for complete thoracic and lumbar vertebrae for each scan. Segmentation masks are also stored in NRRD filed and originally each vertebra had assigned different label. The first vertebra was labelled as 100, the second as 200 and so on. For the semantic binary volumetric segmentation task the masks have been thresholded to greyscale 8bit PNG files and had format where vertebrae tissue was assigned with a value of 255 and non-vertebrae tissue with the value 0 [19].

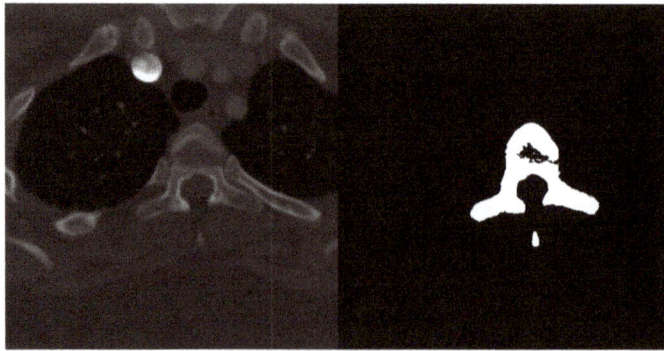

Figure 3. Example transversal image slice from spine dataset (**left**) with according ground truth mask (**right**).

2.1.4. Denoising Data Preparation

After initial segmentation experiments with the spine dataset we found it is a much more challenging task than the brain volumetric segmentation and training the network only on the dataset was not sufficient for highly precise segmentation. Instead of data augmentation, which for 3D algorithms greatly expands computation time, we used denoising autoencoder pre-training. Principle with this method lies in the fact that we add noise to the training data and let the neural network to learn the representation between noised and denoised data and therefore extracting features of the input image and proper learning of deep layers. An example of original training image, image with added noise and denoised image by our network can be seen in Figure 4.

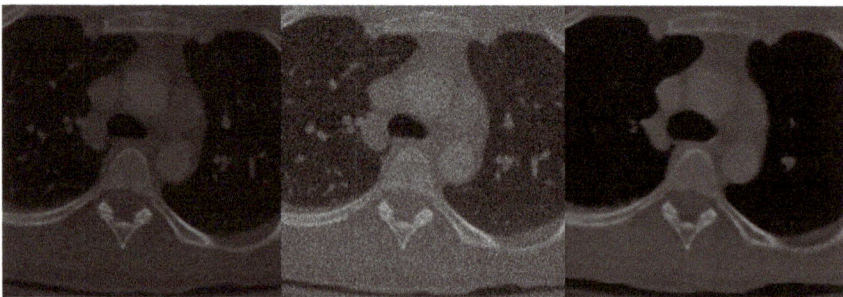

Figure 4. Example of data before adding linear noise (**left**), with added linear noise (**middle**) and after denoising by neural network (**right**).

Training dataset consisted of normal training data with added linear 30 percent noise with normal distribution. The network was trained to denoise data into its original form. As seen in Figure 4 the network learned the representation properly, denoised image only lacks higher level of details [20].

2.1.5. Training and Testing Data Preparation

Brain dataset consisted of 3D images of brain in resolution $400 \times 400 \times 257$ voxels and spine dataset consisted of 3D thoracic images in resolution $512 \times 512 \times 552$ (some scans have up to 600 images in third axis) voxels. Images of this resolution would not fit into GPU RAM memory (Nvidia GTX 1080ti with 11 GB of RAM) and therefore we had to train our network on smaller batches containing 16 images of brain and for spine only 8 images each. To fully utilize information of 3D data we prepared the input data as overlapping batches. The overlapping technique is used to ensure that the algorithm can utilize as much information over the third axis as possible. If we divided the dataset into batches of 16 images without overlapping, the training and predictions on the first and the last slice of every batch would not be sufficiently accurate. The algorithm would not have the information of the surrounding slices. Using the batch overlapping technique is ensured that during training every voxel can be analysed using all its surrounding information. The same principle applies for the output prediction of the algorithm. An example can be seen in Figure 5.

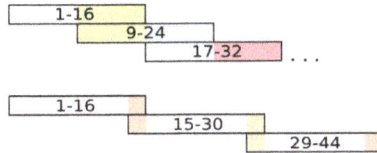

Figure 5. Example of training (**upper**) and testing data batch overlapping (**lower**) for brain dataset. Numbers show which slices of the scan each batch contains.

For brain dataset we used 21 scans as a training data, ten percent of that data served for validation and was automatically separated from training dataset by framework Keras [21]. Last remaining 22nd brain scan was used for testing. Due to the fact that approximately 15 slices on both ends of every subject consisted of only non-brain tissue and therefore contained very little information value, we discarded the last slice of every subject and used only 256 slices from each scan for training so the final number is divisible by 16. After the benchmarking process of evaluation of accuracy of used algorithms we trained the fine-tuned Dense-U-Net network three times to ensure generalization of our results using 3 fold cross validation. Dataset has been every round divided into portions of 21 scans for training and 1 for testing.

We wanted to use the same overlapping data technique for outcome predictions from the system. To shorten the time the systems needs to predict the results, we overlapped only 2 pixels on each end of every batch. Prediction then consists of central 14 slices of each batch. This results in 252 testing images used for prediction. An example can be seen in Figure 5.

The spine dataset consisted of ten 3D scanned thoracic CT images. We used nine for training with ten percent of the training dataset used for validation. We used the last remaining scan for testing. After the benchmarking process of evaluation of accuracy of used algorithms we trained the fine-tuned Dense-U-Net network three times to ensure generalization of our results using 3 fold cross validation. Dataset has been every round divided into portions of 9 scans for training and 1 for testing. As the spine segmentation proved to be a more dificult problem than the brain segmentation, we used the same overlapping technique for training and testing datasets. Therefore the neighbouring batches always overlapped four slice each and for prediction on testing dataset we could use inner four slices in each batch and utilize as much information over the third axis as possible.

2.2. Methodology

2.2.1. Neural Network Architectures

In this paper we propose 3D Dense-U-Net network which is based on original U-Net implementation [3] and on 3D U-Net version [8] but with added interconnections between layers processing the same feature size, its model is in Figure 6.

We wanted to test the idea of residual and dense interconnections to the segmentation U-net type networks. This principle is used commonly in classification deep neural networks. Using this principle we created and tested the Residual-U-Net and Dense-U-Net networks for 3D data processing which are based on original U-Net and 3D U-Net architectures. Results proved that using residual and dense interconnections can help achieve much better results, but is also computationally expensive both in the terms of time and required GPU memory. The resulting architecture is optimised for high resolution image processing and can be used on GPU devices with 11 GB of RAM or more. Other work using interconnections in the U-Net [9] use different architecture and use low resolution image processing only. Difference of this work is in the fact that it is capable of processing medical image data in original resolution and achieve higher accuracy than the standard U-Net or 3D U-Net. The interconnections help the network to achieve faster learning curve and obtain higher level of details.

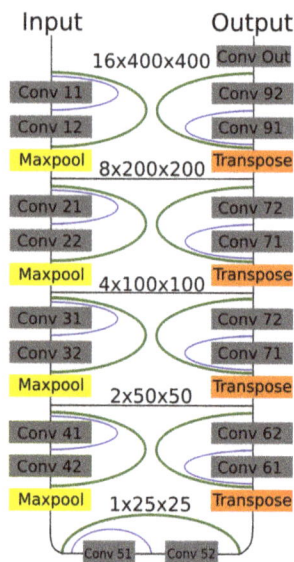

Figure 6. Dense-U-Net network model. Residual interconnections are in green color, dense interconnections in blue. Feature sizes are valid for batches of images from brain dataset.

It is an autoencoder type architecture [3] with 4 down-sampling and 4 up-sampling blocks which are connected by a bridge block (in the most lower part of the network). Feature size is halved after passing through every down-sampling block by a maxpooling layer at its bottom. A number of neurons in each layer can be found in Table 1. At the beginning of every up-sampling block the feature size is doubled using transposed convolution layer as described by [22] with stride of size 2 for each dimension. The feature downsampling and upsampling change for four times can only be done fully for the brain dataset, because it has batch containing 16 images. For spine data we have to change the pooling and strides in the most lower parts of the network to the core size of [1,2,2] as can be seen in Table 2. Passing information through interconnections is possible by using zero padding. After every convolution operation we fill the newly computed vector with zeros to its original length and therefore both convolution layers use the same size of the input vector.

Table 1. Number of neurons in each layers.

Input		Output	
Layers	**Neurons**	**Layers**	**Neurons**
11,12	32	91,92	32
21,22	64	81,82	64
31,32	128	71,72	128
41,42	256	61,62	256
51,52	512	51,52	512
Bridge block—Convolutions 51,52—512			

Our results in the next section prove, that the 3D Dense-U-Net network exceeds the results of other used architectures. In our experiment we compared the results of original U-Net and Residual-U-Net with added only residual interconnection. On brain dataset we used all three architectures in 3D and their corresponding 2D version to examine whether the 3D models will have better performance. Basic U-Net in 3D and in 2D is an implementation of [3] and Residual-U-Net adds residual interconnections to it (green in Figure 6). Especially Residual-U-Net network had very good results taking into account its number of parameters and can be recommended for application with less computation resources.

Table 2. Parameters of network layers—their convolution kernel size, strides and activation function.

Downsampling Block			
Type	Kernel (Pool) Size	Strides	Activation
Convolution 3D	3,3,3	0,0,0	Relu
Convolution 3D	3,3,3	0,0,0	Relu
Maxpooling 3D	2(1),2,2	2,2,2	-

Upsampling Block Block			
Type	Kernel (Pool) Size	Strides	Activation
Transpose Conv.3D	2,2,2	2(1),2,2	-
Convolution 3D	3,3,3	0,0,0	Relu
Convolution 3D	3,3,3	0,0,0	Relu

Bridge Block			
Type	Kernel (Pool) Size	Strides	Activation
Convolution 3D	3,3,3	0,0,0	Relu
Convolution 3D	3,3,3	0,0,0	Relu

Output Convolution			
Type	Kernel (Pool) Size	Strides	Activation
Convolution 3D	1,1,1	0,0,0	Sigmoid

2.2.2. Supervised Denoise Pretraining

Rather than to use random values for neural network weights initialization, we pretrained the network first as a denoising autoencoder. The motivation behind this is to help to propagate gradient during the training, help to not get stuck in local minima and increase overall network stability. First some additional noise was added to the input image and then the network was trained to reconstruct the original image. Preparation of the data for this process is described in more detail in Section 2.1.4. In order to make the experiment better comparable over all the examined architectures, 50 epochs limit was used for all the trainings. It resulted in 84.47 percent pixel accuracy in reconstructing the original image (see Figure 4—the network was able to extract the input image features quite successfully). Resulting weights were then used for initialization the fine-tuning phase where 3D Dense-U-Net network was trained using spine dataset. Not only the network started first epoch with validation accuracy over 80 percent (randomly initialized weights have under 30 percent accuracy during first epoch) but also helped the deeper layers to be trained correctly and prevent so called gradient vanishing, which would have significant impact on neural network training capabilities.

2.2.3. Training Process

For both datasets all the considered architectures were evaluated based on their computation time and accuracy (see Results in Tables 3–7). They were trained for 50 epochs with the same setting of hyper-parameters. Details regarding hyper-parameter settings can be found in the next Section 2.2.4.

The training of the networks was done in two phases—(1) benchmark phase and (2) fine-tuning phase. First we wanted to compare all the used networks for benchmark purposes to examine whether the Dense-U-Net can achieve the highest accuracy among all the tested architectures. Due to the computation time needed for training we decided to evaluate the tested architectures after 50 epochs of training. When we verified that the Dense-U-Net network performs the best, the limit of 50 epochs was removed and during the fine tuning phase we tried to achieve the best possible accuracy.

Thus the fine-tuned Dense-U-Net network was then trained for 99 epochs using brain dataset. The hyper parameters used were exactly the same as used for the benchmarking phase. This was repeated also for spine dataset using 3D Dense-U-Net network with weights initialized from denoise pre-training. This network was trained as well for 99 epochs. The learning rate was set higher to 10^{-4} and the decay to 1.99×10^{-6}. The reason of that was we wanted to give the network higher learning rate at the beginning, which helped to escape from pre-trained auto-encoder local minimum and to continue on with searching for global minimum.

2.2.4. Implementation Details

Complete source code used in this paper is available at [23]. All neural networks architectures were trained using Keras framework [21] using Tensorflow backend. Training was done on Nvidia GTX 1080ti graphics card with 11 GB of memory using CUDA 8.0. As an inspiration for the first U-Net model we used [24] open-source project. Our basic U-Net architecture uses 3D data processing layers. All networks use a binary cross-entropy as a loss function and Adam optimizer with parameters learning_rate equal to 10^{-5}, *beta*1 to 0.9, *beta*2 to 0.999, *epsilon* to 10^{-8} and *decay* to 1.99×10^{-7}. Using decay parameter we lower the learning rate parameter each epoch by constant value, which helps to fine-tune the network.

Residual-U-Net and Dense-U-Net architectures were designed by adding interconnections to the basic 3D U-Net architecture. Residual-U-Net uses only interconnection over whole down or up-sampling blocks as can be seen in Figure 6 and residual interconnection is denoted as green links. Dense-U-Net uses both residual and dense interconnections (blue in Figure 6) which pass unprocessed information to the middle layer of down and up-sampling blocks.

Network uses as input data values which are in range between 0 and 1. For this reason the input pixel or voxel values, which were encoded as 8-bit monochromatic images in png format, i.e., in range [0,255], were normalized to the range 0.0 to 1.0.

As we used sigmoid activation on output network layer, the output of the network is not labeled by just discrete values but with continuous values in range from 0 to 1. For brain dataset we used as only post-processing on predicted data thresholding. All pixels with value lesser than 0.5 were labelled as 0 and with value greater than 1. However the results on spine dataset required thresholding level to be set to 0.9 and above labelled positive, under the level negative. This has to be done to ensure least amount of artifacts in the output segmentation. Even after thresholding usually one middle sized artifact in the upper thoracic region stayed in the segmented image. To ensure clean segmentation suitable for 3D modeling we had to threshold the model to remove all stand alone objects smaller than 7500 voxels. This ensured the quality output without any artifacts in the segmented image.

3. Results

Five different metrics for evaluation were used so the results can be easily compared to related works. All metrics measured are compared to masks labelled by human expert which are used as ground truth. The first used metrics is pixel accuracy, its equation is (1) where N_{TP} stands for true positive pixels or voxels, N_{TN} true negative, N_{FP} false positive and N_{FN} false negative.

$$A_P(X,Y) = \frac{N_{TP} + N_{TN}}{N_{TP} + N_{TN} + N_{FP} + N_{FN}} \tag{1}$$

$$D(X,Y) = \frac{2*|X \cap Y|}{|X| + |Y|} \tag{2}$$

$$I_{oU}(X,Y) = \frac{|X \cap Y|}{|X \cup Y|} = \frac{|X \cap Y|}{|X| + |Y| - |X \cap Y|} \tag{3}$$

Dice coefficient [25] is expressed by Equation (2) and intersection over union [26], also known as Jaccard index, is expressed by Equation (3). X and Y stand for a set of positive pixels/voxels on first and second compared mask. We used Visceral segmentation tool [26] for computing the results in all metrics.

The training of the segmentation neural networks used in this paper is relatively computationally demanding. For this reason, we compare the resulting architectures also from the point of view of computational time needed for training and computational time needed for prediction. Results can be seen in Table 3. As it is obvious from the table, the computational time increases with the complexity of the model and image data resolution, on the other hand it is in all the cases below one minute during application of the model to data.

Table 3. Segmentation alogorithms comparison based on their computation time when ran on gtx 1080ti GPU.

Segmentation Algorithm	Prediction Time	Training Time [50 epochs]
U-Net—brain dataset	8 s	6 h
U-Net—spine dataset.	23 s	9 h
Res-U-Net—brain dataset	12 s	10 h
Res-U-Net—spine dataset	32 s	16 h
Dense-U-Net—brain dataset	21 s	12 h
Dense-U-Net—spine dataset	43 s	21 h
FSL—brain dataset	3 s	-

3.1. Brain Dataset Results

First all the architectures were evaluated on brain dataset. The considered architectures are Dense-U-Net, Residual-U-Net and basic U-Net network, all were tested in 2D and 3D mode, i.e., six neural network architectures. This phase verified, that 3D Dense-U-Net architecture performs better than the remaining architectures (see Table 4).

Table 4. Comparison of tested U-Net versions on brain dataset in benchmark training phase versus Human expert and FMRIB's Automated Segmentation Tool. Used metrics—pixel accuracy, Dice coefficient, intersection over union, average Hausdorff distance [voxel] and area under ROC curve.

Metric	Dense-U-Net	Res-U-Net	U-Net	Human	FSL
3D Networks					
P.A.	0.99703	0.99662	0.99619	0.99489	0.94289
Dice c.	0.98843	0.98686	0.98514	0.98033	0.79698
I.o.U.	0.97713	0.97407	0.97072	0.96141	0.66248
A.H.D. [voxel]	0.01334	0.01911	0.02427	0.02479	4.58848
A.u.R.C.	0.99439	0.99353	0.99205	0.98325	0.96696
2D Networks					
P.A.	0.99576	0.99639	0.99636	0.99489	0.94289
Dice c.	0.98344	0.98357	0.98574	0.98033	0.79698
I.o.U.	0.96743	0.96768	0.97189	0.96141	0.66248
A.H.D. [voxel]	0.09477	0.09130	0.05632	0.02479	4.58848
A.u.R.C.	0.99647	0.99639	0.99663	0.98325	0.96696

As stated in Section 2.1.5, the performance was evaluated on the last brain scan which was not used for training and therefore our results show reliability of our method on unseen data. We compared the predicted data to mask labelled by a human expert as in standard medical praxis, which are in Table 4 in column labelled "Human" and also to output of the FSL segmentation tool.

Dense-U-Net network proved to give the best performance in all considered metrics in 3D processing mode. There should be stated, that even segmentation using basic U-Net network gave better accuracy than a human expert. When compared to other 2D networks, the Dense-U-Net had better results in all metrics except for average Hausdorff distance. This is most probably caused by the fact that 50 epochs of training was not enough for a network with so many parameters and therefore the network generated more artifacts in the output segmentation than the simpler versions. As can be seen in Figure 7 the methods had problems segmenting the area around nasal cartilage. FSL method results in this case were unusable for medical praxis. Dense-U-Net network had the best results also thanks to the fact, that it was able to successfully segment this area on the MRI scan.

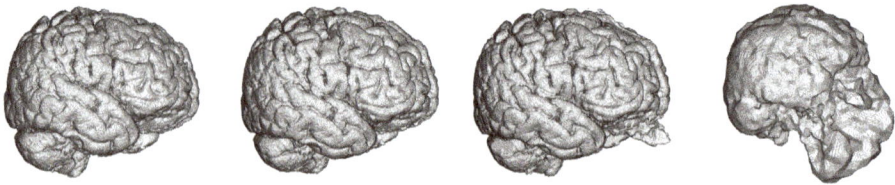

Figure 7. Visualisation of predictions from 3D benchmark brain models and FSL segmentation. From left to right—Dense-U-Net, Residual-U-Net, U-Net, FSL.

After the evaluation in the benchmark phase we trained the 3D Dense-U-Net again, now with 99 epochs, to get the final fine-tuned model. Obtained results are in Table 5.

Table 5. Fine-tuned model trained using 3-folds cross-validation, results includes standard deviation. Used neural network is Dense-U-Net network. Achieved results are compared to another human expert results. For evaluation, several metrics were used: pixel accuracy, Dice coefficient, intersection over union, average Hausdorff distance [voxel] and area under ROC curve.

Metric	Dense-U-Net	Human
P.A.	0.99721 ± 0.00026	0.99510 ± 0.00032
Dice c.	0.98870 ± 0.00134	0.98087 ± 0.00167
I.o.U.	0.97873 ± 0.00165	0.96032 ± 0.00189
A.H.D. [voxel]	0.01302 ± 0.00183	0.02519 ± 0.00342
A.u.R.C.	0.99463 ± 0.00022	0.98413 ± 0.00093

The Dense-U-Net network was trained in the fine-tuning phase (99 epochs) and validated using 3-folds cross-validation. In all five metrics the results are more accurate than a human expert. Output of the segmentation is visualised in Figure 8. It is clearly visible, that the proposed segmentation output overcame the human results and also results of older methods not based on deep neural networks such as the FSL. The network was able to learn the input image features as well as to generalize the brain segmentation problem. All the evaluations were made using data independent to training and the validation data.

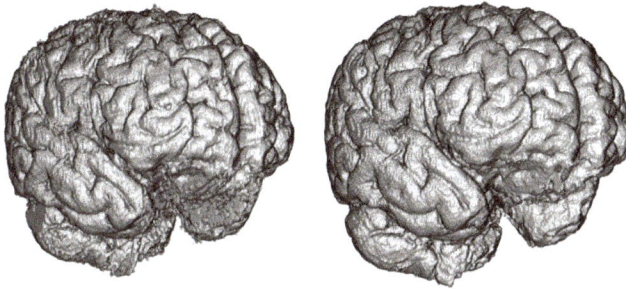

Figure 8. Comparison of ground truth brain model (**left**) and brain model segmented by Dense-U-Net (**right**) after the final phase of training.

3.2. Spine Dataset Results

As the benchmark comparison of tested architectures in 2D vs. 3D versions is done on brain dataset, we trained and evaluated only 3D versions of the networks. A comparison of results achieved during benchmark phase using all three architectures can be seen in Table 6 and visualisation of the segmented spine models can be seen in Figure 9. This dataset does not contain second set of labels done by a human expert and therefore the results cannot be compared with human precision.

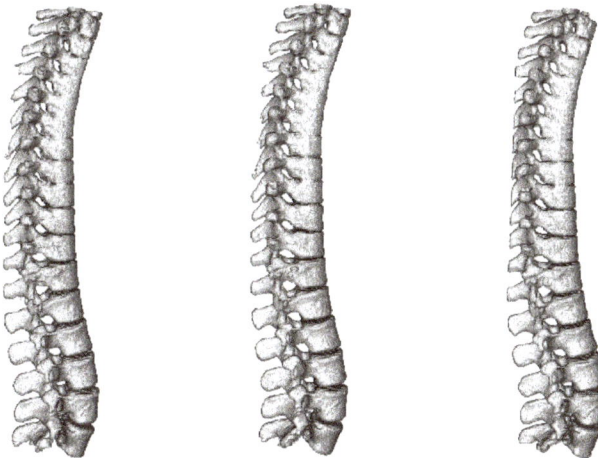

Figure 9. Visualisation of segmented spine models from the benchmark training phase. From left to right—Dense-U-Net, Residual-U-Net, U-Net. Please notice that the abnormal vertebrae adhesions exist also in original ground truth masks as can be seen in Figure 10.

Table 6. Comparison of tested 3D U-Net versions on spine dataset in benchmark training phase. The training was limited by 50 epochs for each. Used metrics for evaluation are: pixel accuracy, Dice coefficient, intersection over union, average Hausdorff distance [voxel] and area under ROC curve.

Metric	Dense-U-Net	Res-U-Net	U-Net
P.A.	0.99732	0.99727	0.99721
Dice c.	0.96784	0.96733	0.96635
I.o.U.	0.93770	0.93672	0.93490
A.H.D. [voxel]	0.21982	0.08754	0.09226
A.u.R.C.	0.98589	0.98539	0.98262

Dense-U-Net network has achieved the highest results in all metrics except for average Hausdorff distance. Reason for this is that the model does not perform well on borders of the image. In the dataset there were labelled only thoracic and lumbar vertebrae, but the CT scans contained also the first cervical vertebra and the network did include it in its segmentation results. You can see a part of the first cervical vertebra on top of the segmented spine in Figure 10.

After accuracy of Dense-U-Net was verified to outperform the other architectures in the benchmark phase, we trained the network in the fine-tuning phase to achieve the best results possible, now using 99 epochs with weights initialized using pre-trained model. 3-folds cross-validation was used for evaluation. Results of the Dense-U-Net network in fine-tuning phase is depicted in Table 7. The visualisation of fine-tuned Dense-U-Net network result can be seen in Figure 10. Please notice that the abnormal vertebrae adhesions exist also in ground truth masks on the model in Figure 10 and therefore it is not a failure of the segmentation algorithm.

Results on testing data which the network had not seen during the training process show, that the algorithm is capable of precise segmentation of human spine on CT images. The network has achieved on the spine dataset even better results in metrics pixel accuracy and in dice coefficient in comparison with volumetric segmentation of MRI brain images. This is a very good result because of much higher difficulty of the volumetric CT spine segmentation problem.

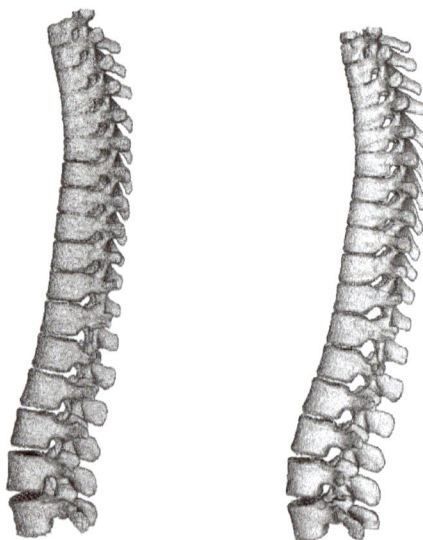

Figure 10. Comparison of ground truth model (**left**) and model segmented by Dense-U-Net network (**right**) after the fine-tuning phase of training. Notice difference on top of the figure—the first cervical vertebra and abnormal vertebrae adhesions that exist also in the original ground truth mask.

Table 7. Segmentation results using Dense-U-Net and spine dataset in the fine-tuning phase. Results includes standard deviation of 3-folds cross-validation. Used metrics are: pixel accuracy, Dice coefficient, intersection over union, average Hausdorff distance [voxel] and area under ROC curve.

Metric	Dense-U-Net
P.A.	0.99805 ± 0.00014
Dice c.	0.97082 ± 0.00292
I.o.U.	0.94332 ± 0.00553
A.H.D.	0.16711 ± 0.11010
A.u.R.C.	0.98627 ± 0.00220

4. Conclusions

In this paper we have proposed 3D Dense-U-Net: a new upgraded U-Net architecture with densely connected layers optimized for high resolution 3D medical image data analysis. We evaluated the performance of the network on two independent datasets. The first dataset is the MRI T1 brain dataset and this network achieved pixel accuracy on testing data 99.72 ± 0.02 percent which exceeded human expert performance done as in standard medical praxis (99.51 ± 0.03 percent). On the second spine dataset the network achieved 99.80 ± 0.01 percent accuracy, which surpassed its results on brain dataset. The network can segment high resolution 3D data in less than one minute using standard PC equipped with Nvidia GTX 1080ti. Using data preparation technique described in Section 2.1.5, we were able to analyse data with a deep neural network for 3D data segmentation using GPU with 11 GB RAM in its original resolution. Many other related segmentation algorithms are designed for data in smaller resolution (because of time or memory demands) and therefore the results are often not usable in practice.

Our approach can easily be applied to any segmentation method already using U-Net architecture. Resulting source-code was released as an open-source and its link is provided in Section 2.2.2. In Appendix A. we provide a manual how to use the source code for other data.

In future we plan to further upgrade data preparation technique so we will be able to train more densely interconnected architectures, plan to evaluate our method on other publicly available datasets and to design a more universal data preparation technique. Also we plan to learn the network using denoising training in a semi-supervised manner to detect novelty in image dataset.

Author Contributions: Data curation, K.Ř.; Formal analysis, M.K.D.; Methodology, M.K., R.B., V.U. and M.K.D.; Software, M.K. and V.U.; Supervision, R.B.; Visualization, K.Ř.; Writing—original draft, M.K.

Funding: This research was funded the Ministry of the Interior of the Czech Republic by the grant VI20172019086, Ministry of Health of the Czech Republic by the grant NV18-08-00459 and the National Sustainability Program under grant LO1401.

Acknowledgments: Research described in this paper was financed by the Ministry of the Interior of the Czech Republic by the grant VI20172019086 and the National Sustainability Program under grant LO1401. For the research, infrastructure of the SIX Center was used. This work was also supported by Ministry of Health of the Czech Republic, grant nr. NV18-08-00459.

Conflicts of Interest: The authors declare no conflict of interest.

Abbreviations

The following abbreviations are used in this manuscript:

GPU	graphic processing unit
MRI	magnetic resonance imaging
CT	computed tomography
NRRD	nearly raw raster data medical image format
FSL	FMRIB's Automated Segmentation Tool

Appendix A

Source code consists of two main files—"data.py" for data loading and preprocessing and "dense-unet.py", which trains the network and segment data and evaluate performance. Depending on the size of the input data you should accordingly modify dimension for resolution and batch depth in data.py. This information has to be changed in dense-unet.py as well. Code is aimed to load slices of .png images located in folders /train, /masks, /test and to create numpy files with encoded images. These files can be used to train the network and make prediction using functions train() and predict() in the dense-unet.py. We recommend using anaconda virtual python environment for execution, the source file for its creation is also included in the github repository.

References

1. Garcia-Garcia, A.; Orts-Escolano, S.; Oprea, S.; Villena-Martinez, V.; Garcia-Rodriguez, J. A review on deep learning techniques applied to semantic segmentation. *arXiv* **2017**, arXiv:1704.06857.
2. Long, J.; Shelhamer, E.; Darrell, T. Fully convolutional networks for semantic segmentation. In Proceedings of the IEEE Conference on Computer Vision and Pattern Recognition, Boston, MA, USA, 7–12 June 2015; pp. 3431–3440.
3. Ronneberger, O.; Fischer, P.; Brox, T. U-net: Convolutional networks for biomedical image segmentation. In *International Conference on Medical Image Computing And Computer-Assisted Intervention*; Springer: Cham, Switzerland, 2015; pp. 234–241.
4. He, K.; Zhang, X.; Ren, S.; Sun, J. Deep residual learning for image recognition. In Proceedings of the IEEE Conference on Computer Vision and Pattern Recognition, Las Vegas, NV, USA, 27–30 June 2016; pp. 770–778.
5. Huang, G.; Liu, Z.; Van Der Maaten, L.; Weinberger, K.Q. Densely Connected Convolutional Networks. In Proceedings of the Computer Vision and Pattern Recognition, Honolulu, HI, USA, 21–26 July 2017; Voume 1, p. 3.
6. Jégou, S.; Drozdzal, M.; Vazquez, D.; Romero, A.; Bengio, Y. The one hundred layers tiramisu: Fully convolutional densenets for semantic segmentation. In Proceedings of the 2017 IEEE Conference on Computer Vision and Pattern Recognition Workshops (CVPRW), Honolulu, HI, USA, 21–26 July 2017; pp. 1175–1183.
7. Milletari, F.; Navab, N.; Ahmadi, S. V-Net: Fully Convolutional Neural Networks for Volumetric Medical Image Segmentation. In Proceedings of the 2016 Fourth International Conference on 3D Vision (3DV), Stanford, CA, USA, 25–28 October 2016; pp. 565–571.
8. Çiçek, Ö.; Abdulkadir, A.; Lienkamp, S.S.; Brox, T.; Ronneberger, O. 3D U-Net: Learning Dense Volumetric Segmentation from Sparse Annotation. *arXiv* **2016**, arXiv:1606.06650.
9. Li, X.; Chen, H.; Qi, X.; Dou, Q.; Fu, C.W.; Heng, P.A. H-DenseUNet: Hybrid densely connected UNet for liver and liver tumor segmentation from CT volumes. *arXiv* **2017**, arXiv:1709.07330.
10. Despotović, I.; Goossens, B.; Philips, W. MRI segmentation of the human brain: challenges, methods, and applications. *Comput. Math. Methods Med.* **2015**, *2015*, 450341. [CrossRef] [PubMed]
11. Uher, V.; Burget, R. Automatic 3D segmentation of human brain images using data-mining techniques. In Proceedings of the 2012 35th International Conference on Telecommunications and Signal Processing (TSP), Prague, Czech Republic, 3–4 July 2012; pp. 578–580.
12. Uher, V.; Burget, R.; Masek, J.; Dutta, M.K. 3D brain tissue selection and segmentation from MRI. In Proceedings of the 2013 36th International Conference on Telecommunications and Signal Processing (TSP), Rome, Italy, 2–4 July 2013; pp. 839–842.
13. Zhang, Y.; Brady, M.; Smith, S. Segmentation of brain MR images through a hidden Markov random field model and the expectation-maximization algorithm. *IEEE Trans. Med. Imaging* **2001**, *20*, 45–57. [CrossRef] [PubMed]
14. Akkus, Z.; Galimzianova, A.; Hoogi, A.; Rubin, D.L.; Erickson, B.J. Deep learning for brain MRI segmentation: State of the art and future directions. *J. Dig. Imaging* **2017**, *30*, 449–459. [CrossRef] [PubMed]
15. Kolařík, M.; Burget, R.; Uher, V.; Dutta, M.K. 3D Dense-U-Net for MRI Brain Tissue Segmentation. In Proceedings of the IEEE 2018 41st International Conference on Telecommunications and Signal Processing (TSP), Athens, Greece, 4–6 July 2018; pp. 1–4.
16. Lu, J.T.; Pedemonte, S.; Bizzo, B.; Doyle, S.; Andriole, K.P.; Michalski, M.H.; Gonzalez, R.G.; Pomerantz, S.R. DeepSPINE: Automated Lumbar Vertebral Segmentation, Disc-level Designation, and Spinal Stenosis Grading Using Deep Learning. *arXiv* **2018**, arXiv:1807.10215.
17. Whitehead, W.; Moran, S.; Gaonkar, B.; Macyszyn, L.; Iyer, S. A deep learning approach to spine segmentation using a feed-forward chain of pixel-wise convolutional networks. In Proceedings of the 2018 IEEE 15th International Symposium on Biomedical Imaging (ISBI 2018), Washington, DC, USA, 4–7 April 2018; pp. 868–871.
18. Larobina, M.; Murino, L. Medical image file formats. *J. Dig. Imaging* **2014**, *27*, 200–206. [CrossRef]
19. Yao, J.; Burns, J.E.; Munoz, H.; Summers, R.M. Detection of vertebral body fractures based on cortical shell unwrapping. In *International Conference on Medical Image Computing and Computer-Assisted Intervention*; Springer: Berlin/Heidelberg, Germnay, 2012; pp. 509–516.

20. Vincent, P.; Larochelle, H.; Lajoie, I.; Bengio, Y.; Manzagol, P.A. Stacked denoising autoencoders: Learning useful representations in a deep network with a local denoising criterion. *J. Mach. Learn. Res.* **2010**, *11*, 3371–3408.

21. Chollet, F. Keras. 2015. Available online: https://keras.io (accessed on 8 November 2018).

22. Noh, H.; Hong, S.; Han, B. Learning deconvolution network for semantic segmentation. In Proceedings of the Proceedings of the IEEE International Conference on Computer Vision, Washington, DC, USA, 7–13 December 2015; pp. 1520–1528.

23. Kolařík, M. mrkolarik/3D-brain-segmentation. Available online: https://github.com/mrkolarik/3D-brain-segmentation (accessed on 8 November 2018)

24. Jocić, M. Deep Learning Tutorial for Kaggle Ultrasound Nerve Segmentation Competition, Using Keras. 2017. Available online: https://github.com/jocicmarko/ultrasound-nerve-segmentation (accessed on 1 May 2018).

25. Zou, K.H.; Warfield, S.K.; Bharatha, A.; Tempany, C.M.; Kaus, M.R.; Haker, S.J.; Wells, W.M., III; Jolesz, F.A.; Kikinis, R. Statistical validation of image segmentation quality based on a spatial overlap index1: Scientific reports. *Acad. Radiol.* **2004**, *11*, 178–189. [CrossRef]

26. Taha, A.A.; Hanbury, A. Metrics for evaluating 3D medical image segmentation: Analysis, selection, and tool. *BMC Med. Imaging* **2015**, *15*, 29. [CrossRef] [PubMed]

applied
sciences

MDPI

Article

Retrieval of Similar Evolution Patterns from Satellite Image Time Series [†]

Anamaria Radoi * and Corneliu Burileanu

Faculty of Electronics, Telecommunications and Information Technology, University Politehnica of Bucharest, Bulevardul Iuliu Maniu 1-3, Bucharest 061071, Romania; corneliu.burileanu@upb.ro
* Correspondence: anamaria.radoi@upb.ro
† This paper is an extended version of our paper published in 2018 41st International Conference on Telecommunications and Signal Processing (TSP) Held in Athens, Greece, 4–6 July 2018.

Received: 29 October 2018; Accepted: 22 November 2018; Published: 1 December 2018

Abstract: Technological evolution in the remote sensing domain has allowed the acquisition of large archives of satellite image time series (SITS) for Earth Observation. In this context, the need to interpret Earth Observation image time series is continuously increasing and the extraction of information from these archives has become difficult without adequate tools. In this paper, we propose a fast and effective two-step technique for the retrieval of spatio-temporal patterns that are similar to a given query. The method is based on a query-by-example procedure whose inputs are evolution patterns provided by the end-user and outputs are other similar spatio-temporal patterns. The comparison between the temporal sequences and the queries is performed using the Dynamic Time Warping alignment method, whereas the separation between similar and non-similar patterns is determined via Expectation-Maximization. The experiments, which are assessed on both short and long SITS, prove the effectiveness of the proposed SITS retrieval method for different application scenarios. For the short SITS, we considered two application scenarios, namely the construction of two accumulation lakes and flooding caused by heavy rain. For the long SITS, we used a database formed of 88 Landsat images, and we showed that the proposed method is able to retrieve similar patterns of land cover and land use.

Keywords: pattern recognition; dynamic time warping; maximum likelihood criterion; similarity measure; multitemporal data; multispectral information

1. Introduction

Over the years, the remote sensing domain has been characterized by numerous technological improvements (e.g., increased spatial resolution, shorter revisit time, increased number of spectral bands). These improvements were made possible through several Earth Observation missions, e.g., the Landsat program sustained by NASA and the United Stated Geological Survey (USGS), the Sentinel program financed by European Space Agency, Envisat's ASAR mission. In addition, many of the recent missions have functioned under an open data access policy for research purposes. This fact has a direct impact on the interest manifested in using this type of data in many realms, such as agriculture, land cover and land use planning, resource management, urbanization, and sustainable development.

One possible method for the analysis of satellite image time series (SITS) is to compare two satellite images captured at two successive moments of time, over the same area of interest [1–4]. These methods are generally called change detection methods. However, although they can successfully detect abrupt changes (e.g., deforestation, natural disasters, building construction), these methods are not able to identify complex spatio-temporal structures that evolve in a defined time-frame (e.g., measuring the effects of floods, urban expansion, land cover and land use modifications). In the later cases, SITS analysis methods that involve information extraction from multi-temporal data are usually preferred.

The main difficulties when dealing with SITS are the irregular time sampling, the missing samples (e.g., due to cloud cluttering and haze that affect images captured by optical sensors, technical artifacts) and, also, the high spatial and temporal data dimensionality. In general, multi-temporal approaches require the processing of large amounts of data characterized by a high degree of variability in terms of temporal, spatial and multi-spectral information.

The spatio-temporal evolution patterns may span different periods of time and may affect small to large regions. For example, urban transformations affect small regions and have multiple construction phases spanning several years. In contrast, flood events may affect larger areas then in the previous case, but, depending on their intensity, the consequences may be observed over shorter or longer periods of time (e.g., land slides are irreversible). In this context, the characterization of temporal evolutions is strictly related to the fact that the processes that occur in dynamic scenes have different time scales and affect small to large areas.

Therefore, the requirements that SITS analysis methods must fulfill are: (1) the ability to capture complex spatio-temporal changes and (2) the potential to emphasize both short-term and long-term modifications. Developing an algorithm that can deal with the spatial, temporal and spectral diversity of Earth Observation data is a challenging task to accomplish and more so when the amount of knowledge that a human expert transfers to the system is limited.

Many multi-temporal analysis techniques use a SITS classification that incorporates various dissimilarity measures to compare two temporal sequences. In SITS, by similar evolution, we understand a group of scene points that share the same behavior for the entire period of time or for most of it. Each point in a SITS is characterized by a spectro-temporal signature. An example of a spectro-temporal signature is provided in Figure 1 for a SITS acquired by the Landsat sensor with six spectral bands.

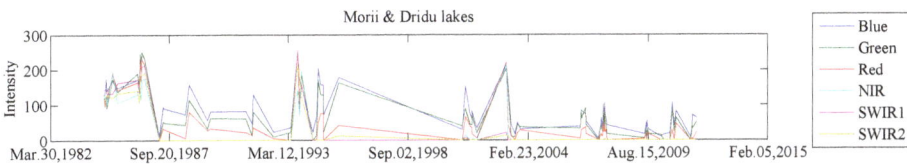

Figure 1. Spectro-temporal signature in a Landsat satellite image time series (SITS).

Some of the methods used to assess the degree of similarity between temporal sequences have emerged directly from text classification, speech recognition, or genomic data classification [5,6]. Dynamic Time Warping (DTW) has proved to be an essential tool not only for spoken-word recognition, but also for understanding similarities between evolution profiles in SITS [7,8]. DTW is able of handling comparisons between sequences characterized by irregular time sampling or missing data issues [7]. In this regard, K-means with DTW as the distance measure the yield land cover and land use maps of SITS [7]. However, performing an unsupervised clustering over an entire multi-spectral SITS data is time-consuming, and the temporal characteristics of the data may be missed. Other versions of DTW include a weighting procedure to mark the seasonality characteristics of temporal evolutions [9,10]. This weighting procedure works well for cropland mapping, but it may not be helpful for other applications that do not account for a periodicity effect (e.g., construction of buildings). In addition, recently, a tree-based structure formed of multiple K-means clustering algorithms that use the DTW distance measure was developed in [11] for the indexing and classification of SITS.

The methodology described in [12] for time series analysis is based on computing change maps between pairs of consecutive images using different similarity measures (e.g., correlation coefficient, first order Kullback–Liebler Divergence, Conditional Information, Normalized Compression Distance). The change maps are transformed into a collection of words through the K-means clustering algorithm. All the words form a dictionary. The transformed change maps are processed afterwards using the Latent Dirichlet Allocation (LDA) model to discover classes (or evolution-topics) based on the latent information from the scene. Thus, the final results of the analysis provide an unsupervised classification of the spatio-temporal evolutions into classes.

A study of several general purpose supervised classifiers, like Random Forest and Support Vector Machine (SVM), was presented in [13] as a solution for land cover mapping. The features extracted from the images were spectral features (e.g., color, normalized difference vegetation, water and build-up indexes, brightness, greenness, wetness, brilliance) and temporal features (e.g., statistical values from normalized difference vegetation index profile, phenological parameters marking important events in a season like sowing, threshing, cropping). However, SVM and Random Forest classifiers are known for having large training times, which may last even several days [13]. In this sense, this type of classification system cannot be used for online retrieval of spatio-temporal evolutions, where a fast response is needed. In addition, supervised classifiers need large amounts of training data. This requirement is not easy to fulfill in SITS classification tasks.

Frequent sequential pattern analysis was introduced in [14,15] for unsupervised SITS mining. This approach performs a quantization of the images in the SITS using a fixed number of gray levels (or, labels) and forms a sequence of T labels for each pixel location, where T is the number of images in the SITS. In this context, an ordered list of labels is called a frequent sequential pattern if the number of its occurrences in the SITS is greater than a fixed threshold. Inferring spatial connectivity constraints when determining the frequent sequential patterns proved to be efficient in monitoring crop or ground deformations [14]. However, the determination of the frequency of sequential patterns in SITS is sensitive to the number of levels used for image quantization. In addition, the method extracts a large number of patterns which may be difficult to interpret in terms of changes that appear in an area. Moreover, missing data may raise difficulties when interpreting the results for SITS analysis.

Apart from the irregular time sampling, missing data, and high data variability in the spatial, temporal, and spectral domains that were mentioned above, another difficult challenge that is met in SITS analysis is the assignment of semantic meaning to patterns of spatio-temporal evolution. This challenge is most frequently met when unsupervised techniques are used for SITS analysis. In these cases, a method to determine the optimal number of evolution classes is usually used [16]. A query-by-example retrieval method would alleviate this issue by associating semantic meaning to each query that a user addresses to the retrieval system. In this regard, we hereby present a query-by-example retrieval method whose goal is to separate spatio-temporal evolutions that are similar to a given query from non-similar ones. After the user performs a specific query, the retrieval method consists of two main steps. Firstly, the system performs the computation of distance measures between the query and the rest of spatio-temporal evolutions. Secondly, an optimal threshold is determined through the Expectation-Maximization technique in order to obtain a binary map displaying similar patterns with respect to the given query. The rest of the paper is structured as follows. The two-step retrieval method is presented in Section 2. Sections 3 and 4 present the experiments conducted over several SITS datasets, discuss the results obtained for different application scenarios, and compare the proposed method to other SITS analysis methods. Finally, the last section concludes the paper.

2. Materials and Methods

In this paper, the SITS analysis is approached through time series information retrieval, which is the process of searching inside a collection of data. The proposed search process starts with a query provided by the end-user, whose aim is to identify evolutions that are similar (i.e., have similar behavior) to the selected query. Firstly, the proposed method determines the degree of similarity between a specific query and the rest of the spatio-temporal evolutions. This step is performed by computing the DTW measure between the query and the rest of the evolutions. The DTW score is expected to be small for similar evolutions and large for non-similar ones. The retrieval process consists of deciding whether a certain evolution is similar to the given query or not. In this sense, the retrieval process can be seen as a problem of separating the spatio-temporal evolutions into two classes (i.e., *similar* versus *non-similar* with respect to a given query). The class identification is performed via Bayesian decision theory and it is based on determining the optimal threshold that separates relevant from irrelevant evolution patterns with respect to a given query.

Specifically, the proposed approach is a two-step process: (1) compute DTW distances and (2) find the optimal threshold. The process is summarized in Figure 2 and starts by computing a DTW distance image containing all the DTW dissimilarity scores with respect to a fixed query. After applying the optimal threshold over the DTW distance image, the final result is a binary map which aids in delimiting similar spatio-temporal evolutions.

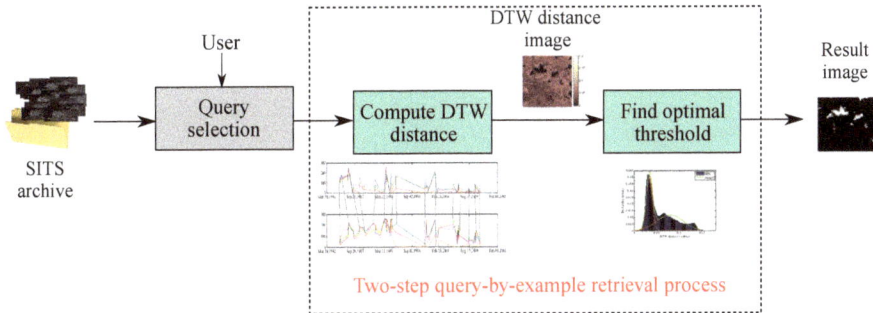

Figure 2. The flowchart of the proposed retrieval of similar patterns in SITS.

2.1. Dynamic Time Warping (DTW)

The first step in retrieving spatio-temporal patterns that are similar to a given query is to measure the degree of similarity between the query sequence and the rest of the sequences. This can be achieved through Dynamic Time Warping, which has been widely used in many applications related to speech recognition [5] and DNA analysis [6]. The method is usually applied to find the optimal alignment path between two sequences, $U_1^T = (\mathbf{u}_1, \ldots, \mathbf{u}_T)$ and $V_1^{T'} = (\mathbf{v}_1, \ldots, \mathbf{v}_{T'})$, which may have different lengths (i.e., $T \neq T'$) [17]. This is a frequently met situation in SITS analysis due to technological artifacts or cloud cluttering problems that may occur in images captured by optical remote sensing sensors. Mathematically, the DTW distance measure between two sequences, $U_1^i = (\mathbf{u}_1, \ldots, \mathbf{u}_i)$ and $V_1^j = (\mathbf{v}_1, \ldots, \mathbf{v}_j)$, can be recurrently defined as

$$D_{i,j} = \delta(\mathbf{u}_i, \mathbf{v}_j) + \min(D_{i-1,j-1}, D_{i-1,j}, D_{i,j-1}) \tag{1}$$

where δ is the Euclidean distance between two current elements $\mathbf{u}_i$ and $\mathbf{v}_j$ of the sequences, $D_{k,l}$ is the partial similarity score between U_1^k and V_1^l for any $k \leq T$ and $l \leq T'$. As can be observed from the above equation, the optimization problem aims at finding the best alignment between subsequences of U_1^T and $V_1^{T'}$.

In the above formulation, the following initial conditions are considered:

$$D_{1,1} = \delta(\mathbf{u}_1, \mathbf{v}_1) \tag{2}$$

$$D_{1,j} = \sum_{p=1}^{j} \delta(\mathbf{u}_1, \mathbf{v}_p) \tag{3}$$

$$D_{i,1} = \sum_{q=1}^{i} \delta(\mathbf{u}_q, \mathbf{v}_1). \tag{4}$$

The overall DTW similarity between U_1^T and $V_1^{T'}$ is given by the score $D_{T,T'}$. Computing the overall score is equivalent to determining the elements of a $T \times T'$ distance matrix where the element on row i and column j is $D_{i,j}$.

The recurrence relation used in the computation of the DTW similarity score implies the computation of a $T \times T'$ matrix, whose elements are the distances $D_{i,j}$ ($1 \le i \le T$, $1 \le j \le T'$). Each distance $D_{i,j}$ uses the smallest value between the closest previous neighbors in the matrix. The DTW matrix is computed from left to right and top to bottom. The last element $D_{T,T'}$ represents the overall DTW score of similarity, whilst the path that is followed to compute the score of similarity provides the optimal alignment. An example of DTW-based optimal alignment together with the DTW distance matrix D is shown in Table 1 for two sequences of one-dimensional elements. In this example, we considered that δ is the difference in absolute values of the elements. The element on the lower-right of the matrix is the total score of the similarity between the two sequences.

Table 1. Example of computation of the Dynamic Time Warping (DTW) distance matrix for two sequences, '5463545' and '0102130'.

	0	1	0	2	1	3	0
5	5	9	14	17	21	23	28
4	9	8	12	14	17	18	22
6	15	13	14	16	19	20	24
3	18	15	16	15	16	16	19
5	23	19	20	18	20	18	21
4	27	22	23	20	21	19	22
5	32	26	27	23	24	21	24

If the similarity scores are computed between spatio-temporal sequences in a multispectral SITS, the sequences are formed by the vector elements $\mathbf{u}_i$ and $\mathbf{v}_j$ whose dimensions are given by the number of frequency bands used by the remote sensing sensors. Let us denote by c the number of spectral bands. Then, δ is the Euclidean distance between two vector elements in a space with c dimensions:

$$\delta(\mathbf{u}_i, \mathbf{v}_j) = \sqrt{\sum_{k=1}^{c}(u_{i,k} - v_{j,k})^2} \tag{5}$$

where $\mathbf{u}_i = [u_{i,1}, \dots, u_{i,c}]^T$ and $\mathbf{v}_j = [v_{j,1}, \dots, v_{j,c}]^T$.

The DTW similarity score is computed between the query provided by the user and each spatio-temporal sequence in the SITS. The result is a DTW distance image **DI** that assesses the scores of similarity between the given query and the evolutions for each pixel location.

As already mentioned, the DTW distance is expected to be small for evolutions that are similar to the query and large for non-similar evolutions. This is due to the fact that, if the spatio-temporal evolutions are similar, the DTW algorithm finds an alignment with a smaller cost between the corresponding temporal sequences and the given query than in the case of non-similar evolutions. Therefore, the separation of similar from non-similar evolution patterns is equivalent to finding an optimal threshold that can be applied over DTW distances in order to retrieve the specific pattern queried by the end-user.

2.2. Determining the Optimal Threshold Using Expectation-Maximization

In the previous subsection, we described the process of determining the DTW distance image **DI** containing DTW similarity scores with respect to a query selected by an end-user. The following step in the automatic extraction of evolutions that have similar behavior to a given query is the optimal thresholding step. More precisely, we aim to separate the DTW distance scores between two classes, a class of *similar* evolutions and a class of *non-similar* spatio-temporal evolutions. Separating these two classes of evolutions translates into finding the optimal threshold to be applied on the DTW similarity scores. In this sense, we formulate the problem of query-by-example SITS retrieval in terms of the Bayesian decision theory.

The DTW similarity scores computed for a given query follow a multimodal distribution. This can be observed from Figure 3, where the query selected for retrieval is from an agricultural area in a long SITS (88 temporal samples). As expected, the first lobe corresponds to similar evolutions. This is argued by the fact that DTW similarity scores are smaller for similar evolutions than for non-similar evolutions. Furthermore, the histogram profile shown in Figure 3 confirms the idea of separating similar from non-similar evolutions using the Maximum Likelihood (ML) technique.

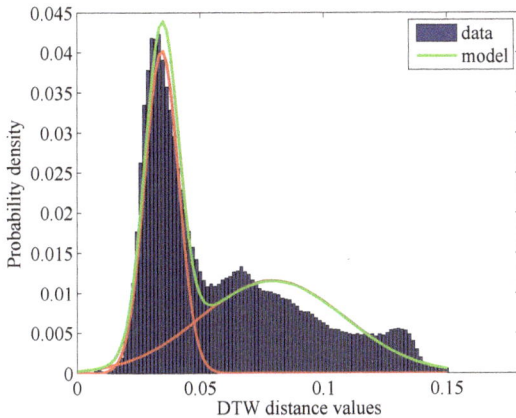

Figure 3. Histogram over DTW similarity scores and a mixture of two Gaussian distributions fitted to these scores.

In the following text, we denote the random variable associated to the computed DTW similarity scores by X, whilst x represents a particular DTW score. Following the formalism initially presented in [18], let us assume that the scores are drawn from a mixture of two Gaussian distributions, $\mathcal{N}(\mu_s, \sigma_s)$ and $\mathcal{N}(\mu_n, \sigma_n)$. The first Gaussian distribution corresponds to evolutions that are *similar* to the given query, whereas the second component refers to *non-similar* evolutions.

The Maximum Likelihood Estimation (MLE) framework determines the parameter values that make the observed data most likely, i.e., that maximize the likelihood function and fit the model to the data. Following the above considerations, the overall posterior probability can be decomposed as

$$p_\theta(x) = \pi_s \mathcal{N}(x|\mu_s, \sigma_s) + \pi_n \mathcal{N}(x|\mu_n, \sigma_n) \tag{6}$$

where $\theta = \{\pi_s, \mu_s, \sigma_s, \pi_n, \mu_n, \sigma_n\}$ is the set of model parameters, whilst π_s and π_n are the mixture probabilities [19] such that $\pi_s + \pi_n = 1$. The mixture components are modeled as Gaussian densities:

$$\mathcal{N}(x|\mu_k, \sigma_k) = \frac{1}{\sqrt{2\pi\sigma_k^2}} \exp\left[-\frac{(x - \mu_k)^2}{2\sigma_k^2}\right] \tag{7}$$

with $k \in \{s, n\}$, whilst (μ_s, σ_s) and (μ_n, σ_n) are the mean and standard deviation pairs corresponding to similar and non-similar evolutions, respectively. The set of parameters θ is determined iteratively using the Expectation-Maximization (EM) algorithm. Specifically, each iteration j is decomposed in two basic steps:

1. *Expectation step (E-step)*. Compute the log-likelihood (i.e., the logarithm of the posterior probability) with respect to the current values of the parameters $\theta^{(t)}$:

$$L(\theta^{(t)}) = \ln p_{\theta^{(t)}}(x) \tag{8}$$

2. *Maximization step (M-step)*. Update the model parameters such that the log-likelihood approaches its maximum, i.e., the convergence of the EM algorithm guarantees that the log-likelihood value is increased with each iteration [19]. It is usually convenient to introduce a mapping between the new model parameters $\theta^{(t+1)}$ and the previous ones $\theta^{(t)}$. According to [18] and following the notations in [19], the means, squared standard deviations, and mixture probabilities for each class $k \in \{s, n\}$ can be re-estimated using the following set of relations:

$$\pi_k^{(t+1)} = \frac{\sum\limits_x \zeta_k^{(t)}(x)}{N} \tag{9}$$

$$\mu_k^{(t+1)} = \frac{\sum\limits_x \zeta_k^{(t)}(x) \cdot x}{\sum\limits_x \zeta_k^{(t)}(x)} \tag{10}$$

$$\sigma_k^{(t+1)} = \left(\frac{\sum\limits_x \zeta_k^{(t)}(x) \cdot \left(x - \mu_k^{(t)}\right)^2}{\sum\limits_x \zeta_k^{(t)}(x)} \right)^{\frac{1}{2}} \tag{11}$$

where $N = W \times H$ is the total number of pixels in the distance image (i.e., the number of spatio-temporal evolutions) and $\zeta_k^{(t)}(x)$ values are derived as

$$\zeta_k^{(t)}(x) = \frac{\pi_k^{(t)} \mathcal{N}(x | \mu_k^{(t)}, \sigma_k^{(t)})}{\sum_{k'} \pi_{k'}^{(t)} \mathcal{N}(x | \mu_{k'}^{(t)}, \sigma_{k'}^{(t)})}. \tag{12}$$

The initialization of the EM algorithm is done by separating the similarity scores into two subsets, $\mathcal{S}_n^{(0)}$ and $\mathcal{S}_c^{(0)}$. Any unsupervised clustering method can be applied, but we chose the K-means ($K = 2$) clustering method due to its fast response time when the number of clusters is small. However, being an unsupervised clustering method, a constraint must still be fulfilled. Namely, the cluster of similar evolutions must be characterized by DTW similarity scores that are smaller than the DTW similarity scores obtained for the temporal sequences pertaining to the cluster of non-similar evolutions with respect to the query. The initial values for the prior probabilities, means, and squared standard deviations can be easily computed as statistics over the two subsets, $\mathcal{S}_n^{(0)}$ and $\mathcal{S}_c^{(0)}$. Compared to the initialization proposed in [1], the K-means initialization speeds up the convergence of the EM algorithm [19], and thus, the speed of the whole algorithm.

After the estimation of the θ model parameters, the optimal threshold T_o that separates the two classes (i.e., *similar* and *non-similar*) is determined from the equality

$$\pi_s \mathcal{N}(T_o | \mu_s, \sigma_s) = \pi_n \mathcal{N}(T_o | \mu_n, \sigma_n) \tag{13}$$

that naturally follows from the Maximum Likelihood rule

$$\pi_s \mathcal{N}(x | \mu_s, \sigma_s) \gtrless_n^s \pi_n \mathcal{N}(x | \mu_n, \sigma_n). \tag{14}$$

Taking the logarithm of both sides of Equation (13) yields a quadratic equation in T_o

$$(\sigma_n^2 - \sigma_s^2)T_o^2 + 2\left(\mu_n\sigma_s^2 - \mu_s\sigma_n^2\right)T_o + \mu_s^2\sigma_n^2 - \mu_n^2\sigma_s^2 - 2\sigma_s^2\sigma_n^2 \ln\left[\frac{\sigma_n\pi_s}{\sigma_s\pi_n}\right] = 0 \tag{15}$$

that has two possible solutions. However, only one of these solutions lies between μ_s and μ_n and this solution represents the optimal threshold value.

After determining the optimal threshold value T_o, the DTW distance image **DI** is transformed into a binary image **S**, delimiting the locations of evolutions that are similar to the given query:

$$\mathbf{S}(w,h) = \begin{cases} 1, & \mathbf{DI}(w,h) \leq T_o \\ 0, & \text{otherwise} \end{cases} \tag{16}$$

where $1 \leq w \leq W$ and $1 \leq h \leq H$, W and H being the width and height of the images in SITS.

3. Experiments

The proposed method for the retrieval of specific spatio-temporal patterns in SITS was tested on series of Landsat images captured at different periods of time and over different locations. In all application scenarios described hereafter, all of the available spectral bands were taken into consideration, namely green, red, blue, Near Infrared and two Short-Wave Infrared bands. Moreover, the spatial resolution of the Landsat sensor was 30 m.

The first series consists of a short time series of 10 images acquired between 1984 and 1993 (i.e., an image is acquired each year), over Bucharest, Romania, and regions surrounding the city. The size of the images is 1702 × 1975 pixels and the captures were taken in different seasons of the years. Several samples of the time series, along with their acquisition moments, are shown in Figure 4. In this case, the task was to determine similar changes that occurred during the formation of three accumulation lakes near Bucharest, namely Dridu, Mihailesti, and Morii. Two of these lakes evolved similarly, whereas the third one went through several modifications over the time period mentioned above.

The second dataset is formed of 13 Landsat images of 1250 × 400 pixels capturing the Dobrogea region, Romania, between 6 May 2000 and 9 September 2001. Some images from the dataset, together with the corresponding timespan, are shown in Figure 5. In May 2000, the heavy rain led to the swelling of Danube river which caused floods in this region (Figure 5a). A rapid assessment of the area affected by floods is necessary in this type of situation. Therefore, the application considered in this case was oriented towards the delimitation of areas affected by floods.

The third time series spans a longer period of time than the previous time series, almost 28 years between 14 September 1984 and 27 October 2011, and contains 88 multispectral images of 700 × 700 pixels. The location is still Bucharest, Romania and surrounding regions, but the area captured by the SITS is smaller than in the previous case. The first and last images from the long-term SITS, along with the corresponding temporal distribution of the acquisitions, are shown in Figure 6. In general, long-term SITS acquisition is characterized by irregular temporal sampling with data captured under different meteorological (e.g., precipitation, clouds, season) and illumination conditions. These issues make the query-by-example retrieval in long-term SITS a challenging task to accomplish if other types of distance measure (e.g., Euclidean distance) are used to assess the degree of similarity between the temporal sequences. There are two main applications where the information extracted from long-term SITS shows its potential, namely land cover and land use mapping. These two applications impact the sustainable management of the natural resources, urban planning, and agriculture.

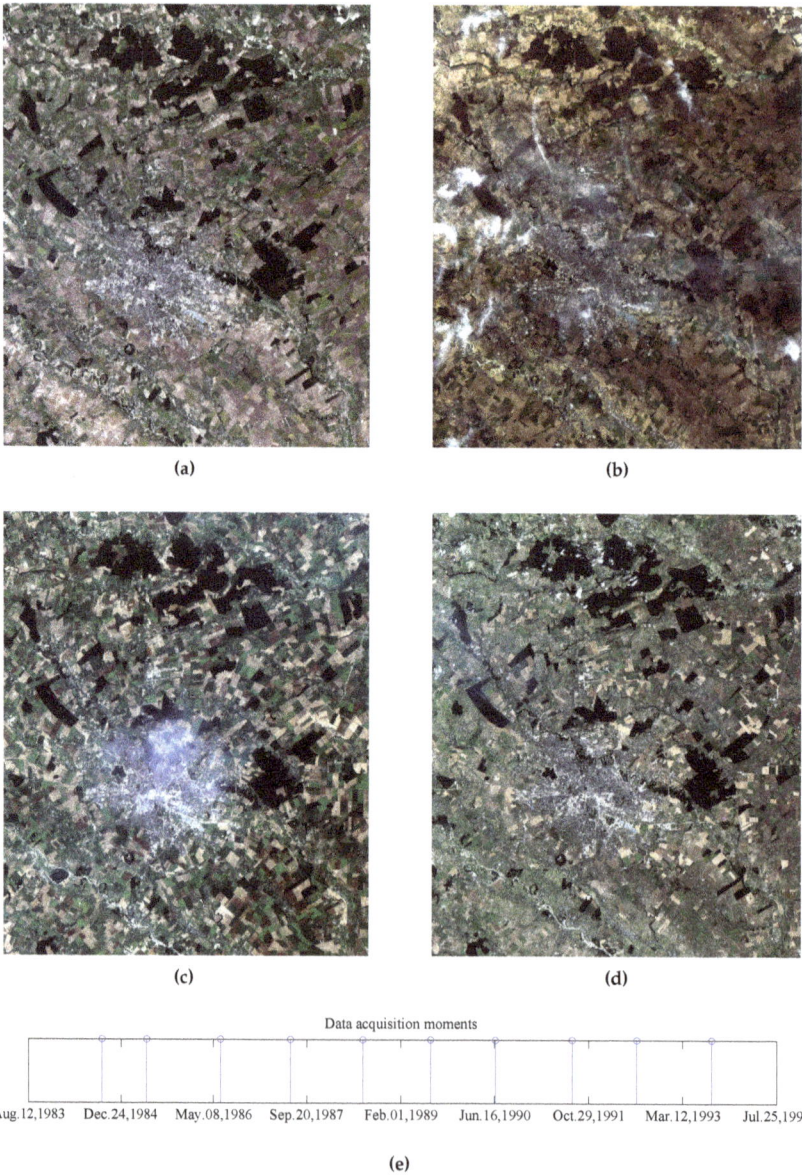

(a)   (b)

(c)   (d)

Data acquisition moments

Aug.12,1983   Dec.24,1984   May.08,1986   Sep.20,1987   Feb.01,1989   Jun.16,1990   Oct.29,1991   Mar.12,1993   Jul.25,1994

(e)

Figure 4. Short Landsat SITS comprised of 10 images captured between 1984–1993. Only four representative images (i.e., containing specific changes) of the series are shown: (**a**) 1984, (**b**) 1987, (**c**) 1988, (**d**) 1992. The distribution of the acquisition moments is shown in (**e**).

(a)

(b)

(c)

(d)

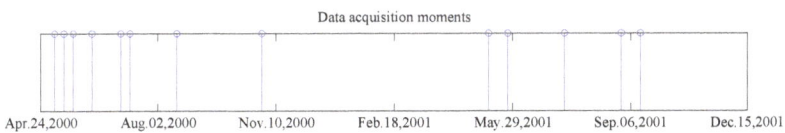

(e) Capture moments of the Dobrogea Landsat SITS

Figure 5. Dobrogea Landsat SITS comprised of 13 images captured between 6 May 2000 and 14 September 2001, in the Dobrogea region. Four images of the series are shown, namely: (**a**) 6 May 2000, (**b**) 22 May 2000, (**c**) 9 July 2000, and (**d**) 29 October 2000. The distribution of the acquisition moments is shown in (**e**).

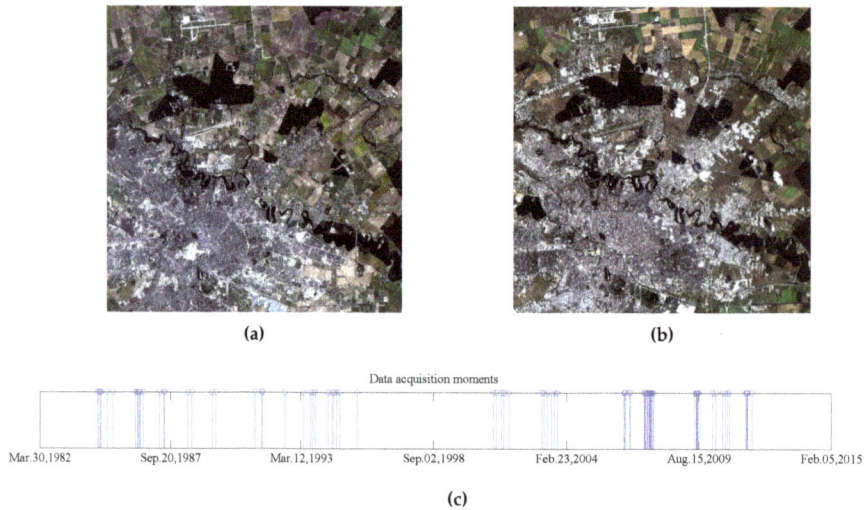

Data acquisition moments

Mar.30,1982   Sep.20,1987   Mar.12,1993   Sep.02,1998   Feb.23,2004   Aug.15,2009   Feb.05,2015

(c)

Figure 6. Long Landsat SITS comprised of 88 images captured in 1984–2011. Only the first and the last images of the series are shown, namely (**a**) 1984 and (**b**) 2011. The distribution of the acquisition moments is shown in (**c**).

In all cases, the user was asked to select a single evolution in the area of interest and then the proposed algorithm was applied to identify other spatio-temporal evolutions with similar spectro-temporal signatures.

4. Discussion

4.1. Discovery of Similar Patterns in SITS

In the first setup, we aimed to discover two types of spatio-temporal evolution related to the formation of three accumulation lakes in the Bucharest region, namely Morii, Dridu and Mihailesti. If compared to the Morii and Dridu lakes, Lake Mihailesti has a distinct spatio-temporal evolution because this accumulation lake was emptied several times (in 1987 and in 1992) during its construction. The spectro-temporal differences between the two queries can also be observed from the spectro-temporal signatures shown in Figure 7. In this sense, the user performs two independent queries, each selected from the regions of interest. The analyzed period of time was between 1984 and 1993 and the dataset considered for retrieval was the short Landsat SITS composed of 10 images captured in the mentioned period (i.e., one image per year).

The results of the proposed query-by-example retrieval method are shown in Figure 8, whereas the ground truth is presented in Figure 9. The first row of Figure 8 represents all the DTW distances measured between the query and the rest of the spatio-temporal sequences, whereas the second row shows the output of the retrieval system (i.e., after applying the optimal threshold over the DTW distance image). For a numerical evaluation of the performances reached by the system, we measured the overall accuracy (OA), the missed alarm rate (MAR) and the false alarm rate (FAR), which are defined as

$$OA = \frac{TP + TN}{TP + TN + FP + FN} \tag{17}$$

$$MAR = \frac{FN}{TP + FN} \tag{18}$$

$$FAR = \frac{FP}{TN + FP} \tag{19}$$

where TP indicates the number of similar spatio-temporal evolutions that were correctly identified, TN indicates the number of non-similar spatio-temporal evolutions that were correctly identified, FP is the number of non-similar spatio-temporal evolutions that were determined as similar, and FN is the number of similar spatio-temporal evolutions that were determined to be non-similar to the given query.

The results, reported in Table 2, show that the system is able to accurately determine similar spatio-temporal patterns with respect to the given queries in term of overall accuracy and false alarm rate. The missed alarm rate can still be decreased if spatial constraints are imposed (i.e., similar sequences are more likely to be close to the query location, and, conversely, non-similar spatio-temporal evolutions are likely to be surrounded by other non-similar spatio-temporal evolutions). In this case, Markov Random Fields models can be employed [1], but they are difficult to use in SITS analysis due to their computational complexity. In addition, if the searched spatio-temporal evolutions are not compact (e.g., construction of new buildings), the solution may not achieve satisfactory results.

Table 2. Performance evaluation on short SITS.

Query	Overall Accuracy	Missed Alarm Rate	False Alarm Rate
Morii and Dridu	99.68%	26.84%	0.23%
Mihailesti	99.36%	30.36%	0.56%

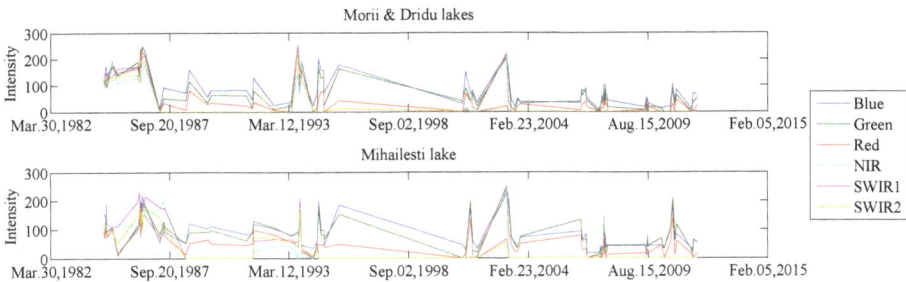

Figure 7. Spectro-temporal signatures during the construction of the accumulation lakes. The two spectro-temporal signatures are characterized by different temporal evolutions in the period that corresponds to their construction, namely 1984–1993.

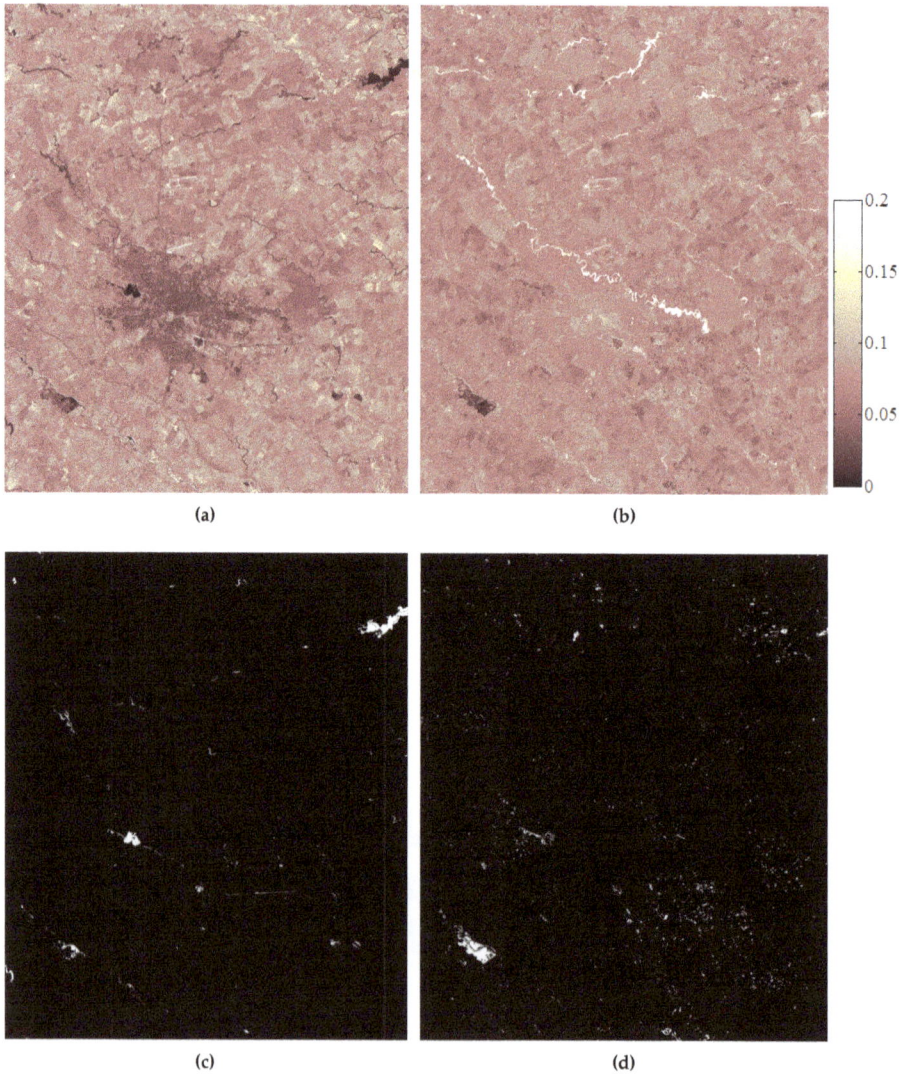

Figure 8. Pattern discovery in short Landsat SITS. (**a**) DTW distance image for a query marked inside the Morii & Dridu accumulation lakes. (**b**) DTW distance image for a query marked inside Mihailesti accumulation lake. (**c**) Pattern discovery using the proposed method for Morii & Dridu accumulation lakes. (**d**) Pattern discovery using the proposed method for Mihailesti accumulation lake. In the case of DTW distance images presented in (**a,b**), dark color represents similar evolutions and bright color represents non-similar evolutions, whereas in (**c,d**), white pixels correspond to spatio-temporal patterns that are similar to the query, which was selected from the region of interest.

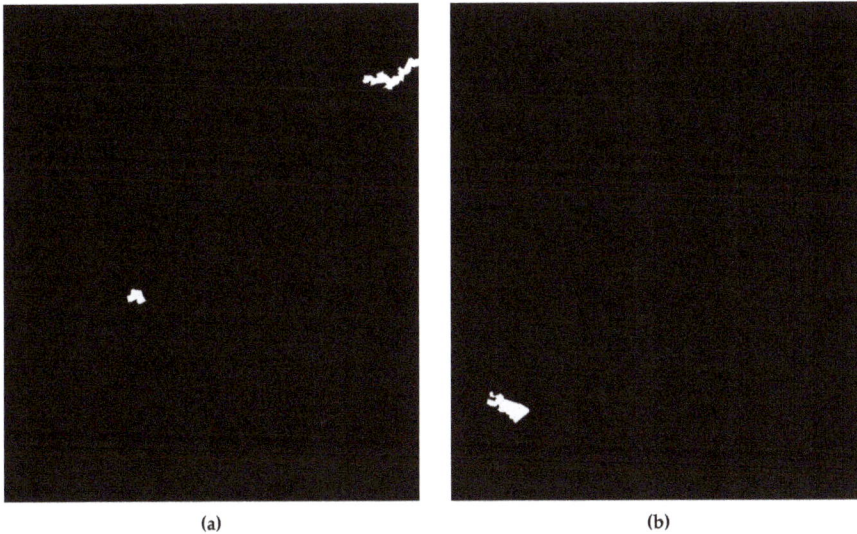

(a)   (b)

Figure 9. Ground truth for short Landsat SITS. White pixels delimit (**a**) Morii & Dridu accumulation lakes and (**b**) Mihailesti accumulation lake.

4.2. Damage Assessment

Undoubtedly, the retrieval of spatio-temporal patterns is extremely important when assessing the spatial and temporal extent of the transformations suffered by an area. In this experiment, we aimed to identify the area affected by the floods that occurred in 2000 in the Dobrogea region, Romania, and to identify their harmful effects over several months. As shown in Figure 5, the first image of the series was captured immediately after the Danube river swelled, whereas the rest of the images were acquired after the river began to retreat. The result of the query-by-example retrieval proposed method is shown Figure 10, along with the corresponding ground truth. The numerical evaluation of the performance achieved by the proposed algorithm is presented in Table 3 and confirms the effectiveness of the method for delimiting specific areas, e.g., flooded areas in this particular case. An interesting aspect is the fact that DTW managed to bypass the problem of distorted data—the region of interest is partially occluded by clouds in the last image from the series, but the proposed algorithm was still able to correctly retrieve the spatio-temporal evolutions in the flooded area.

Table 3. Performance evaluation for delimitation of flooded areas in Dobrogea.

Query	Overall Accuracy	Missed Alarm Rate	False Alarm Rate
Flooded area	99.96%	2.36%	0.13%

(a)

(b)

(c)

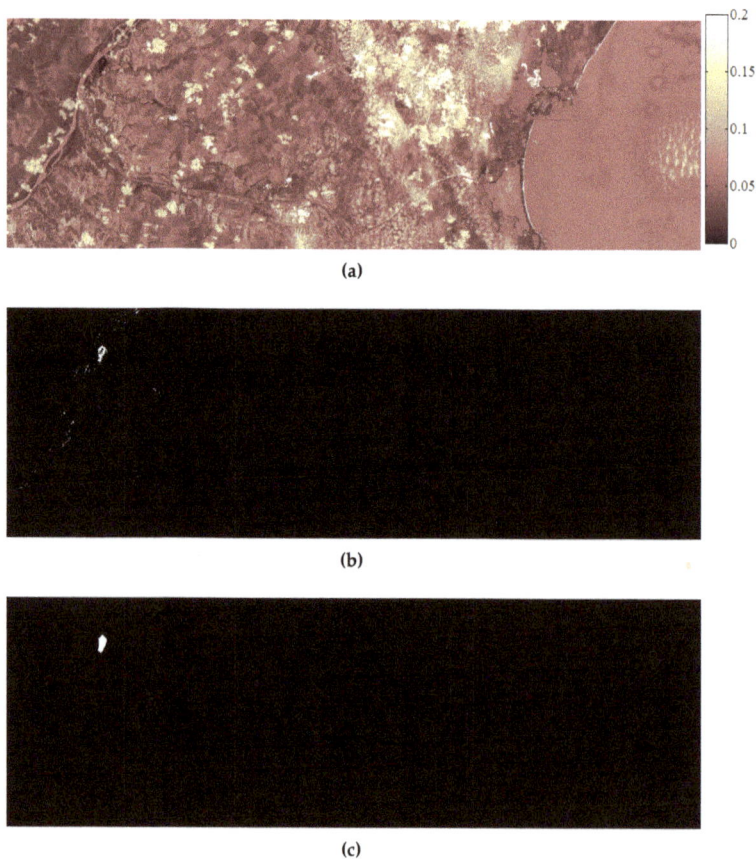

Figure 10. Delimitation of areas affected by floods in Dobrogea. (**a**) DTW distance image for a query in the flooded area. (**b**) Delimitation of the affected area using the proposed algorithm. (**c**) Ground truth marking the flooded areas. In the case of DTW distance image presented in (**a**), dark color represents similar evolutions and bright color represents non-similar evolutions, whereas in (**b**) white pixels correspond to spatio-temporal patterns that are similar to the query, which was selected from the region of interest.

4.3. Land Cover and Land Use Mapping over Long Time Series

In the third setup, the application scenarios were related to the land cover and land use mapping using archives of SITS. Among the queries that we experimented with, we recalled the identification of extra-urban expansion, the demarcation of the urban area, the delimitation of forest areas, the discovery of agricultural areas, and the search for water bodies that remained unmodified over the years. However, due to the complexity of the SITS (i.e., containing 88 images captured over 27 years) and of the queries, the performance of the retrieval algorithm was assessed by visual inspection. The results shown in Figures 11 and 12 with Figure 6 show that the retrieval system was able to discover the queried spatio-temporal patterns for land cover and land use mapping even when the data was irregularly sampled and images were captured under different meteorological and illumination conditions.

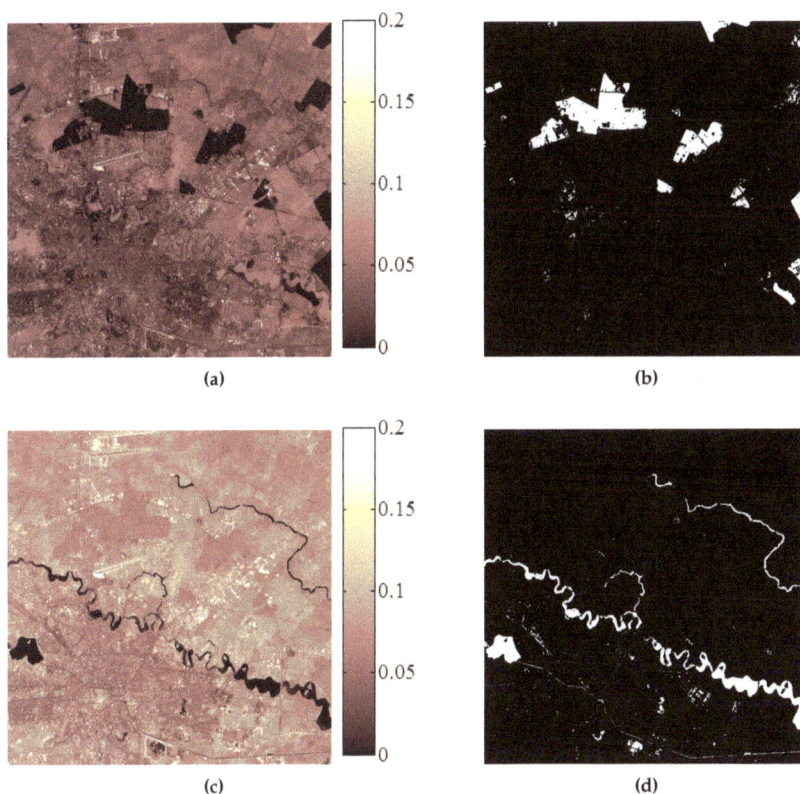

Figure 11. Land cover mapping in long Landsat SITS. (**a**) DTW distance image for a query made inside the forestry area. (**b**) Forestry area delimitation using the proposed query-by-example retrieval method. (**c**) DTW distance image for a query made in an area covered by water. (**d**) Water delimitation using the proposed query-by-example retrieval method. In the case of DTW distance images presented in (**a**,**c**), dark color represents similar evolutions and bright color represents non-similar evolutions, whereas in (**b**,**d**), white pixels correspond to spatio-temporal patterns that are similar to the query, which was selected from the region of interest.

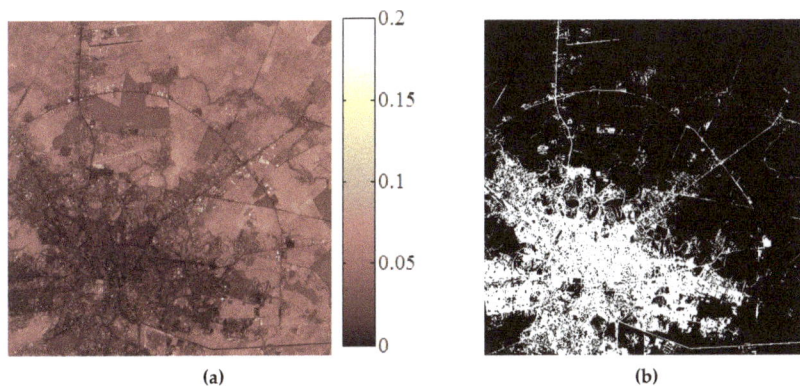

Figure 12. *Cont.*

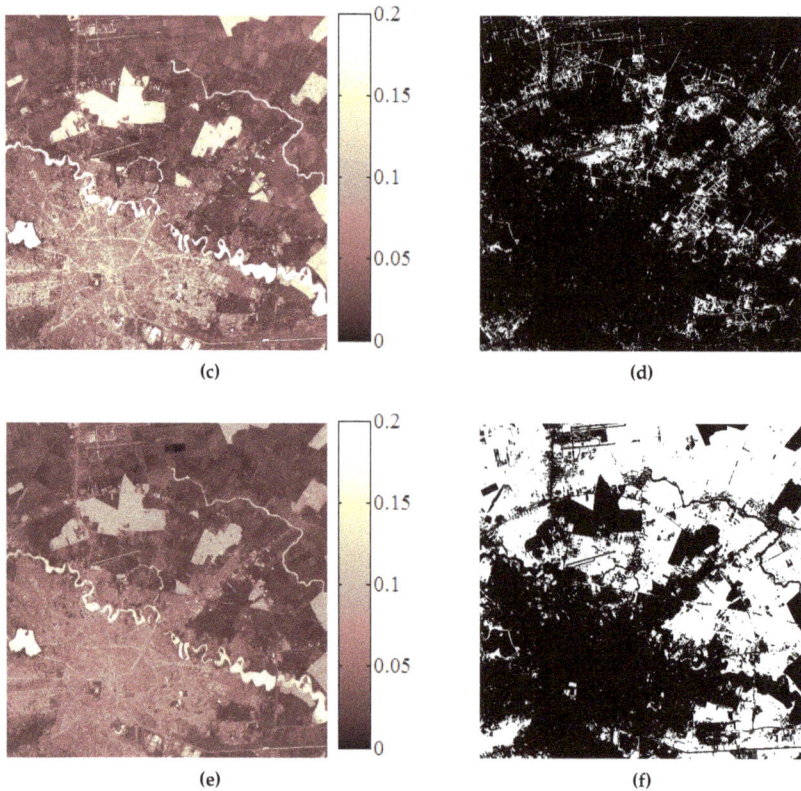

Figure 12. Land use mapping in long Landsat SITS. (**a**) DTW distance image for a query made in the urban area. (**b**) Urban area delimitation using the proposed query-by-example retrieval method. (**c**) DTW distance image for a query made in the extra-urban area. (**d**) Extra-urban area delimitation using the proposed query-by-example retrieval method. (**d**) DTW distance image for a query made inside the agricultural area. (**e**) Agricultural area delimitation using the proposed query-by-example retrieval method. In the case of DTW distance images presented in (**a,c,e**), dark color represents similar evolutions and bright color represents non-similar evolutions, whereas in (**b,d,f**), white pixels correspond to spatio-temporal patterns that are similar to the query, which was selected from the region of interest.

4.4. Comparison with Other State-of-the-Art SITS Analysis Methods

As mentioned in the Introduction, the DTW-based K-means algorithm [7] and the LDA-based method described in [12] are unsupervised clustering methods for SITS analysis. We show a series of results obtained by applying the above mentioned methods over the experimental datasets in Figures 13–15. Following the recommendations in [12], the number of clusters considered was 15 and the dictionary was formed of 150 words. In the case of the DTW-based K-means method, we used the same number of clusters as for the LDA-based method.

As it can be observed in Figure 15, both algorithms produce land use and land cover maps, for which the assignation of a temporal meaning (i.e., action name or evolution name) to the clusters is a difficult task to accomplish. Moreover, the extraction of particular temporal evolutions is not easy, since, most often, the spectral and spatial characteristics inhibit the temporal properties of spatio-temporal evolutions. This happens mostly when the number of clusters, K, is small. However, in both cases, using a greater K leads to inconsistent results and to oversegmentation of the images caused by nonhomogeneus spatio-temporal evolutions, cloud cluttering, noise, different illumination conditions, or seasonal changes. Our proposed approach aims to overcome these drawbacks by retrieving specific spatio-temporal evolutions that, at the end of the retrieval process, will receive the semantic label of the query.

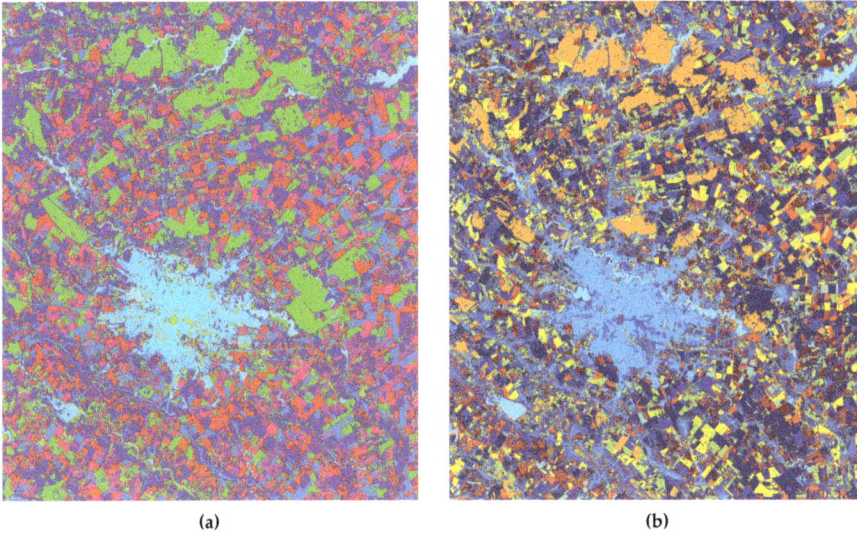

(a)   (b)

Figure 13. Analysis of short Landsat SITS: (**a**) DTW-based K-means clustering [7], (**b**) LDA-based clustering [12].

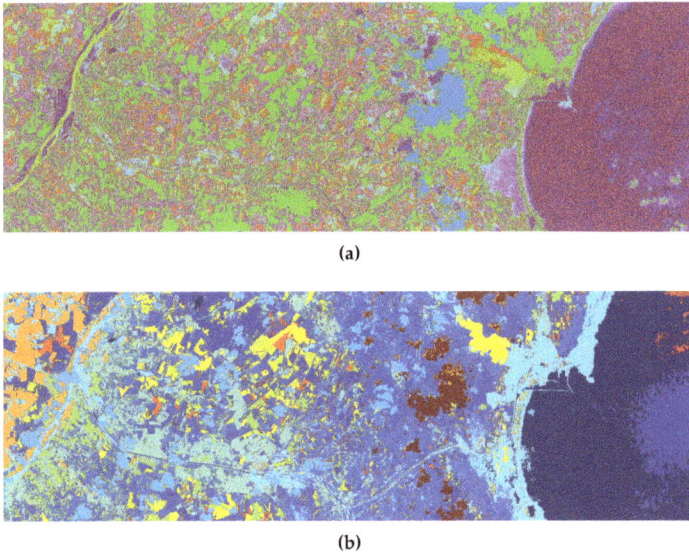

(a)

(b)

Figure 14. Analysis of Dobrogea SITS: (**a**) DTW-based K-means clustering [7], (**b**) Latent Dirichlet Allocation (LDA)-based clustering [12].

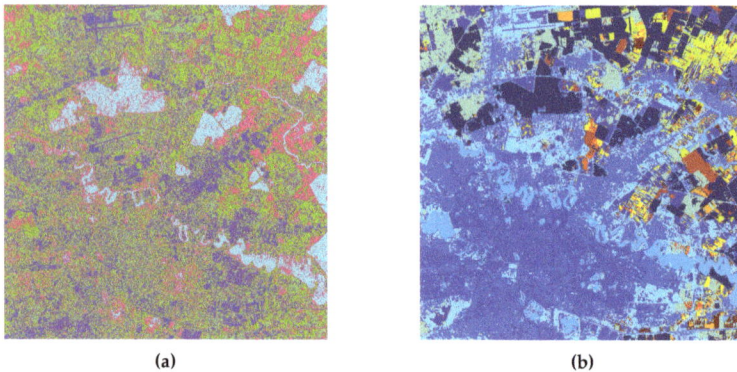

Figure 15. Analysis of long Landsat SITS: (**a**) DTW-based K-means clustering [7], (**b**) LDA-based clustering [12].

The LDA-based method performs an analysis of change map time series that is determined by applying different similarity measures. However, these similarity measures do not take the temporal extent of a transformation into account. This explains the insertion of the two different evolutions related to the building of accumulation lakes Morii & Dridu and Mihailesti into the same cluster as can be observed in Figure 13b. The DTW-based clustering divides the evolution of Mihailesti lake into two classes, one that corresponds to Morii & Dridu evolution and one that captures the evolution of Bucharest city. The SITS analysis method proposed in this paper is able to determine an optimal threshold that separates the two types of evolution related to the construction of the accumulation lakes to separate them from other types of evolutions. In contrast, the proposed query-by-example distinguishes between the two different spatio-temporal patterns, whose dissimilar evolutions can be observed in the corresponding spatio-temporal signatures shown in Figure 7.

In the second use-case scenario, the DTW-based K-means method includes the flooded region in a category containing a portion of the Black Sea captured on the right hand side of Figure 14a, whereas the LDA-based analysis does not distinguish this particular evolution from the rest of evolution patterns. Therefore, this event is not marked as a separate cluster of spatio-temporal evolutions on the resulting maps shown in Figure 14. On the contrary, the proposed query-by-example retrieval method also shows its potential in this use-case scenario by clearly delimiting the affected area and allowing estimation of the damaged surface.

4.5. Final Remarks

Finally, the running time for performing a query-by-example retrieval over the short Landsat SITS is 1 min 18 s/query, whereas over the long Landsat SITS, the running time is 10 min/query. The running times are considerably smaller than those required to perform unsupervised clustering of the spatio-temporal evolutions using the DTW-based K-means algorithm with $K = 15$ classes, as presented in [7] (i.e., 38 min 17 s for the short SITS and approximately 16 h for the long SITS). A shorter running time was registered for the LDA-based method—almost 25 min for short SITS and approximately 2 h for the long SITS.

5. Conclusions

In this paper, we described an effective query-by-example retrieval system that can be used for the exploitation of Earth Observation SITS. The strategy is based on a two-step procedure. The first step consists of measuring the degree of similarity between a query provided by the user and other spatio-temporal evolutions in SITS. The second step consists of separating similar and non-similar patterns via an optimal thresholding technique applied over the DTW distance image obtained in the first step.

The experiments showed that, even if the SITS is irregularly sampled, affected by clouds or haze, or if the acquisitions are performed in different seasons and under different illumination conditions, the method is able to recognize specific patterns in SITS. In order to show the effectiveness and applicability of our proposed retrieval method, we considered several application scenarios. First, we considered the problem of retrieving similar spatio-temporal patterns from the SITS by analyzing a specific case, namely the construction of two accumulation lakes with different histories. In the second scenario, we exploited the proposed method to assess the damage that floods produce over a region. This is a typical scenario in emergency situations when a fast and effective evaluation of the affected areas is mandatory. In the last scenario, we considered long SITS and aimed to build land cover and land use maps that provide fundamental information for many applications, including change analysis, crop estimation, sustainable management of natural resources, and urbanization planning.

Being characterized by a low computational complexity, the proposed method for retrieving similar spatio-temporal evolutions based on a specific query represents a good candidate for the online analysis and annotation of SITS. Moreover, the method is designed to use the entire information provided by a multispectral remote sensing sensor, regardless of the number of spectral bands.

Author Contributions: The contribution to this work is as follows: conceptualization, A.R. and C.B.; methodology, A.R.; software, A.R.; validation, A.R.; writing—review and editing, A.R.; review and supervision, C.B.; funding acquisition, C.B.

Funding: This work has been funded by University Politehnica of Bucharest, through the Excellence Research Grants Program, UPB GEX 2017. Identifier: UPBGEX2017, Ctr. No. 32/2017.

Acknowledgments: The authors thank NASA USGS for making freely available the datasets for research purposes.

Conflicts of Interest: The authors declare no conflict of interest.

Abbreviations

The following abbreviations are used in this manuscript:

SITS	Satellite Image Time Series
USGS	United States Geological Survey
DTW	Dynamic Time Warping
LDA	Latent Dirichlet Allocation
SVM	Support Vector Machine
ML	Maximum Likelihood
MLE	Maximum Likelihood Estimation
EM	Expectation-Maximization
OA	Overall Accuracy
MAR	Missed Alarm Rate
FAR	False Alarm Rate

References

1. Bruzzone, L.; Prieto, D.F. Automatic analysis of the difference image for unsupervised change detection. *IEEE Trans. Geosci. Remote Sens.* **2000**, *38*, 1171–1182. [CrossRef]
2. Bovolo, F.; Bruzzone, L. A Theoretical Framework for Unsupervised Change Detection Based on Change Vector Analysis in the Polar Domain. *IEEE Trans. Geosci. Remote Sens.* **2007**, *45*, 218–236. [CrossRef]
3. Celik, T. Unsupervised Change Detection in Satellite Images Using Principal Component Analysis and k-Means Clustering. *IEEE Geosci. Remote Sens. Lett.* **2009**, *6*, 772–776. [CrossRef]
4. Radoi, A.; Datcu, M. Automatic Change Analysis in Satellite Images Using Binary Descriptors and Lloyd-Max Quantization. *IEEE Geosci. Remote Sens. Lett.* **2015**, *12*, 1223–1227. [CrossRef]
5. Rabiner, L.; Juang, B.H. *Fundamentals of Speech Recognition*; Prentice-Hall, Inc.: Upper Saddle River, NJ, USA, 1993.

6. Skutkova, H.; Vitek, M.; Babula, P.; Kizek, R.; Provaznik, I. Classification of genomic signals using dynamic time warping. *BMC Bioinform.* **2013**, *14*, S1. [CrossRef] [PubMed]

7. Petitjean, F.; Inglada, J.; Gancarski, P. Satellite Image Time Series Analysis Under Time Warping. *IEEE Trans. Geosci. Remote Sens.* **2012**, *50*, 3081–3095. [CrossRef]

8. Radoi, A.; Burileanu, C. Query-by-Example Retrieval in Satellite Image Time Series. In Proceedings of the 2018 41st International Conference on Telecommunications and Signal Processing (TSP), Athens, Greece, 4–6 July 2018; pp. 1–5. [CrossRef]

9. Maus, V.; Camara, G.; Cartaxo, R.; Sanchez, A.; Ramos, F.M.; de Queiroz, G.R. A Time-Weighted Dynamic Time Warping Method for Land-Use and Land-Cover Mapping. *IEEE J. Sel. Top. Appl. Earth Obs. Remote Sens.* **2016**, *9*, 3729–3739. [CrossRef]

10. Belgiu, M.; Csillik, O. Sentinel-2 cropland mapping using pixel-based and object-based time-weighted dynamic time warping analysis. *Remote Sens. Environ.* **2018**, *204*, 509–523. [CrossRef]

11. Tan, C.W.; Webb, G.; Petitjean, F. Indexing and classifying gigabytes of time series under time warping. In Proceedings of the 2017 SIAM International Conference on Data Mining, Houston, TX, USA, 27–29 April 2017.

12. Vaduva, C.; Costachioiu, T.; Patrascu, C.; Gavat, I.; Lazarescu, V.; Datcu, M. A Latent Analysis of Earth Surface Dynamic Evolution Using Change Map Time Series. *IEEE Trans. Geosci. Remote Sens.* **2013**, *51*, 2105–2118. [CrossRef]

13. Pelletier, C.; Valero, S.; Inglada, J.; Champion, N.; Dedieu, G. Assessing the robustness of Random Forests to map land cover with high resolution satellite image time series over large areas. *Remote Sens. Environ.* **2016**, *187*, 156–168. [CrossRef]

14. Julea, A.; Meger, N.; Bolon, P.; Rigotti, C.; Doin, M.P.; Lasserre, C.; Trouve, E.; Lazarescu, V.N. Unsupervised Spatiotemporal Mining of Satellite Image Time Series Using Grouped Frequent Sequential Patterns. *IEEE Trans. Geosci. Remote Sens.* **2011**, *49*, 1417–1430. [CrossRef]

15. Petitjean, F.; Gançarski, P.; Masseglia, F.; Forestier, G. Analysing Satellite Image Time Series by Means of Pattern Mining. In *Intelligent Data Engineering and Automated Learning*; Lecture Notes in Computer Science; Springer: Berlin/Heidelberg, Germany, 2010; Volume 6283, pp. 45–52.

16. Radoi, A.; Datcu, M. Spatio-temporal characterization in satellite image time series. In Proceedings of the 8th International Workshop on the Analysis of Multitemporal Remote Sensing Images, MultiTemp 2015, Annecy, France, 22–24 July 2015; pp. 1–4. [CrossRef]

17. Muller, M. Dynamic Time Warping. In *Information Retrieval for Music and Motion*; Springer: Berlin/Heidelberg, Germany, 2007; pp. 69–84. [CrossRef]

18. Redner, R.A.; Walker, H.F. Mixture Densities, Maximum Likelihood and the EM Algorithm. *SIAM Rev.* **1984**, *26*, 195–239. [CrossRef]

19. Bishop, C.M. *Pattern Recognition and Machine Learning (Information Science and Statistics)*; Springer-Verlag New York, Inc.: Secaucus, NJ, USA, 2006.

applied
sciences

MDPI

Article

Calibration for Sample-And-Hold Mismatches in *M*-Channel TIADCs Based on Statistics

Xiangyu Liu [1], Hui Xu [1], Yinan Wang [1], Yingqiang Dai [2], Nan Li [1,*] and Guiqing Liu [1]

[1]   College of Electronic Science, National University of Defense Technology, Changsha 410073, China; liuxiangyu15@nudt.edu.cn (X.L.); xuhui@nudt.edu.cn (H.X.); wangyinan@nudt.edu.cn (Y.W.); nswe123456789@gmail.com (G.L.)
[2]   Technical Maintenance Office, Logistical Support Battalion, The PLA 77156 Unit, Wuzhong 751100, China; malarky@126.com
*   Correspondence: linan@nudt.edu.cn; Tel.: +86-0731-87003132

Received: 26 October 2018; Accepted: 3 January 2019; Published: 8 Janurary 2019

Abstract: Time-interleaved analog-to-digital converter (TIADC) is a good option for high sampling rate applications. However, the inevitable sample-and-hold (S/H) mismatches between channels incur undesirable error and then affect the TIADC's dynamic performance. Several calibration methods have been proposed for S/H mismatches which either need training signals or have less extensive applicability for different input signals and different numbers of channels. This paper proposes a statistics-based calibration algorithm for S/H mismatches in *M*-channel TIADCs. Initially, the mismatch coefficients are identified by eliminating the statistical differences between channels. Subsequently, the mismatch-induced error is approximated by employing variable multipliers and differentiators in several Richardson iterations. Finally, the error is subtracted from the original output signal to approximate the expected signal. Simulation results illustrate the effectiveness of the proposed method, the selection of key parameters and the advantage to other methods.

Keywords: time-interleaved analog-to-digital converter (TIADC); sample-and-hold (S/H) mismatch; modulo *M* quasi-stationary; Taylor series; Richardson iteration

---

1. Introduction

Time-interleaved analog-to-digital converters (TIADCs) perform high-throughput A/D conversions without loss of dynamic performance if all channels have identical electronic characteristics [1]. In reality, electronic mismatches, which periodically modulate the input signal and degrade the output signal's dynamic performance, are inevitable. In recent years, methods have been put forward to mitigate timing mismatches [2–17], bandwidth mismatches [18,19], frequency response mismatches [20–30] (frequency response mismatch contains the gain, timing and bandwidth mismatches altogether), and nonlinearity mismatches [31–34]. To consider all kinds of mismatches, ref. [35,36] propose the joint calibration methods. This paper is an extended version of our paper published in 2018 41st International Conference on Telecommunications and Signal Processing (TSP) [37].

1.1. Review of Literature

Foreground methods. In [18,36], the mismatch coefficients are identified with the help of training signals, which is more accurate than the background methods but requires interruption of the normal operation. An additional channel is needed to assist the calibration in [36], which brings about additional hardware cost.

Background methods. In [19], a test tone is injected near the Nyquist frequency for the coefficient estimation. The drawback of this semi-blind method is that the input signal can only occupy the middle part of a Nyquist band, causing a low utilization of the frequency band. In [27], the coefficients are tracked with the help of low-pass filters and fractional delay filters, whose bandwidth utilization efficiency is a little higher than [19]. An input-free band (IFB) is utilized in [23,35] for the coefficient identification in 2-channel TIADCs. This method consumes fewer resources than [27], but it fails for some narrow-band signals, which will be further explained in Section 5.3. The in-phase/quadrature (I/Q) mismatch calibration technique is borrowed for dual- and quad-channel TIADC's frequency response mismatch calibration in [24,25]. However, when the input signal contains mainly sinusoidal components, the calibration is less satisfactory, which will be further explained in Section 5.3.

Among others, ref. [3,5] propose timing mismatch identification algorithms based on wide sense stationary (WSS) property and modulo M quasi-stationary property of the input signal, but they do not mention the calibration of the S/H mismatch.

1.2. Contribution of the Paper

This paper proposes a statistics-based calibration method for S/H mismatches in M-channel TIADCs. First of all, a cost function is established assuming the WSS and modulo M quasi-stationary properties of the input signal, and the mismatch coefficients are identified by eliminating the cost function in a least-mean-square (LMS) sense. Next, the error is approximated with the aid of multipliers and differentiators, and Richardson iteration is employed to achieve higher precision. At last, the signal is calibrated by suppressing the error. The resource consumption of the proposed method is higher compared with the IFB-based method and the I/Q-based method, but the proposed method can apply to extensive types of signals and numbers of channels whereas the other two methods cannot.

1.3. Outline

The remaining paper is organized as follows. The model of the TIADC with bandwidth mismatches is illustrated in Section 2. Since the compensation structure is utilized in the identification, the compensation is considered first in Section 3. Then the identification algorithm is treated in Section 4. The simulations and the comparisons with the IFB-based method and the I/Q-based method are presented in Section 5. Finally, the conclusion is given in Section 6.

2. The Model

The sampling rate of the TIADC is denoted as f_s. The analog input signal $x(t)$, is assumed to be band-limited to $f_s/2$, indicating that $x(t)$ can be perfectly recovered from the uniform-sampling samples $x[n] = x(nT)$, $T_s = 1/f_s$. As the focus of this paper is the bandwidth mismatch, it is assumed that other types of mismatches have been removed by the methods mentioned in Section 1.

The S/H circuit is usually modeled by a first-order RC filter which is illustrated in Figure 1 [18,38]. The circuit is essentially a low-pass filter with the -3 dB bandwidth $\Omega_c = 1/RC$. ("R" is the equivalent resistance of the circuit, and "C" is the equivalent capacitance.) In reality, the values for R and C differ between channels owing to variations in the manufacture, and they change slowly due to the fluctuation in temperature and voltage.

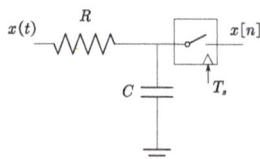

Figure 1. RC model of sample-and-hold circuits.

The transfer function of the m-th channel's S/H circuit can be expressed as (1)

$$H_m\left(j\Omega\right) = \frac{1}{1 + j\Omega\frac{1+\beta_m}{\Omega_c}}, \tag{1}$$

where β_m is the mismatch coefficient and Ω_c is the -3 dB bandwidth of the reference channel. The M-channel TIADC's model can be described as Figure 2.

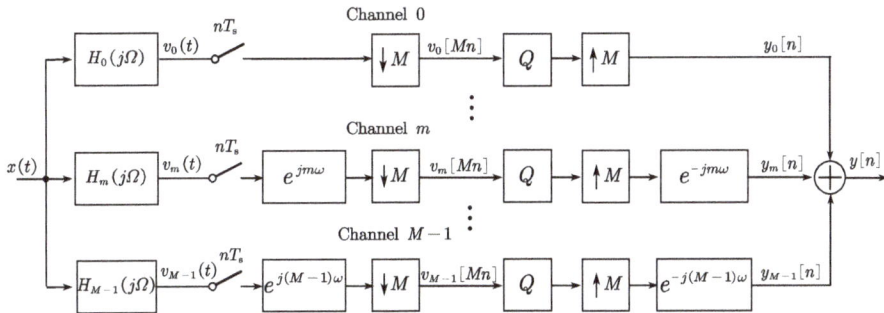

Figure 2. The model of an M-channel time-interleaved analog-to-digital converters (TIADC) with bandwidth mismatches. (The "↓ M" module is a down-sampler of factor M, while the "↑ M" module is an up-sampler of factor M. The "Q" module is the quantizer.)

To investigate the influence which bandwidth mismatch has on the dynamic performance of the TIADC, the discrete-time Fourier transform (DTFT) of downsampled signals $v_m[Mn]$ (defined in Figure 2) are

$$V_m\left(e^{jM\omega}\right) = \frac{1}{M}\sum_{k=0}^{M-1} H_m\left(e^{j\left(\omega-\frac{2k\pi}{M}\right)}\right)e^{j\left(\omega-\frac{2k\pi}{M}\right)}X\left(e^{j\left(\omega-\frac{2k\pi}{M}\right)}\right), \tag{2}$$

where

$$H_m(e^{j\omega}) = H_m(j\Omega T_s), \quad -\pi < \omega < \pi, \tag{3}$$

and $X\left(e^{j\omega}\right)$ is the DTFT of signal $x[n]$. Then the DTFT of the TIADC's output signal is

$$Y\left(e^{j\omega}\right) = \sum_{m=0}^{M-1} Y_m\left(e^{j\omega}\right) = \sum_{m=0}^{M-1} V_m\left(e^{jM\omega}\right)e^{-jm\omega}, \tag{4}$$

where $Y_m\left(e^{j\omega}\right)$ is the DTFT of $y_m[n]$ defined in Figure 2.

Taking 2-channel TIADC for instance, (4) is reduced to

$$\begin{aligned}
Y\left(e^{j\omega}\right) &= Y_0\left(e^{j\omega}\right) + Y_1\left(e^{j\omega}\right) = V_0\left(e^{j2\omega}\right) + V_1\left(e^{j2\omega}\right)e^{-j\omega} \\
&= \frac{1}{2}\left[H_0\left(e^{j\omega}\right) + H_1\left(e^{j\omega}\right)\right]X\left(e^{j\omega}\right) + \frac{1}{2}\left[H_0\left(e^{j(\omega-\pi)}\right) - H_1\left(e^{j(\omega-\pi)}\right)\right]X\left(e^{j(\omega-\pi)}\right),
\end{aligned} \tag{5}$$

where the first term is the linear distortion of the input spectrum, and the second term is the error spectrum. The two spectra are symmetric around frequency $\pi/2$, which is shown in Figure 3.

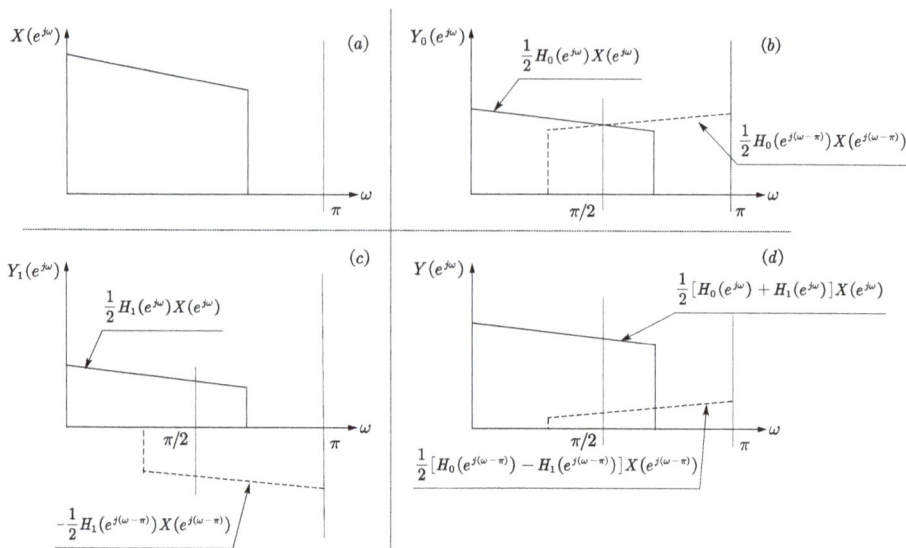

Figure 3. The spectra of a 2-channel TIADC with bandwidth mismatch. (**a**) The input spectrum. (**b**) The output spectrum of Channel 0. (**c**) The output spectrum of Channel 1. (**d**) The output spectrum of the TIADC.

Like in a single ADC, there's no need to explicitly equalize the frequency response, and it is not obligatory here for a TIADC [21,22]. So Channel 0 can be chosen as the reference channel ($\beta_0 = 0$), and the filter $H_m(j\Omega)$ ($m \neq 0$) can be divided into two cascaded filters as is depicted in Figure 4. The bandwidth mismatch error can be regarded as the effect of an error generator whose transfer function is $H_m(j\Omega)/H_0(j\Omega)$.

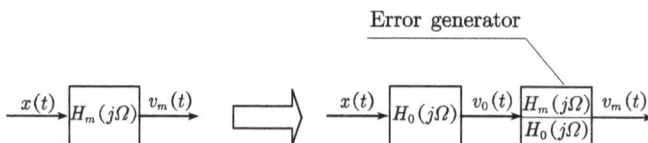

Figure 4. The filter $H_m(j\Omega)$ can be divided into two cascaded filters.

To simplify the compensation, we try to use polynomials to approximate the transfer function of the error generator. Then the transfer function can be expressed in 2-order Taylor series as [39]

$$\frac{H_m(j\Omega)}{H_0(j\Omega)} = \frac{1 + j\Omega\frac{1}{\Omega_c}}{1 + j\Omega\frac{1+\beta_m}{\Omega_c}} \approx 1 - j\Omega\frac{\beta_m}{\Omega_c} + (j\Omega)^2\frac{\beta_m(1+\beta_m)}{\Omega_c^2}, \tag{6}$$

where $j\Omega$ is the frequency response of a differentiator. Usually, the bandwidth mismatch is small enough that 2-order Taylor series are sufficient to approximate the transfer function [21,22,29].(When the mismatch is larger, higher-order terms of Taylor series are needed. For example, in [22] where $\Omega_c = 4\pi f_s$ and β_m is around 10^{-3}, the third order terms of Taylor series is needed.)

Then $v_m[Mn]$ (defined in Figure 2) can be expressed as (7)

$$v_m[Mn] \approx \tilde{v}_m[Mn] - \frac{\beta_m}{\omega_c}\tilde{y}'_m[Mn] + \frac{\beta_m(1+\beta_m)}{\omega_c^2}\tilde{y}''_m[Mn] = \tilde{v}_m[Mn] + e_m[Mn]. \tag{7}$$

The meanings of $\tilde{v}_m[Mn]$, $\tilde{y}'_m[Mn]$ and $\tilde{y}''_m[Mn]$ are given in Figure 5. Now our task is to cancel the error by identifying the mismatch coefficient β_m and approximating the error $e_m[Mn]$.

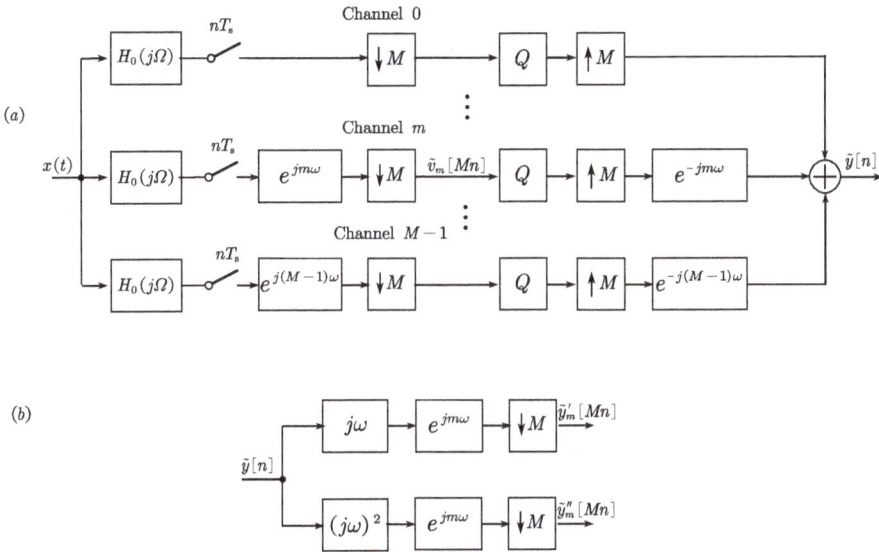

Figure 5. (a) $\tilde{v}_m[Mn]$ is obtained through a TIADC without mismatches. (b) $\tilde{y}'_m[Mn]$ and $\tilde{y}''_m[Mn]$ are the subsequences of signal $\tilde{y}$'s first differential and second differential, respectively. The $j\omega$ is a first-order differentiator, and $(j\omega)^2$ is a second-order differentiator.

3. Mismatch Compensation

Since the compensation structure is utilized in the identification, the compensation is firstly considered. When the mismatch coefficient $\hat{\beta}_m$ is identified (which is described in detail in Section 4), the error $e_m[Mn]$ in (7) can be generated in theory. In practice, however, no prior knowledge about the desirable signal $\tilde{y}[n]$ (defined in Figure 5) is available. One widely acceptable solution is to approximate it by $y[n]$ [6,7,22,28].

The Richardson iteration structure illuminated in Figure 6 is introduced for high precision compensation [40]. In Stage 1, we try to generate the 1st-order term of Taylor series in (7) using the output signal $y[n]$ and then eliminate it from the original output signal. Then the compensated down-sampled signal of Channel m is

$$
\begin{aligned}
p_m[Mn] &= v_m[Mn] + \frac{\hat{\beta}_{m1}}{\Omega_c}y'_m[Mn] \\
&= \tilde{v}_m[Mn] + \frac{\hat{\beta}_{m1} - \beta_m}{\Omega_c}\tilde{y}'_m[Mn] + \frac{\beta_m(1+\beta_m-\hat{\beta}_{m1})}{\Omega_c^2}\tilde{y}''_m[Mn] + o\left(\tilde{y}''_m[Mn]\right),
\end{aligned}
\tag{8}
$$

where $\hat{\beta}_{m1}$ is the identified coefficient for β_m in Stage 1, and $o\left(\tilde{y}''_m[Mn]\right)$ is higher order differential of $\tilde{y}_m[Mn]$ which is too weak to consider. Comparing (8) with (7), $p_m[Mn]$ is closer to $\tilde{v}_m[Mn]$ than $v_m[Mn]$, where the detailed derivation is given in [22,40]. Such being the case, we can use $p_m[Mn]$ instead of $v_m[Mn]$ for further approximation to $\tilde{v}_m[Mn]$ in the error cancellation of Stage 2.

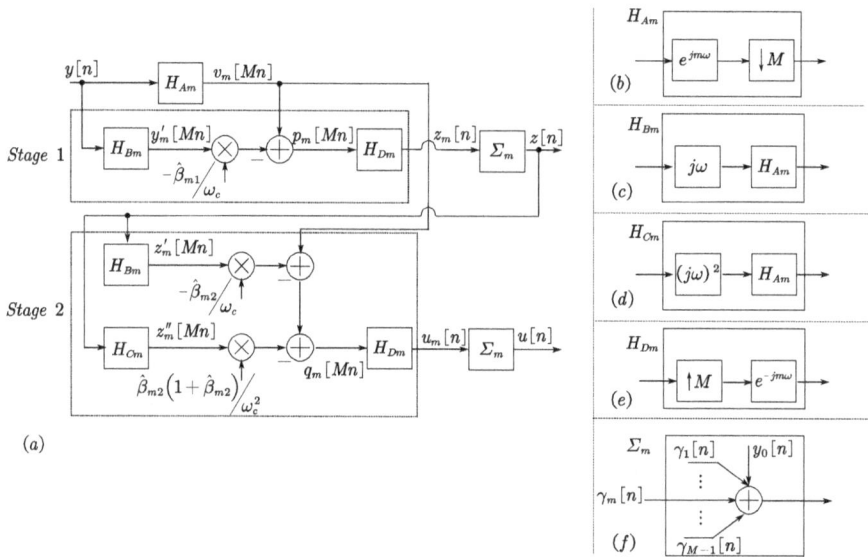

Figure 6. (**a**) The compensation structure using Richardson iteration. (**b**) The details of block H_{Am}. (**c**) The details of block H_{Bm}. (**d**) The details of block H_{Cm}. (**e**) The details of block H_{Dm}. (**f**) The details of block Σ_m, where γ represents for z or u in (**a**).

In Stage 2, we try to approximate the 1st- and 2nd-order terms of Taylor series in (7). Note that the generated error should be subtracted from the original signal $v_m[Mn]$ to further suppress the distortion. Then the compensated down-sampled signal of Channel m is denoted as

$$
\begin{aligned}
q_m[Mn] &= v_m[Mn] + \frac{\hat{\beta}_{m2}}{\Omega_c} z'_m[Mn] - \frac{\hat{\beta}_{m2}(1+\hat{\beta}_{m2})}{\Omega_c^2} z''_m[Mn] \\
&= \tilde{v}_m[Mn] + \frac{\hat{\beta}_{m2} - \beta_m}{\Omega_c} \tilde{y}'_m[Mn] + \frac{(1+\beta_m-\hat{\beta}_{m1})(\beta_m-\hat{\beta}_{m2}) - \hat{\beta}_{m2}^2}{\Omega_c^2} \tilde{y}''_m[Mn] + o\left(\tilde{y}''_m[Mn]\right),
\end{aligned}
$$
(9)

where $\hat{\beta}_{m2}$ is the identified coefficient for β_m in Stage 2. The error is further suppressed in Stage 2 compared with Stage 1.

More stages are needed if the mismatch is larger or higher precision is required. For instance, in Stage 3, the 1st-, 2nd-, and 3rd-order terms of the Taylor's series can be approximated using $u[n]$, and are then subtracted from $y[n]$ to get a signal closer to the expected one.

4. Coefficient Identification

The coefficient identification is based on the statistical properties of the input signals. Most real-life signals are WSS and modulo M quasi-stationary, whose definitions are given below [3,5].

Definition 1. *Wide-Sense Stationary*

A discrete-time signal $u[k]$ is said to be WSS if its 1st and 2nd moments are time-invariant. That is

$$
m_u = \lim_{N \to \infty} \frac{1}{N} \sum_{k=1}^{N} E(u[k]),
$$
(10)

and

$$R_u[n] = \lim_{N \to \infty} \frac{1}{N} \sum_{k=1}^{N} E(u[k+n]u[k]), \tag{11}$$

where $E(\cdot)$ is the expectation.

Definition 2. *Modulo M Quasi-Stationary*

Assume

$$\bar{f}_{u_{i_1}, u_{i_2}, \cdots} = \lim_{N \to \infty} \frac{1}{N} \sum_{t=1}^{N} f(u_{i_1}[t], u_{i_2}[t], \cdots), \quad i_1, i_2, \cdots = 0, 1, \cdots, M-1 \tag{12}$$

exists for a function $f(\cdot, \cdot, \cdots)$. Then u is modulo M quasi-stationary with respect to f if

$$\bar{f}_{i_1, i_2, \cdots} = \bar{f}_{\{(i_1+l) \bmod M, (i_2+l) \bmod M, \cdots\}}, \quad l \in Z. \tag{13}$$

("mod" is the remainder operator, and "Z" is the set of integers.) The modulo M quasi-stationary property guarantees that the input signal manifests the same statistical properties for all channels in the time-interleaved system.

Assume $x[n]$ is WSS and modulo M quasi-stationary with respect to the function $f(x_i, x_{i-1}) = (x_i - x_{i-1})^2$. Then the down-sampled signal for Channel m is denoted as

$$v_m[Mn] = A_m x(MnT_s + t_m), \quad m = 0, 1, 2, \cdots, M-1, \tag{14}$$

where

$$A_m = \frac{1}{\sqrt{1 + \left(\frac{(1+\beta_m)\omega}{\Omega_c}\right)^2}} \tag{15}$$

is the magnitude response of the filter $H_m(e^{j\omega})$, and

$$t_m = -\frac{1}{\omega} \arctan\left(\frac{(1+\beta_m)\omega}{\Omega_c}\right), \quad s.t. \ t_m \in \left(-\frac{\pi}{2}, \frac{\pi}{2}\right) \tag{16}$$

is the phase response of the filter $H_m(e^{j\omega})$.

Then the mean squared difference between the down-sampled signals of adjacent channels is

$$
\begin{aligned}
D_{v_m, v_{m-1+\lfloor\lfloor\frac{m-1}{M}\rfloor\rfloor \cdot M}} &= \lim_{N \to \infty} \frac{1}{N} \sum_{n=1}^{N} \left(v_m[Mn] - v_{m-1+\lfloor\lfloor\frac{m-1}{M}\rfloor\rfloor \cdot M} \left[M\left(n + \lfloor\frac{m-1}{M}\rfloor\right)\right]\right)^2 \\
&= \left(A_m + A_{m-1+\lfloor\lfloor\frac{m-1}{M}\rfloor\rfloor \cdot M}\right) \sigma^2 - 2 A_m A_{m-1+\lfloor\lfloor\frac{m-1}{M}\rfloor\rfloor \cdot M} R_x \left(T_s + t_m - t_{m-1+\lfloor\lfloor\frac{m-1}{M}\rfloor\rfloor M}\right), \\
&\qquad m = 0, 1, \cdots, M-1,
\end{aligned} \tag{17}
$$

where $\lfloor \cdot \rfloor$ is the floor operator, and $|\cdot|$ is the absolution operator. In (17), σ^2 stands for the variance of the signal $x[n]$

$$\sigma^2 = \lim_{N \to \infty} \frac{1}{N} \sum_{n=0}^{N} x^2[n], \tag{18}$$

and $R_x(\cdot)$ means the autocorrelation of the signal $x[n]$

$$R_x(\tau) = \lim_{N \to \infty} \frac{1}{N} \sum_{n=0}^{N} x[n] x[n+\tau]. \tag{19}$$

When $m > 1$, (17) becomes

$$D_{v_m,v_{m-1}} = \lim_{N\to\infty} \frac{1}{N} \sum_{n=1}^{N} (v_m[Mn] - v_{m-1}[Mn])^2, \quad m = 1, \cdots, M-1. \tag{20}$$

When $m = 0$, (17) becomes

$$D_{v_0,v_{M-1}} = \lim_{N\to\infty} \frac{1}{N} \sum_{n=1}^{N} (v_0[Mn] - v_{M-1}[M(n-1)])^2. \tag{21}$$

The value of (17) varies with the bandwidth mismatch β_m.

LMS algorithm is adopted to estimate β_m. By considering all the distinctions between the mean squared differences indicated by (17), the loss function can be established as

$$P = \sum_{i=1}^{M-1} \sum_{j=0}^{i-1} \left(D_{v_i,v_{i-1+\lfloor\lfloor\frac{i-1}{M}\rfloor\rfloor\cdot M}} - D_{v_j,v_{j-1+\lfloor\lfloor\frac{i-1}{M}\rfloor\rfloor\cdot M}} \right)^2. \tag{22}$$

Only when all $\beta_m = 0$, the cost function $P = 0$.

In Stage 1, the identified coefficient $\hat{\beta}_1$ is searched as the following four steps.

Step 1: Calculate the mean squared differences between adjacent channels' compensated signals as (In practice, the limit expressed in (17) cannot be realized, and it is approximated by a batch of finite samples instead [3,5].)

$$D_{p_i,p_{i-1+\lfloor\lfloor\frac{i-1}{M}\rfloor\rfloor\cdot M}} = \frac{1}{N} \sum_{n=1}^{N} \left(p_i[Mn] - p_{i-1+\lfloor\lfloor\frac{i-1}{M}\rfloor\rfloor\cdot M} \left[M\cdot\left(n+\lfloor\frac{i-1}{M}\rfloor\right)\right]\right)^2, \tag{23}$$

where $p_0[Mn] = v_0[Mn]$.

Step 2: Calculate the cost function as

$$P = \sum_{i=1}^{M-1} \sum_{j=0}^{i-1} \left(D_{p_i,p_{i-1+\lfloor\lfloor\frac{i-1}{M}\rfloor\rfloor\cdot M}} - D_{p_j,p_{j-1+\lfloor\lfloor\frac{i-1}{M}\rfloor\rfloor\cdot M}} \right)^2, \tag{24}$$

which is also shown in Figure 7.

Step 3: Calculate the partial differential of $\hat{\beta}_{m1}$ as

$$\begin{aligned}\frac{\partial P(k)}{\partial\hat{\beta}_{m1}(k)} =& 2\sum_{i=m+2}^{M-1}\sum_{j=m}^{i-1}\left(D_{p_i,p_{i-1}} - D_{p_j,p_{j-1}}\right)\left(-\frac{\partial D_{p_j,p_{j-1}}}{\partial\hat{\beta}_{m1}(k)}\right)\\ &+2\sum_{i=m}^{m+1}\sum_{j=0}^{m-1}\left(D_{p_i,p_{i-1}} - D_{p_j,p_{j-1+\lfloor\lfloor\frac{i-1}{M}\rfloor\rfloor\cdot M}}\right)\cdot\frac{\partial D_{p_i,p_{i-1}}}{\partial\hat{\beta}_{m1}(k)}\\ &+2\left(D_{p_{m+1},p_m} - D_{p_m,p_{m-1}}\right)\cdot\left(\frac{\partial D_{p_{m+1},p_m}}{\partial\hat{\beta}_{m1}(k)} - \frac{\partial D_{p_m,p_{m-1}}}{\partial\hat{\beta}_{m1}(k)}\right)\end{aligned} \tag{25}$$

$$\frac{\partial D_{p_{m+1},p_m}}{\partial\hat{\beta}_{m1}(k)} = \frac{2}{N}\sum_{n=1}^{N}(p_{m+1}[Mn] - p_m[Mn])\cdot\left(-\frac{\partial p_m}{\partial\hat{\beta}_{m1}(k)}\right), \tag{26}$$

$$\frac{\partial D_{p_m,p_{m-1}}}{\partial\hat{\beta}_{m1}(k)} = \frac{2}{N}\sum_{n=1}^{N}(p_m[Mn] - p_{m-1}[Mn])\cdot\frac{\partial p_m}{\partial\hat{\beta}_{m1}(k)}, \tag{27}$$

$$\frac{\partial p_m}{\partial\hat{\beta}_{m1}(k)} = \frac{y_m'[Mn]}{\Omega_c}, \tag{28}$$

where k means the k-th searching loop and $m = 1, 2, \cdots, M - 1$.

Step 4: Update the coefficients as

$$\hat{\beta}_{m1}(k+1) = \hat{\beta}_{m1}(k) - \mu \frac{dP(k)}{d\hat{\beta}_{m1}(k)}, \tag{29}$$

where μ is the searching step.

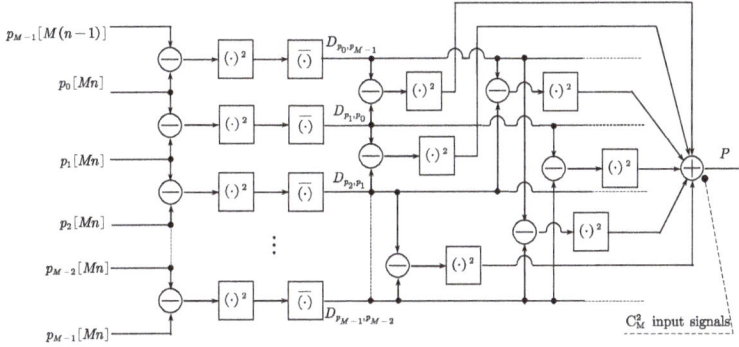

Figure 7. The calculation of the cost function in Stage 1. ("$(\cdot)^2$" is the square operation, and "$\overline{(\cdot)}$" is the averaging operation).

In Stage 2, the identification procedure is similar to the above four steps except that $q_m[Mn]$ is used instead of $p_m[Mn]$. The partial derivative $\partial q_m[Mn] / \partial \hat{\beta}_{m2}(k)$ used in *Step 3* is

$$\frac{\partial q_m[Mn]}{\partial \hat{\beta}_{m2}(k)} = \frac{z'_m[Mn]}{\Omega_c} - \frac{(2\hat{\beta}_{m2}(k) + 1) z''_m[Mn]}{\Omega_c^2}. \tag{30}$$

Moreover, it should be noted that before the identification, notch filters with notch frequencies at $m \cdot f_s / (2M)$ are needed to exclude the error caused by coherent sampling, where $m = 1, 2, \cdots, M - 1$.

5. Simulation and Comparison

A quantizer of 14 bits is utilized in the simulations below. The filters are designed by "firpm" function in MATLAB©, which uses the Parks-McClellan optimal equiripple design algorithm. The -3 dB bandwidth of Channel 0 is $\Omega_c = 6\pi f_s$. We define a term for sinusoidal signals to indicate the dynamic performance of the TIADC: the Largest-Signal-to-largest-Spurious-component-Ratio (LSSR).

$$\text{LSSR (dBc)} = 10 \log_{10} \frac{\max\left(P_{signal,i}\right)}{\max\left(P_{spurious,j}\right)}, \tag{31}$$

where $P_{signal,i}$ is the power of the i-th sinusoidal component in the DFT spectrum, and $P_{spurious,j}$ is the power of the j-th spurious component including harmonic component. For a single-tone sinusoid, the LSSR is equal to the spurious-free dynamic range (SFDR).

5.1. Effectiveness

We evaluate the effectiveness of the proposed approach by simulating two cases.

In the first case, an 8-channel TIADC is simulated. The parameters are set as Tables 1 and 2. The output power spectra are demonstrated in Figure 8. The higher vertical lines up to 0 dB stand for the desirable input signals, while the lower vertical lines are the bandwidth mismatch induced errors which should be suppressed below the noise floor. The signal-to-noise-and-distortion (SINAD) is

62.15 dB and the LSSR is 59.07 dBc before calibration, while they are enhanced to 74.95 dB and 77.53 dBc after Stage 1 calibration, and then to 75.78 dB and 91.39 dBc after Stage 2 calibration. The identified coefficients in Stage 2 calibration plotted in Figure 9 and those in Table 2 are close but not identical, because we use $y[n]$ and $z[n]$ to approximate the error rather than $\tilde{y}[n]$ in (8) and (9). Therefore, the algorithm proposed in this paper can accurately identify the mismatch coefficients without much prior information of the input signal other than its WSS and modulo 8 quasi-stationary properties.

Table 1. Simulation parameters.

Frequency Range	Signal Type	Batch Size	Differentiator Order
$0 \sim 0.4 f_s$	10-tone sinusoid	$N = 512$	$N_d = 20$

Table 2. Mismatch coefficients.

β_1	β_2	β_3	β_4	β_5	β_6	β_7
0.01	−0.005	0.012	−0.014	0.02	−0.008	0.008

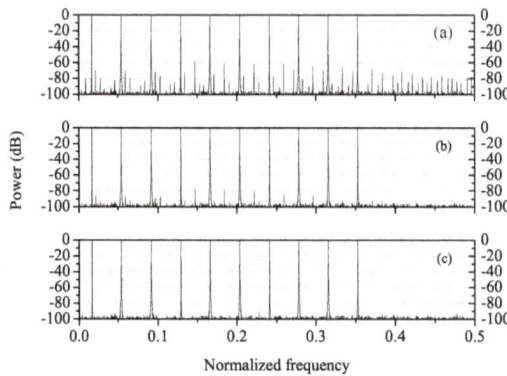

Figure 8. Output power spectra of the 8-channel TIADC (**a**) without calibration, (**b**) after Stage 1 calibration, and (**c**) after Stage 2 calibration.

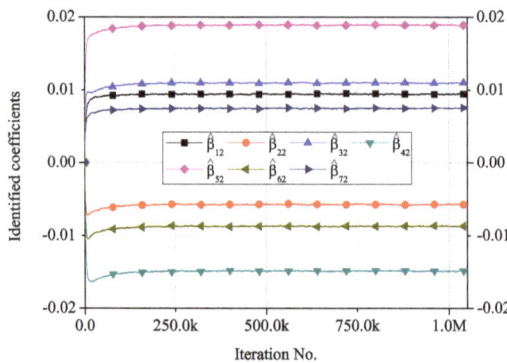

Figure 9. Identified coefficients for the 8-channel TIADC in Stage 2 calibration.

In the second case, the performance enhancement for different mismatch coefficients in a 2-channel TIADC is illustrated. Parameters are set as Table 3. The performance before and after calibration are depicted in Figure 10, showing that our algorithm works well under circumstances of different mismatch levels. Theoretically, larger promotion should have been achieved at smaller mismatch circumstance where the error in the compensation resulted from approximating y or z to $\tilde{y}$ is smaller (ref. (8) and (9)). Whereas actually in Figure 10, the SINAD and the LSSR initially increase with the decreasing mismatch and then saturate at certain values. This is because the quantization bits of the TIADC limit the further improvement. By increasing the quantization bits, a higher promotion can be achieved, and the SINAD and LSSR will saturate at larger values.

Table 3. Simulation parameters.

Frequency Range	Signal Type	Batch Size	Differentiator Order
$0 \sim 0.45 f_s$	10-tone sinusoids	$N = 16{,}384$	$N_d = 60$

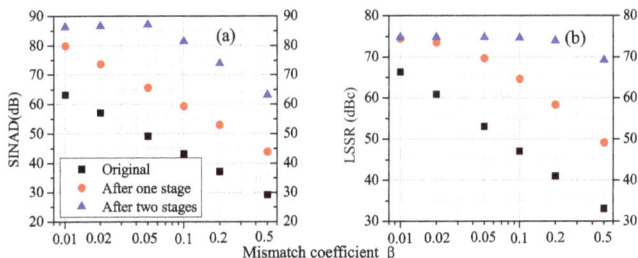

Figure 10. The (**a**) signal-to-noise-and-distortion (SINAD) and (**b**) Largest-Signal-to-largest-Spurious-component-Ratio (LSSR) for different mismatch coefficients before and after calibration.

In this subsection, the calibration method performs well for an 8-channel TIADC and a 2-channel TIADC with different mismatch coefficients.

5.2. Parameter Selection

We show the differentiator order selection and the batch size selection by simulating two cases. A 2-channel TIADC with the mismatch coefficient $\beta = 0.2$ is used for these two cases.

In the first case, the effect which the differentiator order has on the algorithm performance is surveyed. Simulation parameters except the differentiator order are set as Table 3. The results are depicted in Figure 11. One can achieve better SINAD and LSSR along with higher differentiator's order. A 40-order differentiator is enough when the mismatch coefficient is below 0.2 and input spectrum occupies the lower 90% fraction of the Nyquist band. Moreover, the differentiator's order also has a connection with its pass-band (PB) width. For example, a 20-order differentiator is enough if the input signal only occupies the lower 80% fraction of the Nyquist band.

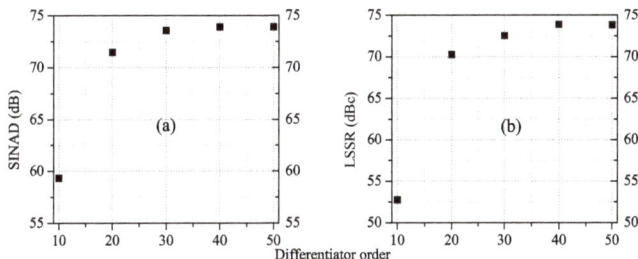

Figure 11. The (**a**) SINAD and (**b**) LSSR after two stages of calibration for different orders of differentiators.

In the second case, the effect which the batch size for coefficient identification has on the calibration precision is surveyed. The input signal is set as Table 3, and the differentiator order is set as 40. The results are depicted in Figure 12. One can achieve better SINAD and LSSR along with larger batch size for identification, because the cost function is established under the circumstance where $N \to \infty$ as (17). However, the resource consumption also increases with the batch size. Batch size of 512 is moderate for both dynamic performance and resource consumption.

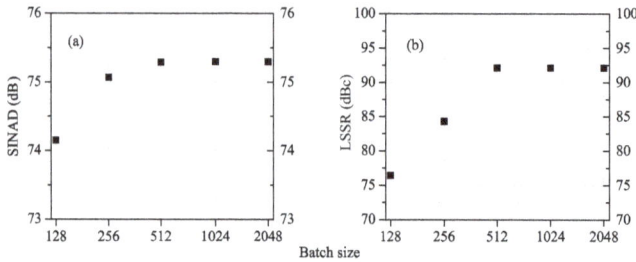

Figure 12. The (**a**) SINAD and (**b**) LSSR after two stages of calibration for different batch sizes.

In this subsection, we show the connection between the calibration precision and the differentiator order or the batch size. Eventually, differentiator order of 20 and batch size of 512 are selected for moderate dynamic performance and resource consumption.

5.3. Comparisons

In this section, we choose the IFB-based method [23,35] and the I/Q-based method [24,25] for comparison which are better than other methods both in bandwidth efficiency and complexity. A 2-channel TIADC is used for the following two cases.

In the first case, the LSSR improvements of the three methods are compared. The parameters are set as Tables 4–7, and the differentiators used in the IFB method and our method are identical. The output power spectra are demonstrated in Figure 13. The SINAD is 59.39 dB and the LSSR is 58.00 dBc before calibration, and they are slightly enhanced to 62.89 dB and 61.77 dBc using the IFB-based method, and they are enhanced to 75.08 dB and 84.95 dBc using the I/Q-based method, and they are enhanced to 77.34 dB and 93.05 dBc employing the method proposed in this paper.

For the IFB-based method, an input-free band is created in the frequency spectrum by oversampling where the mismatch-induced error exists without input signal. By exerting a high-pass filter whose passband coincides with the IFB, the mismatch coefficients can be identified. However, in this case, the error spectrum does not appear in the IFB and the identification fails (ref. Figure 13b). For some other kinds of narrow-band signals, the IFB-based method cannot also work.

For the I/Q-based method, the TIADC's output signal is converted to a complex signal with frequency shift, which is similar to the homodyne receiver's output signal with I/Q mismatch, and the I/Q mismatch calibration technique is used to calibrate the TIADC's mismatch. The compensation filter's coefficients are determined by restoring the complex signal's circularity. However, in this case where the signals are multi-tone sinusoids, there is no remarkable difference between the circularity of the signal without mismatch and that of the signal with mismatch, and therefore the calibration cannot suppress the error to the noise floor. For some other types of signals mainly composed of sinusoids, the I/Q-based method also cannot work. So compared with the other two methods, our method has more extensive applicability for different types of signals.

Table 4. Common simulation parameters for LSSR improvement comparison.

Frequency Range	Signal Type	Mismatch Coefficient
$0.26 \sim 0.38 f_s$	10-tone sinusoids	$\beta = 0.02$

0

Table 5. Simulation parameters for the high-pass filter in the input-free band (IFB) method.

Order	PB Cut-Off Frequency	SB Cut-Off Frequency	PB Attenuation	SB Attenuation
$N_{hp} = 40$	$0.495f_s$	$0.4f_s$	-1 dB	-115 dB

Note: (1) "PB" is the abbreviation for pass-band. (2) "SB" is the abbreviation for stop-band.

Table 6. Simulation parameters for the I/Q method.

Hilbert Filter Order	Hilbert Filter PB	Compensation Filter Order
$N_{hb} = 10$	$0.075f_s \sim 0.425f_s$	$N_{cp} = 2$

Table 7. Simulation parameters for the proposed method.

Batch Size	Differentiator Order
$N = 512$	$N_d = 20$

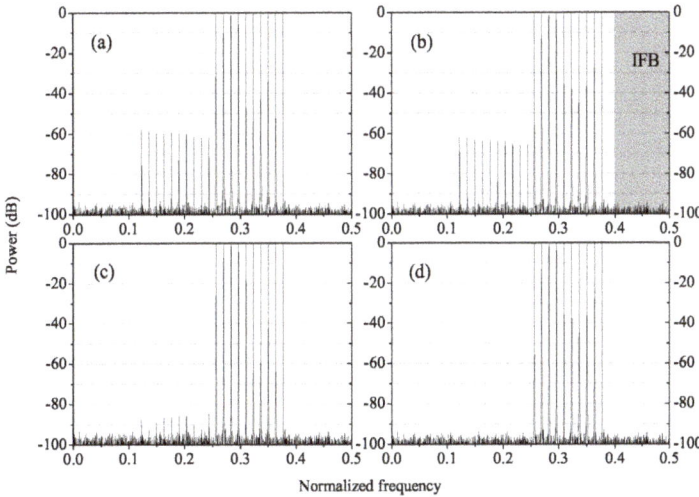

Figure 13. Output power spectra of the 2-channel TIADC (**a**) without calibration, (**b**) calibrated using the IFB-based method, (**c**) calibrated using the I/Q-based method, and (**d**) calibrated using the method in this paper.

In the second case, the resource consumptions are compared among the three methods when the same error attenuation is achieved after calibration. The multiplications used in one loop of calibration is chosen as the indicator. The input signal is set as Table 8 so that all the methods can work effectively. $\beta = 0.08$. The variables N_d, N_{hp}, N_{cp}, N_{hb} and N are defined in Tables 1, 5 and 6.

Table 8. Input signal for resource consumption comparison.

Signal Type	Carrier Frequency	Duration/Sample
16-QAM	$0.05f_s$	10

The resource consumption for the I/Q-based method is calculated as follows. ⟨1⟩ It needs $0.5N_{hb}$ multiplications to pass the output signal through the Hilbert filter because the filter's taps are anti-symmetric with a null center tap. ⟨2⟩ It needs $4(N_{cp} + 1)$ multiplications to compensate the signal, since both the signal and the filter taps are complex, and multiplying two complex numbers actually needs four multiplications of real numbers. $\langle 3 \rangle$ It consumes $4(N_{cp} + 1)$ multiplications to update the compensation filter's taps. Adding $\langle 1 \rangle$ to $\langle 3 \rangle$, the total number of multiplications needed in one calibration iteration is

$$8N_{cp} + 0.5N_{hb} + 8. \tag{32}$$

Here, $N_{hb} = 10$ and N_{cp} is set to be 0, which results in 13 multiplications in total.

The IFB-based method and the proposed method use the common compensation technique and the consumption is calculated as follows. For simplicity, only the first Richardson iteration is considered for both the IFB method and ours. $\langle 4 \rangle$ It needs $0.5N_d$ multiplications to get $v'[2n]$ because the differentiator's taps are anti-symmetric with a null center tap. $\langle 5 \rangle$ It requires 1 multiplication to scale $v'[2n]$ with estimated coefficients as (8).

The consumption of the identification procedure for the IFB-based method is calculated as follows. $\langle 6 \rangle$ It consumes $(0.5N_h + 1)$ multiplications to filter $z[n]$ due to the anti-symmetric property of the high-pass filters' taps. $\langle 7 \rangle$ It requires 2 multiplications to update $\hat{\beta}_1$. Adding $\langle 4 \rangle$ to $\langle 7 \rangle$, the total number of multiplications needed in one calibration loop is

$$0.5\left(N_d + N_{hp}\right) + 3. \tag{33}$$

Here, $N_d = 20$ and $N_{hp} = 40$, which results in 33 multiplications in total.

The consumption of the identification procedure for the proposed method is calculated as follows. $\langle 8 \rangle$ It takes $2N$ multiplications to calculate D_{v_1,v_0} and D_{v_0,v_1} as (17). $\langle 9 \rangle$ It requires $(2N+1)$ multiplications to calculate $\partial P / \partial \hat{\beta}_1$ as (25) to (28). $\langle 10 \rangle$ It requires 1 multiplication to update $\hat{\beta}_1$. (All the constant coefficients used in (17), and (25) to (28) can be combined together into the step μ, so only one multiplication is needed to consider them all.) Adding $\langle 4 \rangle$ to $\langle 5 \rangle$ and $\langle 8 \rangle$ to $\langle 10 \rangle$, the total number of multiplications needed in one calibration loop is

$$4N + 0.5N_d + 2. \tag{34}$$

Here, $N_d = 20$ and N is set to be 32, which results in 140 multiplications in total.

The comparison results of the above two cases are shown in Table 9. The proposed method in this paper is more complex compared with the other two methods when calibrating the same signals, but this method can apply to more types of signals and more channels whereas the other two methods cannot.

Table 9. Comparisons between the proposed method, the IFB-based method, and the I/Q based method.

Method	LSSR Improvement	Resource Consumption	Channel No.
This paper	35.05 dBc	141	M
IFB-based method [23,35]	3.77 dBc	42	2
I/Q-based method [24,25]	26.95 dBc	13	2 or 4

6. Conclusions

This paper proposes a statistics-based calibration method for S/H mismatches in M-channel TIADCs. The mismatch coefficients are identified by eliminating the statistical differences between channels using the LMS algorithm. The mismatch-induced errors are approximated using multipliers and differentiators, and are eliminated from the original output samples afterwards. Although the complexity is higher compared with the IFB-based method and the I/Q-based method, the proposed algorithm in this paper has more extensive applicability for different signals and different numbers of channels.

There are three circumstances where our method should be given priority. When the signal's bandwidth is unknown or is narrow to some extent, it is more reliable to use our method than the

Appl. Sci. **2019**, *9*, 198

IFB-based method. When the signal's components are unknown or the signal is mainly composed of sinusoids, our method can provide higher calibration precision than the I/Q-based method. When the number of channels is more than four, only our method can work.

To realize our method in a real-time way, a full-parallel structure can be employed in FPGAs. For instance in [30], fast FIR algorithm is used to parallelize the differentiators, which can make the FPGA's low clock frequency more compatible with the TIADC's high sampling rate and reduce the power consumption meanwhile. The batch size N can also be reduced since the accumulation in (23) can be shifted into the iteration process of (29) and (30) at the cost of more iterations. Furthermore, for the high sampling rate, the extra iterations do not bring much longer convergence time.

Author Contributions: Conceptualization, X.L., H.X.; Methodology, X.L., Y.W.; Software, Y.D., N.L.; Validation, G.L.; Formal Analysis, X.L.; Investigation, X.L., Y.W.; Resources, X.L.; Data Curation, X.L., N.L.; Writing—Original Draft Preparation, X.L.; Writing—Review & Editing, Y.W.; Visualization, Y.D.; Supervision, H.X.; Project Administration, H.X., Y.W.; Funding Acquisition, Y.W.

Funding: This research was funded by National Natural Science Foundation of China [No. 61701509].

Conflicts of Interest: The authors declare no conflict of interest.

References

1. Black, W.; Hodges, D. Time interleaved converter arrays. *IEEE J. Solid-State Circuits* **1980**, *6*, 1022–1029. [CrossRef]
2. Jamal, S.M.; Fu, D.; Chang, N.; Hurst, P.J.; Lewis, S.H. A 10-b 120-Msample/s time-interleaved analog-to-digital converter with digital background calibration. *IEEE J. Solid-State Circuits* **2002**, *12*, 1618–1627. [CrossRef]
3. Elbornsson, J.; Gustafsson, F.; Eklund, J.E. Blind adaptive equalization of mismatch errors in a time-interleaved A/D converter system. *IEEE Trans. Circuits Syst. I Regul. Pap.* **2003**, *1*, 151–158. [CrossRef]
4. Jamal, S.M.; Fu, D.; Singh, M.P.; Hurst, P.J.; Lewis, S.H. Calibration of sample-time error in a two-channel time-interleaved analog-to-digital converter. *IEEE Trans. Circuits Syst. I Regul. Pap.* **2004**, *1*, 130–139. [CrossRef]
5. Elbornsson, J.; Gustafsson, F.; Eklund, J.E. Blind equalization of time errors in a time-interleaved ADC system. *IEEE Trans. Signal Process.* **2005**, *4*, 1413–1424. [CrossRef]
6. Chi, H.L.; Hurst, P.J.; Lewis, S.H. A Four-Channel Time-Interleaved ADC With Digital Calibration of Interchannel Timing and Memory Errors. *IEEE J. Solid-State Circuits* **2010**, *10*, 2091–2103. [CrossRef]
7. Matsuno, J.; Yamaji, T.; Furuta, M.; Itakura, T. All-Digital Background Calibration Technique for Time-Interleaved ADC Using Pseudo Aliasing Signal. *IEEE Trans. Circuits Syst. I Regul. Pap.* **2013**, *5*, 1113–1121. [CrossRef]
8. Li, D.; Zhu, Z.; Zhang, L.; Yang, Y. A background fast convergence algorithm for timing skew in time-interleaved ADCs. *Microelectron. J.* **2016**, 45–52. [CrossRef]
9. Mafi, H.; Yargholi, M.; Yavari, M. Digital Blind Background Calibration of Imperfections in Time-Interleaved ADCs. *IEEE Trans. Circuits Syst. I, Regul. Pap.* **2017**, *99*, 1–11. [CrossRef]
10. Chen, S.; Wang, L.; Zhang, H.; Murugesu, R.; Dunwell, D.; Carusone, A.C. All-digital calibration of timing mismatch error in time-interleaved analog-to-digital converters. *IEEE Trans. VLSI Syst.* **2017**, *9*, 2552–2560. [CrossRef]
11. Liu, S.; Lv, N.; Ma, H.; Zhu, A. Adaptive semiblind background calibration of timing mismatches in a two-channel time-interleaved analog-to-digital converter. *Analog Integr. Circ. Signal* **2017**, *1*, 1–7. [CrossRef]
12. Liu, S.; Lyu, N.; Cui, J.; Zou, Y. Improved Blind Timing Skew Estimation Based on Spectrum Sparsity and ApFFT in Time-Interleaved ADCs. *IEEE Trans. Instr. Meas.* **2018**. [CrossRef]
13. Lin, C.Y.; Wei, Y.H.; Lee, T.C. A 10-bit 2.6-GS/s Time-Interleaved SAR ADC With a Digital-Mixing Timing-Skew Calibration Technique. *IEEE J. Solid-State Circuits* **2018**, *5*, 1508–1517. [CrossRef]
14. Li, D.; Zhu, Z.; Ding, R.; Liu, M.; Yang, Y.; Sun, N. A 10-bit 600-MS/s Time-Interleaved SAR ADC With Interpolation-Based Timing Skew Calibration. *IEEE Trans. Circuits Syst. II Exp. Briefs* **2018**. [CrossRef]
15. Wang, X.; Li, F.; Jia, W.; Wang, Z. A 14-bit 500MS/s Time-Interleaved ADC with Autocorrelation-Based Time Skew Calibration. *IEEE Trans. Circuits Syst. II Exp. Briefs* **2018**. [CrossRef]

16. Qiu, Y.; Liu, Y.; Zhou, J.; Zhang, G.; Chen, D.; Du, N. All-Digital Blind Background Calibration Technique for Any Channel Time-Interleaved ADC. *IEEE Trans. Circuits Syst. I Regul. Pap.* **2018**, *8*, 2503–2514. [CrossRef]
17. Yang, K.; Wei, W.; Shi, J.; Zhao, Y.; Huang, W. A fast TIADC calibration method for 5GSPS digital storage oscilloscope. *IEICE Electron. Exp.* **2018**, *9*. [CrossRef]
18. Tsai, T.H.; Hurst, P.J.; Lewis, S.H. Bandwidth mismatch and its correction in time-interleaved analog-to-digital converters. *IEEE Trans. Circuits Syst. II Exp. Briefs* **2005**, *10*, 1133–1137. [CrossRef]
19. Satarzadeh, P.; Levy, B.C.; Hurst, P.J. Adaptive semiblind calibration of bandwidth mismatch for two-channel time-interleaved ADCs. *IEEE Trans. Circuits Syst. I Regul. Pap.* **2009**, *9*, 2075–2088. [CrossRef]
20. Johansson, H.; Lowenborg, P. A Least-Squares Filter Design Technique for the Compensation of Frequency Response Mismatch Errors in Time-Interleaved A/D Converters. *IEEE Trans. Microw. Theory Tech.* **2008**, *11*, 1154–1158. [CrossRef]
21. Vogel, C.; Mendel, S. A Flexible and Scalable Structure to Compensate Frequency Response Mismatches in Time-Interleaved ADCs. *IEEE Trans. Circuits Syst. I Regul. Pap.* **2009**, *11*, 2463–2475. [CrossRef]
22. Johansson, H. A polynomial-based time-varying filter structure for the compensation of frequency-response mismatch errors in time-interleaved ADCs. *IEEE J. Sel. Top. Signal Process.* **2009**, *3*, 384–396. [CrossRef]
23. Saleem, S.; Vogel, C. Adaptive blind background calibration of polynomial-represented frequency response mismatches in a two-channel time-interleaved ADC. *IEEE Trans. Circuits Syst. I Regul. Pap.* **2011**, *6*, 1300–1310. [CrossRef]
24. Singh, S.; Anttila, L.; Epp, M.; Schlecker, W.; Valkama, M. Analysis, Blind Identification, and Correction of Frequency Response Mismatch in Two-Channel Time-Interleaved ADCs. *IEEE Trans. Microw. Theory Tech.* **2015**, *5*, 1721–1734. [CrossRef]
25. Singh, S.; Anttila, L.; Epp, M.; Schlecker, W.; Valkama, M. Frequency Response Mismatches in 4-channel Time-Interleaved ADCs: Analysis, Blind Identification, and Correction. *IEEE Trans. Circuits Syst. I Regul. Pap.* **2015**, *9*, 2268–2279. [CrossRef]
26. Bonnetat, A.; Hode, J.M.; Ferre, G.; Dallet, D. Correlation-Based Frequency-Response Mismatch Compensation of Quad-TIADC Using Real Samples. *IEEE Trans. Circuits Syst. II Exp. Briefs* **2015**, *8*, 746–750. [CrossRef]
27. Teyou, G.K.D.; Petit, H.; Loumeau, P. Adaptive and digital blind calibration of transfer function mismatch in time-interleaved ADCs. *Proc. NEWCAS* **2015**, 1–4. [CrossRef]
28. Bonnetat, A.; Hode, J.M.; Ferre, G.; Dallet, D. An Adaptive All-Digital Blind Compensation of Dual-TIADC Frequency-Response Mismatch Based on Complex Signal Correlations. *IEEE Trans. Circuits Syst. II Exp. Briefs* **2016**, *9*, 821–825. [CrossRef]
29. Liu, H.; Wang, Y.; Li, N.; Xu, H. An Adaptive Blind Frequency Response Mismatches Calibration Method for Four-Channel TIADCs Based on Channel Swapping. *IEEE Trans. Circuits Syst. II Exp. Briefs* **2016**, *99*, 625–629. [CrossRef]
30. Liu, G.; Wang, Y.; Liu, X.; Liu, H.; Li, N. Efficient Real-Time Blind Calibration for Frequency Response Mismatches in Two-Channel TI-ADCs. *IEICE Electron. Exp.* **2018**, *12*. [CrossRef]
31. Wang, Y.; Xu, H.; Johansson, H.; Sun, Z.; Wikner, J.J. Digital estimation and compensation method for nonlinearity mismatches in time-interleaved analog-to-digital converters. *Digital Signal Process.* **2015**, 130–141. [CrossRef]
32. Wang, Y.; Johansson, H.; Xu, H.; Diao, J. Bandwidth-efficient calibration method for nonlinear errors in *M*-channel time-interleaved ADCs. *Analog Integr. Circ. Signal* **2015**, *2*, 275–288. [CrossRef]
33. Liu, H.; Wang, Y.; Li, N.; Xu, H. A Calibration Method for Nonlinear Mismatches in *M*-Channel Time-Interleaved Analog-to-Digital Converters Based on Hadamard Sequences. *Appl. Sci.* **2016**, *11*, 362. [CrossRef]
34. Liu, X.; Xu, H.; Liu, H.; Wang, Y. An efficient blind calibration method for nonlinearity mismatches in *M*-channel TIADCs. *IEICE Electron. Exp.* **2017**, *11*. [CrossRef]
35. Wang, Y.; Johansson, H.; Xu, H.; Sun, Z. Joint Blind Calibration for Mixed Mismatches in Two-Channel Time-Interleaved ADCs. *IEEE Trans. Circuits Syst. I Regul. Pap.* **2015**, *6*, 1508–1517. [CrossRef]
36. Huang, G.; Yu, C.; Zhu, A. Analog assisted multichannel digital postcorrection for time-interleaved ADCs. *IEEE Trans. Circuits Syst. II Exp. Briefs* **2017**, *8*, 773–777. [CrossRef]
37. Liu, X.; Xu, H.; Wang, Y.; Li, N.; Liu, G.; Tian, Q. Statistics-Based Correction Method for Sample-and-Hold Mismatch in 2-Channel TIADCs. *Proc. TSP* **2018**, 771–774. [CrossRef]

38. Kurosawa, N.; Kobayashi, H.; Maruyama, K.; Sugawara, H.; Kobayashi, K. Explicit analysis of channel mismatch effects in time-interleaved ADC systems. *IEEE Trans. Circuits Syst. I Fundam. Theory Appl.* **2001**, *3*, 261–271. [CrossRef]

39. Stewart, J. The title of the cited contribution. In *Calculus*, 7th ed.; Thomson Learning: Stamford, CA, USA, 2007; pp. 32–58.

40. Tertinek, S.; Vogel, C. Reconstruction of nonuniformly sampled bandlimited signals using a differentiator-multiplier cascade. *IEEE Trans. Circuits Syst. I Regul. Pap.* **2008**, *8*, 2273–2286. [CrossRef]

MDPI

St. Alban-Anlage 66

4052 Basel

Switzerland

Tel. +41 61 683 77 34

Fax +41 61 302 89 18

www.mdpi.com

Applied Sciences Editorial Office

E-mail: applsci@mdpi.com

www.mdpi.com/journal/applsci

www.ingramcontent.com/pod-product-compliance
Lightning Source LLC
Chambersburg PA
CBHW051854210326

41597CB00033B/5888